Indigenous Ways to the Present:

Native Whaling in the Western Arctic

CCI PRESS

Studies in Whaling No. 6
Canadian Circumpolar Institute (CCI) Press

THE ANTHROPOLOGY
OF PACIFIC
NORTH AMERICA

Herbert D.G Maschner and Katherine L. Reedy-Maschner
series editors

Indigenous Ways to the Present:

Native Whaling in the Western Arctic

edited by Allen P. McCartney

Canadian Circumpolar Institute (CCI) Press
Edmonton

The University of Utah Press
Salt Lake City

Canadian Cataloguing in Publication Data

Main entry under title:

Indigenous Ways to the Present: Native Whaling in the Western Arctic
Allen P. McCartney, editor
(Studies in Whaling No. 6; Occasional Publication / Canadian Circumpolar Institute Press; no. 54)
Includes bibliographical references
ISBN 1-896445-25-X   **ISSN 0069-0303; 54**

1. Whaling—Arctic Regions. 2. Inuit—Hunting—Alaska.
3. Inuit—Alaska—Social life and Customs. I. McCartney, Allen Papen, 1940- II. Canadian Circumpolar Institute. III. Series: Studies in Whaling No. 6. IV. Series: Occasional publication (Canadian Circumpolar Institute); no. 54.

E99.E7145 2003   639.2'8'0916327   C2003-910950-X

Library of Congress Cataloguing in Publication Data

CONTROL NUMBER 2004104754

Cover photos:
Top: courtesy Anchorage Museum of History and Art (Ref. B97.19.23);
Bottom: Barbara Bodenhorn (this volume).

Cover Design by art design printing inc.
Printed in Edmonton by art design printing inc.

Partial funding for this publication was provided by the National Science Foundation (NSF) Office of Polar Programs.

Table of Contents

Editor's Note iii

Contributors iv

Introduction v
    Allen P. McCartney and Roger K. Harritt

Environment and Environmental Change in the Western Arctic and
Subarctic: Implications for Whaling 1
    John C. Dixon

Re-Examining Wales' Role in Bering Strait Prehistory: Some
Preliminary Results of Recent Work 25
    Roger K. Harritt

A Paleo-Geographic Preface to the Origins of Whaling: Cold is Better
....69
    Owen K. Mason and Valerie Barber

Ekven—A Prehistoric Whale Hunters' Settlement on the Asian Shore
of Bering Strait 109
    Yvon Csonka

Secular Dynamics of the Prehistoric Catch and Population Size of
Baleen Whales off the Chukchi Peninsula, Siberia: As Based upon the
Study of Historical Whale Bone from Ancient Coastal Sites
.... 137
    Lev G. Dinesman and Arkady B. Savinetsky

Prehistoric Bowhead Whaling in the Bering Strait and Chukchi Sea
Regions of Alaska: A Zooarchaeological Assessment 167
    James M. Savelle and Allen P. McCartney

Ancient Whaling and the Biogeography of Bowhead and Gray
Whales 185
    Howard W. Braham

The Bowhead Whale Off Chukotka: Integration of Scientific and
Traditional Knowledge...209
    Lyudmila S. Bogoslovskaya

Some Observations on the Influence of Environmental Conditions on
the Success of Hunting Bowhead Whales Off Barrow, Alaska
   255
    John C. George, Stephen Braund, Harry Brower, Jr.,
    Craig Nicolson and Todd M. O'Hara

Fall Whaling in Barrow, Alaska: A Consideration of Strategic
Decision–Making..277
    Barbara Bodenhorn

When Whaling Folks Celebrate: A Comparison of Tradition and
Experience in Two Bering Sea Whaling Communities 307
    Carol Zane Jolles

Festival and Tradition: The Whaling Festival at Point Hope.........341
    Mary A. Larson

'*Story of a Whale Hunt*': Suzanne Rognon Bernardi's Photographs and
Observations of Iñupiaq Whaling, Wales, Alaska, 1901–1902
   357
    Susan W. Fair

Eskimo Laborers: John Kelly's Commercial Shore Whaling Station,
Point Belcher, Alaska, 1891–1892 387
    Mark S. Cassell

Whaling: Indigenous Ways to the Present 427
    Herbert O. Anungazuk

Editor's Note

The majority of the papers included here are the direct outgrowth of research conducted under parallel National Science Foundation (NSF) grants (OPP-9807051, OPP-9806516) entitled *Native Whaling in the Western Arctic: A Regional Integration* made through the University of Arkansas and the Environment and Natural Resources Institute (ENRI) of the University of Alaska Anchorage between 1998 and 2001. Other papers were contributed by ethnographers and archaeologists who have carried out whale research in Alaska for many years. We are fortunate to include three papers that represent Russian research along the Western margin of Bering Strait and the Bering and Chukchi seas. Some of these research results were outlined in a whaling workshop, also sponsored by NSF, in March 2000, held at the Anchorage Museum of Art and History. The workshop was entitled *Native Whaling in the Western Arctic: Development, Spread, and Responses to a Changing Environment*, and those presentations are summarized in a booklet assembled by Roger Harritt, *In Pursuit of Agviq* (available through ENRI, UAA).

Representative of many different bodies of knowledge, academic disciplines, and cultural locales of the Western Arctic, this volume gives an idea of the diversity of interest surrounding the practice of traditional or subsistence whaling in this region, from ancient times to the present.

I wish to acknowledge the contributions of all of the authors, the dedicated efforts of the project participants, the many anonymous reviewers of these papers, and tribal councils, community leaders and elders, whaling captains' and wives' associations, and local consultants. Specifically, we thank ReBecca Hunt (editorial assistant), Sarah Whalen Kraft (technical editor), and Elaine Maloney and her staff at CCI for their respective contributions. John Dixon, James Savelle, and Roger Harritt served with me as the Principal Investigators for the simultaneously awarded NSF grants, and they have all contributed significantly in assisting me as well as managing the grants. I wish to thank them, on behalf of all project participants, for their enduring interest and multiple contributions. Further, all of us are indebted to the kind cooperation expressed by the Alaskan Eskimo Whaling Commission (AEWC), its representatives, and especially Maggie Ahmaogak, Executive Director, and to the Office of Polar Programs, NSF, which underwrote most of the research in this publication. Without the keen interest and insights of all these persons, we could not have engaged in this venture of learning more about Native whaling, past and present.

Allen P. McCartney

Contributors

Anungazuk, Herbert O.—National Park Service, *Anchorage, AK 99503*

Barber, Valerie—Institute of Marine Science, University of Alaska Fairbanks, Fairbanks, AK 99775

Bodenhorn, Barbara—Pembroke College, Cambridge, UK BB2 1RF and Ilisagvik College, Barrow, AK 99723 USA

Bogoslovskaya, Lyudmila S.—Institute of Cultural and National Heritage, 2 Kosmonavtov Street, Moscow, Russia 129366

Bower, Harry Jr.—Department of Wildlife Management, North Slope Borough, Box 69, Barrow, AK 99723

Braham, Howard W.—Department of Biology, Oregon State University, Ashland, OR 97520

Braund, Stephen—Stephen R. Braund and Associates, Anchorage, AK

Cassell, Mark S.—2510 Oakland Rd., *Minnetonka, Minnesota 55305*

Csonka, Yvon—Ilisimatusarfik, The University of Greenland, P.O. Box 279, DK-3900, Nuuk, Greenland

Dinesman, Lev G.—Severtsov Institute of Ecology and Evolution, Russian Academy of Sciences, 33 Leninsky Prospect, Moscow, Russia 117071

Dixon, John C.—Department of Geosciences, University of Arkansas, Fayetteville AR 72701

Fair, Susan W. —at time of submission: The Southwest Center, University of Arizona, Tucson, Arizona 85721-0185

George, John C.—Department of Wildlife Management, North Slope Borough, Box 69, Barrow, AK 99723

Harritt, R.K.—Environment and Natural Resources Institute (ENRI), University of Alaska Anchorage, 707 A Street, Anchorage, AK 99501

Jolles, Carol Zane—University of Washington, Seattle, Washington 98195-3100

Larson, Mary—University of Nevada Oral History Program, Mail Stop 324, Reno, NV

McCartney, Allen P.—Department of Anthropology, University of Arkanas, Fayetteville, Arkansas 72701

Mason, Owen K.—GeoArch Alaska, P.O. Box 91554, Anchorage AK 99509

Nicolson, Craig—University of Massachusetts, Amherst, MA

O'Hara, Todd M.—Department of Wildlife Management, North Slope Borough, Box 69, Barrow, AK 99723

Savelle, James M. —Department of Anthropology, McGill University, 855 Sherbrooke St West., Montreal PQ H3A 2T7 Canada

Savinetsky, Arkady B.—Severtsov Institute of Ecology and Evolution, Russian Academy of Sciences, 33 Leninsky Prospect, Moscow, Russia 117071

Introduction

Allen P. McCartney
Roger K. Harritt

This volume includes contributions from participants of a multidisciplinary Native whaling project carried out in the Western Arctic and others who have focused on similar subjects. This is the second CCI volume devoted to this topic, the first appeared in 1995 (*Hunting the Largest Animals: Native Whaling in the Western Arctic and Subarctic*).

The current papers stress three themes: different variations of whaling practices, history of Yupik and Iñupiat whaling traditions over time, and interactions with changing environmental conditions. While these are, indeed, common elements to traditional whaling found among the ten Alaskan Eskimo communities that engage in modern subsistence whale hunting, each community is at the same time unique in its traditions of interconnected practices that make up whaling and whale use. These variations on the whaling theme are a clear reflection of the uniqueness of local ecologies and cultural patterns for each community. Western Arctic Native whaling was not and is not practiced in the same way by all participants. While most whaling targets bowhead whales grays, minkes, and other species were also taken. Every community had local ceremonies surrounding whaling, but these specifics differed from place to place. While butchering had to be carried out according to prescribe methods to harvest the meat and *muktuk*, different communities had their own plan for dividing the shares of the products. By calling attention to local flexible adaptations, these papers emphasize the commonness while at the same time stressing how societies lived in real time and space.

In tracing whaling traditions of the region back through time, we find that there are important examples of diversity in the conditions and materials of different prehistoric and historic whaling cultures of the past 2000 years. It is clear that changes in climate and other types of environmental change have always had a crucial influence on the development and practice of Native whaling. Thus, some of the papers in this volume amplify and clarify these differences and changes and thereby demonstrate that there are many and varied strands that compose the fabric of Eskimo whaling.

Secondly, some papers directly or indirectly compare the expressions of whaling over time, stressing how the past influences the present. The fact that modern whalers use Global Positioning Systems (GPS) units and marine-band radios and that they acknowledge a great resurgence in subsistence whaling in the early 1970s might give the reader an impression that modern whaling is far removed from subsistence whaling of the 19th century or before. Each generation of Yupik and Iñupiat learns whaling from its elders and, therefore, skills, knowledge, and materials span time to shape the present. Knowledge of bowhead behavior and the progression of seasons over the course of the year remain important subjects for young hunters and they give continuity and meaning to whaling communities and their traditions. Suitable and effective technology such as light colored parkas and skin covered frame boats are still used.

Finally, environmental conditions have always influenced Eskimo whaling. Ice thickness, break-up times, winds, currents, and local atmospheric conditions all impact the process of whaling. In modern times, the environmental dynamics include global warming and its representations in local whaling areas. Major climatic episodes have certainly influenced whaling over its 2,000+–year history in northwestern Alaska, as have annual or decadal perturbations. Cultural adjustments were required for these variations, whether measured annually or for longer periods.

In large measure, therefore, changes in the health and behaviors of the great whales correspond directly with the condition of their ocean environment. And, the traditional pursuit of whales by Eskimo hunters remains an area in which humans articulate directly with natural processes and knowledge of animals and the elements. To dwellers of large, modern cities, such direct relations between modern people, wild animals, and their environment may be exotic prospects, but are nevertheless important pursuits to many modern Iñupiat and Yupik peoples.

The ability to adapt to changes in natural processes and the animals is in itself a quality of a successful culture, and the importance of this relationship between a human enterprise and the natural world in modern times cannot be overstated. The continuity of the relationship itself, from prehistoric times to the present, sheds light on the types of difficulties modern Western societies face in times of global warming and changes precipitated by modern industries. Examining this ancient relationship will undoubtedly illuminate ways in which human societies may successfully adapt to changing environments.

Environment and Environmental Change in the Western Arctic and Subarctic: Implications for Whaling

John C. Dixon

Abstract. Patterns of Alaskan native whaling over the past two millennia have been strongly influenced by both the geography of the natural environment and by changing environmental conditions. The principal environmental parameters influencing both patterns of movement of whales as well as access of Native hunters to whales include the nature and distribution of ocean currents, sea surface temperature, storm and sea level histories, climate processes, and sea ice thickness and concentration. Multiscalar climate forcings, including position of the Aleutian Low, El Niño/Southern oscillation patterns, and the Pacific Decadal Oscillation, exert profound influences on regional temperature, ocean storminess, sea ice thickness and concentration, and sea surface temperature. These environmental parameters strongly influence whale availability, patterns and timing of whale movement, accessibility to migrating whales, and whaling success. Variations in the magnitude of environmental parameters occur at multiple scales in response to multiscalar forcings. In examining spatial and temporal variability in environmental parameters and whaling success, it is important to realize that inter-annual and inter-decadal variability is likely being driven by different forcing mechanisms.

INTRODUCTION

The spatial and temporal movement patterns of whales in northwestern Alaska, as well as hunters' access to them, are greatly influenced by a variety of environmental factors. Whale migration from the southern ice front to the Arctic Ocean in the spring is strongly affected by patterns of food availability (Jacoby 2001), which in turn are influenced by the major ocean currents of the Bering Sea and Bering Strait (Schell *et al.* 1998). In addition, both sea ice extent and thickness have a profound influence on the ability of whales to migrate north (Moore and Reeves 1993). Spring migration patterns are also affected by the distribution of leads and polynyas in the sea ice, which vary depending on ice thickness and concentration as well as patterns of storminess (Stringer and Grove 1991).

Access of hunters to whales is influenced both by the distribution of open waters as well as patterns of sea ice distribution and condition, which are influenced by prevailing atmospheric and sea surface temperature and atmospheric circulation patterns at various scales. In addition, patterns of storminess directly impact hunters' ability to pursue whales in open waters.

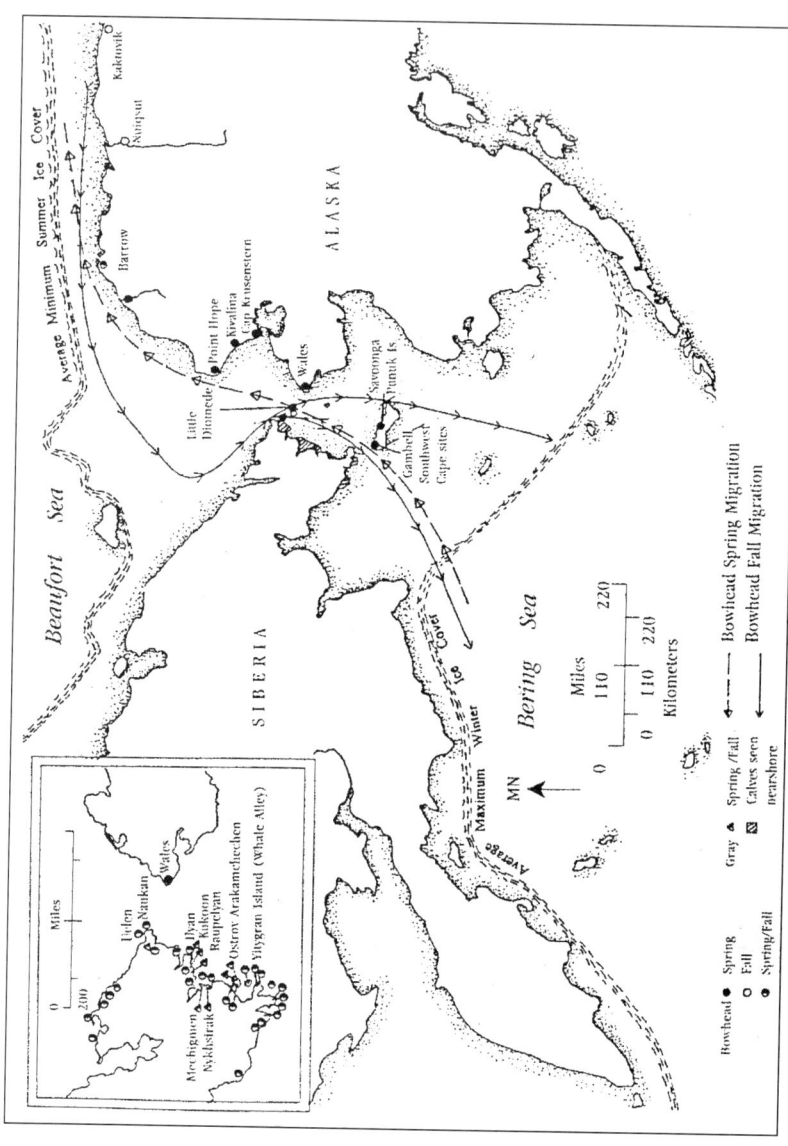

Figure 1. *Location of villages in study area, average position of sea ice front in summer and winter, and migration paths of whales (after Harritt 1995).*

Throughout the two millennia-long history of Native whaling in northwest Alaska, climate change and variability has profoundly influenced patterns and productivity of Native whaling. This variability has displayed both temporal and spatial dimensions, with different parts of the Alaskan Arctic and Subarctic responding differently to climate–forcing mechanisms. Terrestrial responses have also varied depending on the nature of the physical environment, including such factors as oceanographic and coastal configuration, and this has had human consequences.

The purpose of this chapter is twofold. First, it discusses the nature of the physical environment of the western Alaskan Arctic and Subarctic in which the villages that are the focus of the chapters here are located (Fig. 1). Second, it provides some insight into the spatial and temporal variability within those environmental parameters that most strongly influence whale–based subsistence hunting in the Western Arctic. Parameters examined include atmospheric temperature, sea ice extent, concentration, and thickness, storminess, and permafrost. The spatial coverage of this discussion is focused on the Beaufort, Chukchi, and Bering seas and adjacent coastal regions.

Temporal scale varies for each parameter under discussion depending on the length of relevant historic records or the length and resolution of the proxy measure. Millennial–long tree ring records of atmospheric temperature provide calendral scales of resolution. Sea surface temperatures extracted from tree ring records provide similar scales of resolution over a 250–year record. Sea ice records are again annual and are available for a century, but direct measurements of sea ice patterns span a much shorter period. Storm event and recurrence intervals are available only at the resolution of radiocarbon dating. Permafrost extent and temperature characteristics are available for only a very recent portion of the historic record.

LANDSCAPE SETTING

The whaling villages of northwestern Alaska encompass four of Alaska's 20 ecoregions (Gallant *et al.* 1995). The most northerly of these is the arctic coastal plain that extends across the northern portion of the state, from near the international border with Canada's Yukon Territory in the east to the Chukchi Sea in the west, covering an area of some 50,000 sq km. The region is bordered to the north and west by the Arctic Ocean and to the south by the arctic foothills (Wahrhaftig 1965; Gallant *et al.* 1995). The coastal plain is relatively smooth and rises gradually from sea level to approximately 180 m a.s.l. at the foothills of the Brooks Range, with an average slope of less than one degree. The coastal plain is mantled by a variety of Quaternary age deposits principally of fluvial and aeolian origin. The plain was not glaciated during the Pleistocene; consequently, except for glacially–derived sediments from the Brooks Range transported by north–flowing rivers, glacial debris is absent. Extensive areas of dunes occur between the Colville and Kuk rivers (Gallant *et al.* 1995).

The coastal plain is underlain by continuous permafrost, and over much of the region the permafrost table is at or close to the surface. A relatively thin, seasonal active layer develops that varies in thickness depending on a wide variety of influences, but regionally follows air temperature trends (Nelson and Hinkel 2002). The influence of continuous permafrost dominates the landscape, with extensive development of ice wedge polygons giving the landscape a distinctively patterned appearance.

The region also supports extensive occurrence of pingos (ice–cored hills) up to 100 m in height, as well as a variety of other smaller frost–mound features (Nelson *et al.* 1992). In the summer, vast areas of the permafrost thaw, leading to a saturated active layer with extensive ponded areas on the landscape surface. This melting of near–surface permafrost leads to the development of huge areas of oriented, elliptical thaw lakes (Nelson and Hinkel 2002). It has been estimated that as much as 50–70% of the landscape of the North Slope is occupied by these lakes or by mires developed in former lake basins (Hussey and Michelson 1966). In the winter the lakes and rivers freeze, and coupled with the limited snowfall, the landscape takes on the appearance of a continuous blanket of white.

To the south of the coastal plain and extending to the Brooks Range are the arctic foothills, which consist of linear rolling hills and plateaus covering an area of some 124,000 sq km (Wahrhaftig 1965; Gallant *et al.* 1995). The region is divided into two distinct subdivisions. The northern region, with elevations rising to 600 m, consists predominantly of broad, rounded ridges of unconsolidated Quaternary alluvial, glacial, and aeolian debris. The southern section consists of higher elevation buttes, mesas, and ridges with intervening, undulating plains and plateaus (Gallant *et al.* 1995). The area is underlain by continuous permafrost with associated patterned ground; however, thaw lakes are substantially fewer than on the coastal plain. Drainage in the foothills is more strongly integrated than farther north. Like the coastal plain, this region was not glaciated during the Pleistocene.

Forming the eastern side of Bering Strait and the eastern core of the Pleistocene subcontinent of Beringia, Seward Peninsula is a tundra–dominated upland consisting of broad convex hills, flat divides, and steep V–shaped valleys (Wahrhaftig 1965). The margins of the peninsula are occupied by narrow strips of coastal lowland (Gallant *et al.* 1995). Over much of the peninsula, elevations rise to 500 m. The generally flat plateau surface is broken by the presence of isolated groups of rugged glaciated mountains that rise to as much as 1400 m. The peninsula is predominantly underlain by metamorphic rocks that have been intruded by granitic masses. Ancient metamorphosed volcanic rocks occur widely as do more recent volcanics in the northeastern part of the region.

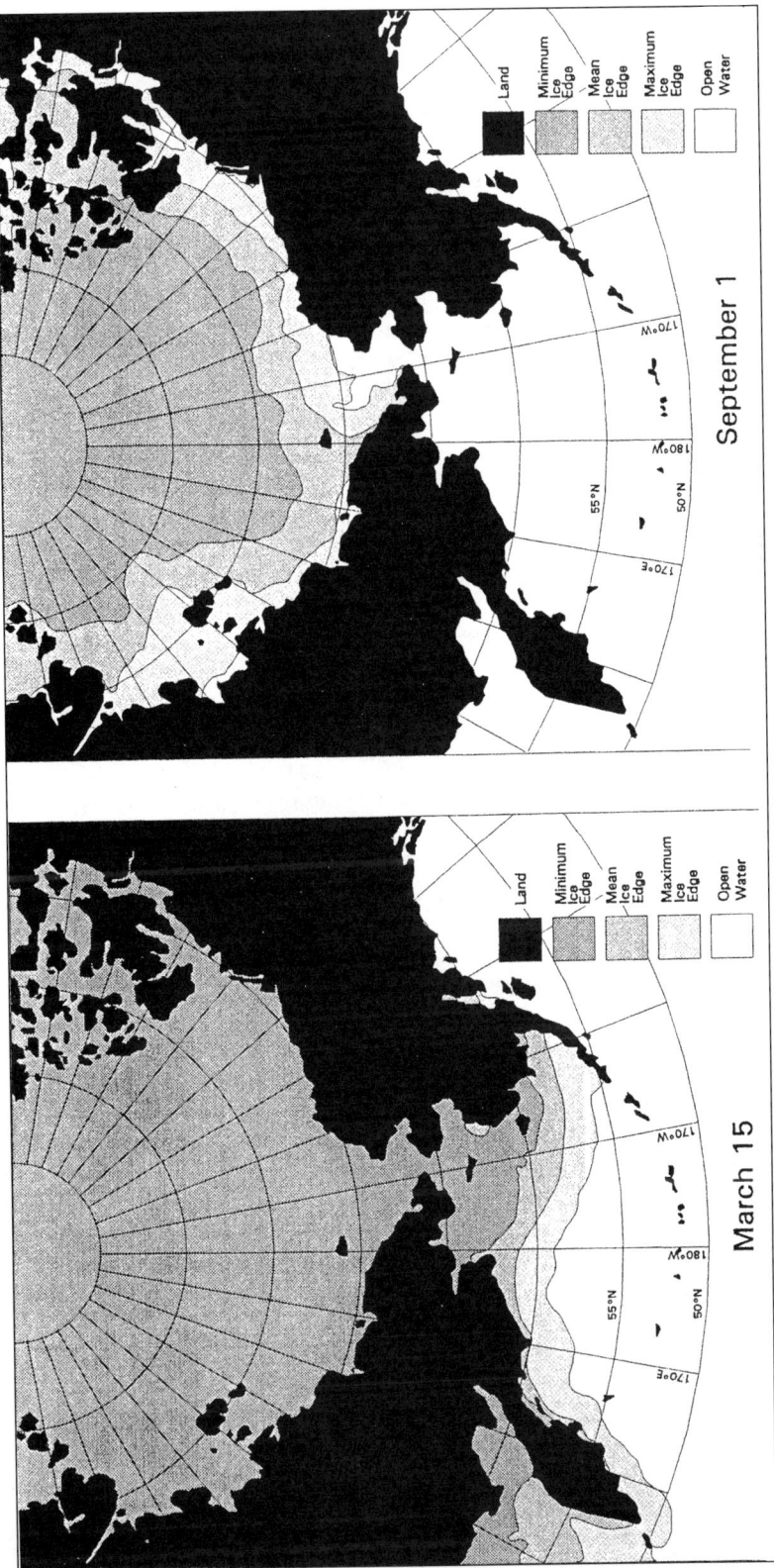

Figure 2. Mean positions of sea ice front in March and September (after Niebauer and Schell 1993).

Continuous permafrost underlies the landscape, and this is reflected in the widespread occurrence of ice wedge polygons, but thaw lakes are largely restricted to the coastal lowlands (Wahrhaftig 1965). Hillslope process outside the mountain massifs are dominated by periglacial gelifluction processes.

Subarctic coastal plains cover approximately 91,000 sq km of the Kotzebue Sound, the Yukon–Kuskokwim River Delta, St. Lawrence Island, and Nunivak Island areas of Alaska (Gallant *et al.* 1995). Typically, these are areas of low relief dominated by coastal plains and river deltas. However, in some areas low basaltic hills with associated cinder cones and shallow volcanic craters occur that rise to as much as 120 m in elevation. The subarctic coastal plains are typically covered by older deposits of interstratified deposits of alluvium and marine deposits. In the Kotzebue Sound area, Pleistocene glaciation occurred and here glacial debris is found. Thin to moderately thick permafrost underlies the region, and thaw lakes are numerous. Pingos also occur in the region, especially around the Salawik River area (Gallant *et al.* 1995). The Yukon–Kuskokwim lowlands represent the emerged portion of the Bering Shelf, which joins Alaska to Siberia. The submerged section of this shallow submarine plain is the Bering Platform (Wahrhaftig 1965).

THE COASTAL ENVIRONMENT

While much of the arctic and subarctic coasts of eastern North America experienced repeated depression and rebound associated with the direct influence of the weight of Quaternary ice sheets, this was not the case for the Western Arctic. Here, beyond the margin of the Wisconsin ice sheet along the coast of the Beaufort Sea, tectonic subsidence, sediment loading, forebulge collapse, and the Flandrian transgression have all combined to favor postglacial submergence (Orme 2002). The coastal environments of the Arctic Ocean and Bering Sea are cold, and as we have seen previously, periglacial landform features extend to the shoreline. Seasonal sea ice is present for seven to nine months of the year, and at the highest latitudes perennial sea ice lies just offshore (Fig. 2). In the Arctic Ocean, cold surface waters descend to 100 m or more before sinking to generate outflowing deep water, but in the Bering Strait the shelf is relatively shallow and this limits exchange of water to the surface. As a result, the low temperature of the Arctic Ocean waters in the Chukchi and Beaufort seas is reinforced (Orme 2002).

The principal ocean currents entering Bering Strait from the south are the Anadyr Current, the Bering Slope Current, and the Alaskan Coastal Current (Coachman *et al.* 1975; Ackerman 1988; Schell *et al.* 1998; Fig. 3). The Anadyr Current follows the Anadyr Basin between the Chukchi Peninsula and St. Lawrence Island. It is an extension of the Bering Slope Current and is high in nutrient content due to deep water mixing (Schell *et al.* 1998), which also imparts low temperature to the current. The Bering Slope Current originates south of St. Lawrence Island and flows north, diverging around the island (Coachman *et al.* 1975; Ackerman 1988). The current is cold and nutrient–rich (Schell *et al.* 1998). Nutrients in these two currents are concentrated at depth until the waters reach the shallower Bering Shelf, where they are able to support abundant phytoplankton growth (Schell *et al.* 1998). Alaskan Coastal Current waters originate in the southeastern Bering Sea and flow north. This current is characteristically of lower salinity due to inputs from large rivers. The northward flow of these three currents is the result of higher sea surface elevations of the Bering Sea compared to the Chukchi Sea (Coachman *et al.* 1975); the flow is additionally enhanced by strong southerly winds that pile up water south of St. Lawrence Island (Ackerman 1988). Flow in the Chukchi Sea is primarily north along the eastern shore, flowing north and east into Kotzebue Sound, then north and west toward Point Hope. The current then splits, with one arm flowing east toward Point Barrow and the other flowing west.

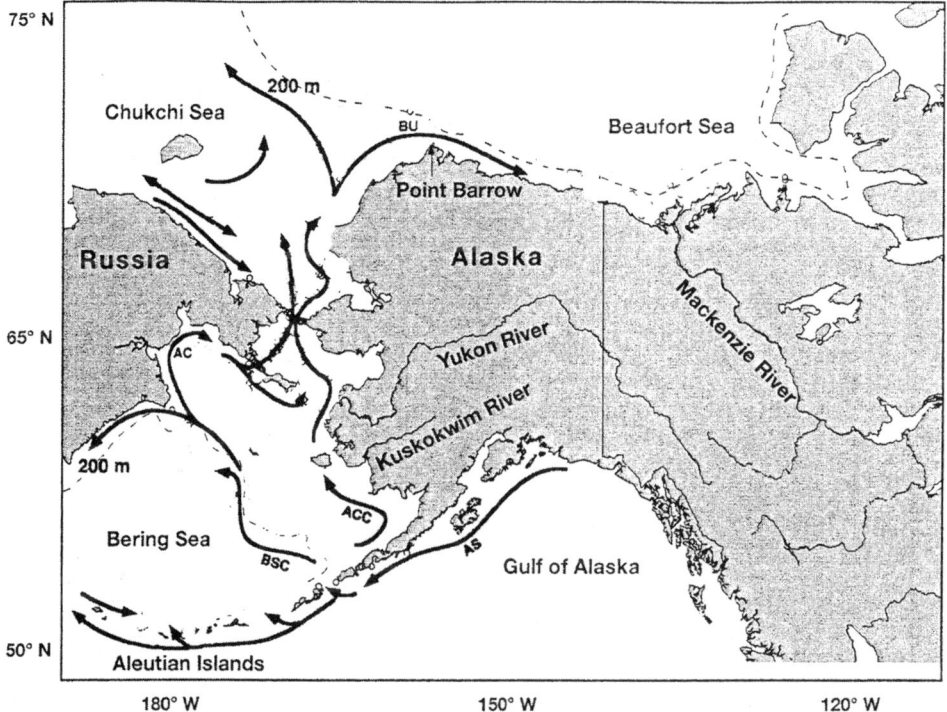

Figure 3. *Dominant ocean currents of Bering Sea region. AS: Alaska Stream; BSC: Bering Slope Current; ACC: Alaska Coastal Current; AC: Anadyr Current; BU: Beaufort Undercurrent (after Schell* et al. *1998).*

Wave action is typically muted by the presence of coastal ice for extended periods of the year but may be significant in late summer and early fall when ice moves offshore. Microtidal conditions also limit the impact of ice, but where tidal ranges increase, boulders may be moved by ice floes. The presence of permafrost and seasonal ground ice results in coastal forms unique to the arctic environment, such as thermoerosional niches and thaw lakes that form arcuate shorelines when breached (Orme 2002).

The lowland coasts of the Beaufort, Chukchi, and Bering seas differ from the shorelines of the Eastern Arctic with respect to the occurrence of coastal dune systems. In the Western Arctic the low coastal gradients favor the formation of cuspate forelands and barrier beaches. Several large, sediment-laden rivers flowing from the Brooks Range have extensive deltas associated with them in a low energy microtidal environment. In the Beaufort Sea, major deltas are associated with the Colville and Sagavanirktok rivers, and in the Bering Sea the Yukon River has at varying times through the Quaternary delivered vast amounts of sediment to the coast along a 600 km stretch from Norton Sound to Kuskokwim Bay (Orme 2002).

Where sediment is available, vast beach ridge-dominated plains and barrier islands have formed. This is particularly the case along the northwestern coast of Seward Peninsula where barrier islands and beach ridge plains extend northeastward from Cape Prince of Wales to Cape Espenberg. In addition, similar sandy complexes are to be found all the way along the coast from Cape Krusenstern to Point Barrow (Mason and Jordan 1993; Jordan and Mason 1999; Mason *et al.* 1995). These complexes derive most of their sediment from coastal cliffs and offshore on the inner shelf (Mason *et al.* 1995; Mason and Jordan 1994). Additionally, however, some sediments are also derived from reworking of marine and lacustrine sediments of the coastal plain (Lea and Waythomas 1990).

Rocky coastlines develop where mountain systems reach the coast. This is particularly the case on the southern coast of the Seward Peninsula, the northwestern part of St. Lawrence Island, where the Brooks Range reaches the Chukchi Sea, and where the Kuskokwim Mountains meet the Bering Sea (Walker 1985).

Regional Climate

At the regional scale, the climate of Alaska is controlled by polar and arctic air masses. In general, these air masses are characterized by low average annual temperatures and small amounts of precipitation, most of which is received in summer. These regions are characterized by strong seasonal fluctuations in solar energy and temperature, resulting in long, severe winters and short, mild summers. Temperature ranges characteristically exhibit greater annual than diurnal ranges. Annual solar radiation intensity is characteristically low (Bailey 2002). The study area is dominated by two climate divisions: the Tundra and the

Subarctic. The Tundra division extends northward from the Arctic Circle to approximately 75ºN and lies within the outer zone of the influence of arctic air masses. This climate is characterized by very short, cool summers and long, severe winters. Average annual temperature of the warmest month is between 0 and 10ºC. Mean daily temperatures rise above 0ºC between 55 and 188 days (Bailey 2002). Annual precipitation is low, although because of extremely low evaporation rates, this climate is regarded as humid. It is dominated by high pressure year round, which accounts for the prevailing low temperatures and lack of moisture, and experiences negligible invasion of cyclonic storms. The whaling villages found in this climate region include Kaktovik, Nuiqsut, Barrow, and Wainwright. The climate data in Table 1 illustrate the relatively low average maximum and minimum annual temperatures and the low overall precipitation and snow received in this climate.

Table 1. Climate of whaling villages showing contrast between arctic (Barter Island, Barrow, Wainwright) and subarctic (Kotzebue, Wales, and Gambell) locations.

	Barter Is.	Barrow	Wainwright	Kotzebue	Wales	Gambell
Average Annual Precipitation (mm)	157.2	116.8	158.8	241.3	291.6	446.0
Average Annual Snowfall (mm)	1061.7	756.9	N/A	1264.9	967.7	1790.7
Average Minimum Temp.ºC	-15.5	-15.3	-15.3	-9.3	-9.1	-6.1
Average Maximum Temp. ºC	-9.2	-9.3	-8.3	-2.3	-3.0	-1.6

Source: *Alaska Climate Research Center (http://climate.gi.alaska.edu).*

The vast area of Alaska is predominantly a subarctic climate, which is dominated by continental polar air masses that form between 50–70ºN.

This climate exhibits a large seasonal temperature range. Winters are severe, and summers are cool and short with only one month of the year having a mean temperature above 10°C. As high pressure systems dominate the winter season, the relatively small amount of annual precipitation is concentrated in the three summer months (Oliver 2002). The transition between summer and winter is abrupt, with an extremely short spring and fall; these are the traditional times of whale hunting, depending on specific location. While the two dominant climate types of Tundra and Subarctic are dominated by arctic and continental polar air masses respectively, in the Bering Sea region there is a strong maritime influence. Here, maritime polar air masses originating in the North Pacific dominate the climate of the region. While these air masses are very similar in their characteristics to continental polar air masses, they differ with respect to more moderate temperature and moisture characteristics. The influence of these air masses is well illustrated in the climate contrasts between Barrow, Barter Island, and Wainwright on the one hand and Gambell, Kotzebue, and Wales on the other. The latter three stations all show substantially warmer average temperatures and greater amounts of average annual precipitation and snow depths (Table 1).

The maritime polar air mass is dominated by low pressure, and this persistent low pressure cell is referred to as the Aleutian Low. This low pressure system is the dominant control on much of Alaska's weather in the winter and spring (Wilson and Overland 1987). Inter-annual variability in Alaskan climate, particularly as it relates to annual temperatures and storm tracks, is strongly tied to variability in the path of movement of the Aleutian Low (Miller *et al.* 1994). Changes in the pathways and intensity of the Aleutian Low, for example, were responsible for a marked warming in temperature in southern Alaska between 1976 and 1988 (Trenberth 1990). These marked changes have been related to the occurrence of ENSO (El Niño/Southern Oscillation) events in the tropical oceans (Niebauer 1998; Mysak 1986). More recently, however, extratropical forcings have been proposed as the dominant influence on the strength and position of the Aleutian Low (Latif and Barnett 1994). These forcings are related to a pattern of Pacific climate variability referred to as the Pacific Decadal Oscillation, which is more strongly tied to subtropical gyre circulation than to the shorter frequency tropical ENSO forcings (Wiles 1997).

ATMOSPHERIC TEMPERATURE TRENDS

Long-term, instrument-based climate records from the Western Arctic are generally lacking, so researchers interested in climate change over centuries are forced to rely on other proxy climate (especially temperature) indicators. One of the most widely used of these is tree rings, as they represent a proxy with annual resolution, a feature essentially unmet by any other proxy indicators except perhaps varves when available. A surprisingly large and growing number of arctic and subarctic Alaskan tree-ring chronologies are now in existence. Based on eleven tree-ring chronologies from across northern North America, Jacoby and D'Arrigo (1989) and D'Arrigo and Jacoby (1993) developed an arctic temperature reconstruction for the period 1671–1973 (Fig. 4) in which they demonstrated that temperature had warmed in the early 1700s following a period of markedly cooler temperatures in the 17th century. This warmer period continued until the early 1800s, when once again a period of marked cooling is recorded through the 1840s. After that a long and sustained warming period began. This pattern of summer temperatures from tree rings is corroborated by the melt history of the Devon Island Icecap (Jacoby and D'Arrigo 1989; Koerner 1977; Patterson *et al.* 1977). A more recent study by Jacoby *et al.* (1996) focuses on Alaskan climate reconstructions for both arctic and subarctic locations using substantially more tree ring sites. This more regionally focused study also corroborates the earlier studies' broad patterns but highlights intervals of marked warming during the Little Ice Age in the 1720s and 1770s. Further, it demonstrates periods of marked cooling during the overall warming of the last 160 years, especially from 1950 to the early 1970s. Reconstructed temperatures subsequently return to pre-1950 levels. The climatic warming so widely identified in the instrumented record since the late 1970s is not reflected in the tree–ring reconstruction, and this may be due to moisture stress, especially in interior sites (Jacoby *et al.* 1996).

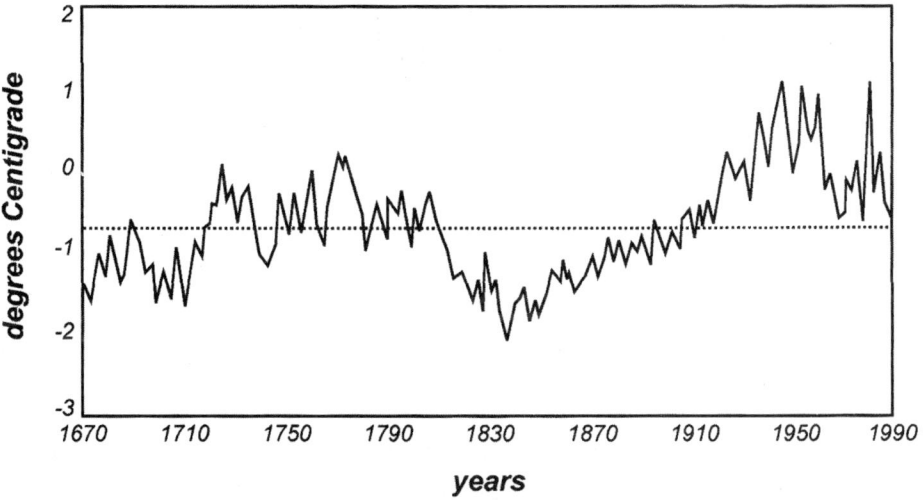

Figure 4. *Reconstructed Arctic temperature departures from tree rings, 1671–1973 (after Jacoby and D'Arrigo 1989).*

Jacoby and D'Arrigo (1995) attribute the temperature trends observed in their three–century reconstruction to the influence of three principal forcing mechanisms: volcanic activity, changes in insolation, and atmospheric CO_2 levels. In addition, they acknowledge that changes in ocean circulation associated with ENSO events, as well as intrinsic variability in the climate system, may also be important in explaining observed climate variability in recent centuries. However, they find no overwhelming support for a significant influence of CO_2 in the pre-1980 data sets.

A recent temperature reconstruction from southern Alaska for the past millennium (Fig. 5), based on tree-ring analyses of trees exposed from beneath retreating glaciers shows two major periods of ice expansion (cooling) between A.D. 1250 and 1300 and A.D. 1650 and 1750 (Mann *et al.* 1999; Wiles *et al.* 1999). Beginning just prior to this latter cooling episode, decade-long changes in the climate system become apparent, especially with respect to the occurrence of cold periods. The warmest decades recorded in this tree-ring chronology appear in the 20th century. These climate variations are attributed to decadal oscillations in the North Pacific climate system.

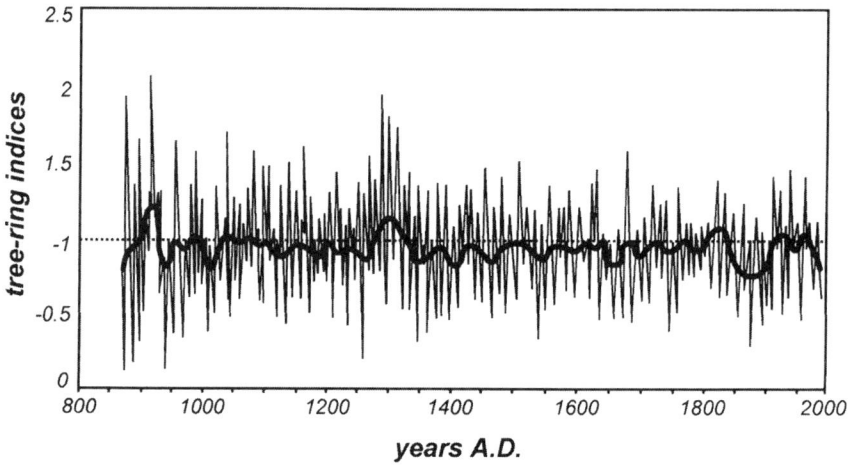

Figure 5. *Reconstructed southern Alaska millennial temperature record from tree rings, 873–1991 (after Barclay et al. 1999).*

While tree-ring chronologies provide the best annual, decadal, century, and millennial proxy measures of climate variability, several studies of arctic temperature trends, based on instrumental records over decades have been undertaken. These studies generally confirm tree- ring reconstructions for the late 20th century. Stone (1997) reports a winter and spring warming trend for the period of 1965–1995 from the

Barrow instrumented record, and Morison *et al.* (2000) note that for the very latest part of their record (since 1989), there has also been a fall season warming. Atmospheric temperature clearly has direct impact on other environmental parameters such as sea surface and deep-water temperatures, sea ice extent and thickness, open–water distribution patterns, and food availability patterns.

SEA SURFACE TEMPERATURES (SST)

Sea surface temperature (SST) trends, which have a significant influence on patterns of sea ice distribution, thickness, and concentration, have recently been reconstructed for the North Pacific over the period 1750–1983 (D'Arrigo *et al.* 1999). These reconstructions are from tree-ring chronologies in southern and central Alaska, and closely parallel the temperature reconstructions discussed previously. Tree-ring derived sea surface temperature reconstructions show cooling in the early and mid-1800s and an overall warming through the 20th century. Within the 20th century warming trend, however, marked periods of cooling are identified in the 1960s and 1970s. The changes in SST are attributed to variability in the Pacific Decadal Oscillation, with which SST reconstructions are positively correlated.

STORMINESS

Storm and sea level history appears to have had a profound influence on the development of whaling in the Western Arctic (Mason, this volume). Work conducted over the past decade has established a complex history of beach ridge and dune building, beginning about 4,000 yrs B.P., interspersed with marked periods of storminess and beach ridge degradation (Mason and Jordan 1993). While detailed histories appear to vary from one ridge complex to another, there is an amazing amount of synchroneity between north and south Chukchi Sea sites (including Bering Strait) and for western Beaufort Sea sites (Mason and Jordan 1993; Mason *et al.* 1995). Two major periods of beach ridge truncation (erosion) are widely recognized. Separated by an intervening period of beach ridge complex progradation, the former occurring between 3300–1700 and 1200–900 yrs B.P. and the latter falling between these two periods (Mason and Jordan 1993; Mason *et al.* 1995; Jordan and Mason 1999). In addition, century-long periods of increased storminess are recorded over the last millennium. Within these periods of enhanced storminess, Mason (this volume) recognizes five widespread periods of storm erosion throughout the region. These periods occur at 1600–1200 B.C., 800–200 B.C., A.D. 750–950, A.D.1030–1200, and A.D.1450–1800. Mason and Jordan (1993) link these periods of enhanced storminess to northerly shifts in North Pacific storm tracks associated with changing positions of the Aleutian Low. Mason (this volume) links these periods of enhanced storminess to the expansion of whaling at several locations along the northwest coast of Alaska.

A recent detailed analysis by Mason *et al.* (1996) of approximately the last century of storm activity in the Bering Sea, using newspaper reports, reveals four periods of enhanced storminess and two periods of marked quiescence. Recurrence intervals appeared to fall into 3–,5–, to 7–, and 11–year periodicities. Enhanced storminess occurs at 1900–1913, 1936–1946, 1974–1976, and 1992, while quiet periods occur between 1916–1928 and 1947–1959. These periodicities are also correlated with climate anomalies such as ENSO and accompanying shifts in the position of the Aleutian Low (Mason *et al.* 1996).

Periods of enhanced storminess are also correlated with other climate proxies including alpine glacier expansions (Calkin 1988; Denton and Karlen 1977), widespread alluviation (Hamilton 1981; Mason and Beget 1991), ice wedge polygon development (Hamilton *et al.* 1983), and activation of Brooks Range slope deposits (Ellis and Calkin 1984). These other climate anomalies are commonly associated with cold events and increases in moisture availability (Mason and Jordan 1993). Storminess has the potential to limit access to the marine environment as well as to marine mammal resources, both in terms of the ability to launch boats and the influence of storms on the distribution of shore fast-ice. The presence/absence of shore-fast ice has a direct impact on whalers' ability to gain close access to whale populations. In any given season, this impact may be directly opposite for adjacent communities.

SEA ICE EXTENT AND VARIABILITY

Arctic Sea Ice Extent and Thickness
Arctic sea ice extent over the past 30 years has been widely reported to be diminishing (Vinnikov *et al.* 1999). This decrease has also been accompanied by a thinning of the ice.

The reductions in sea ice extent that have been so widely reported are regionally averaged determinations (Parkinson 2000a, 2000b), which apply to combined analyses from the Eastern, Western, and Central Arctic. Examination of individual regional trends (Parkinson 2000b) reveals that sea ice extent histories over the past 20 years vary markedly from one part of the arctic region to another. The research presented here concerns the Western Arctic, so discussion of sea ice behavior over the past 50 years will focus on two areas: the Bering Sea and Arctic Ocean.

One of the longest sea ice extent records for the Bering and Chukchi seas is that of Niebauer (1998), who examined sea ice extent for the period 1947–1996 (Fig. 6). This record displays considerable inter-annual variability over its length but most significantly displays a step decrease of some 5% in the late 1970s. While both periods are marked by considerable inter-annual variability in ice extent (Cavalieri *et al.* 1999; Parkinson 2000a) related locally to trends in the Aleutian

Low and regionally in El Niño patterns, the last 20 years of data show a weak, overall trend toward very slightly increasing ice extent (Parkinson 2000b). This trend, however, is not statistically significant. Whereas some parts of the Arctic as a whole also show statistically significant changes in sea ice extent patterns over the last decade, no such changes are noted for the Bering Sea. In the Arctic Ocean, the last 20 years reveal a slight, overall trend of diminishing ice extent (Johannessen *et al.* 1995; Bjorgo *et al.* 1995). Examination of the last 10 years reveals a slight increase in ice extent. However, this apparent reversal in trend is not statistically significant (Parkinson 2000b:Fig. 7). It is apparent that generalized statements about sea ice extent need to be made with reference to specific areas of the arctic and subarctic realms, as considerable temporal and spatial variability is to be found in the region. Clearly, this spatial variability in ice extent impacts whaling communities in different ways within the same season.

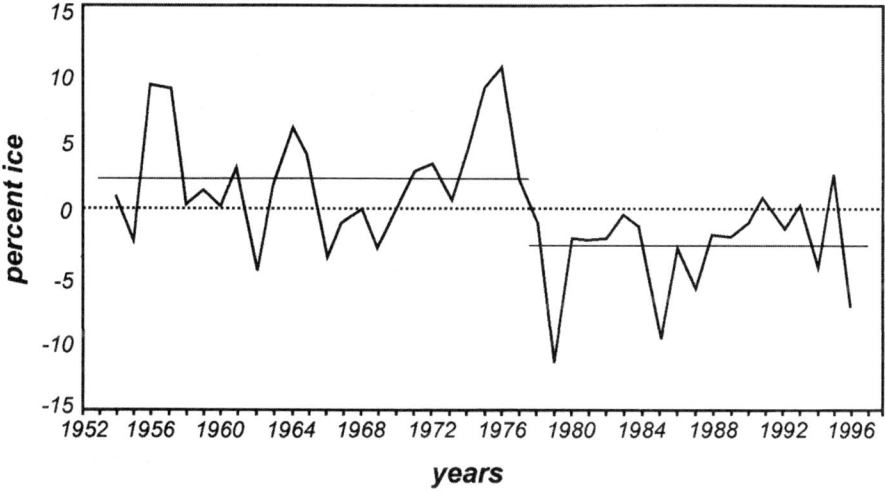

Figure 6. *Percent ice cover anomalies from Bering and Chukchi Seas 1950–1996 (after Niebauer 1998)*

The length of the sea ice season, defined by Parkinson (1992) as the number of days during the year when sea ice concentration exceeds 15%, also shows considerable variability both spatially and temporally. Over the past 20 years, for the Arctic as a whole, sea ice season length has decreased, with the number of arctic regions showing season shortening exceeding those with season lengthening, while in the Western Arctic the trend is considerably more variable. Parkinson (2000a) finds that in the Chukchi Sea there is a shortening of the sea ice season over the past two decades; however, in the Bering Sea and the eastern Beaufort Sea the sea ice season has lengthened. Again, length of ice season displays considerable temporal and spatial variability over a relatively small geographic area. This clearly has quite variable implications for whale hunting communities of northwestern Alaska.

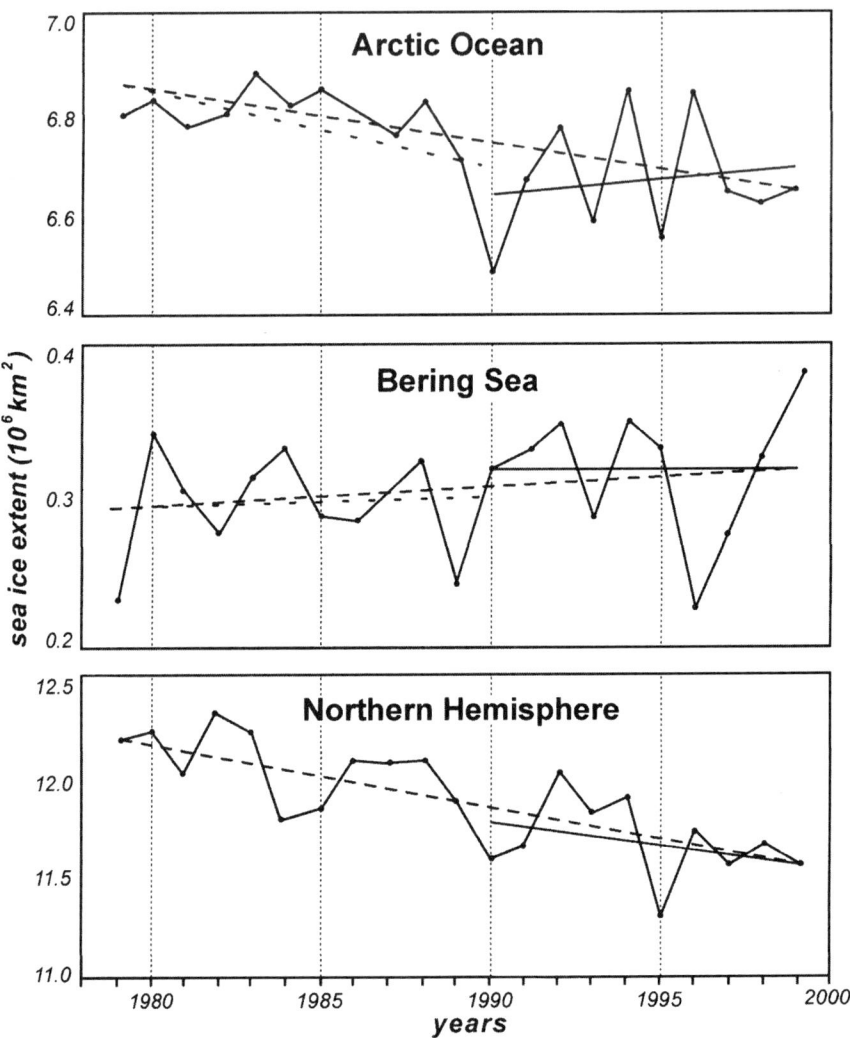

Figure 7. *Time series of yearly averaged sea ice extents, 1979-1999 (after Parkinson 2000b).*

Driving Mechanisms

Ice variability in the Bering Sea is strongly tied to atmospheric circulation. The ice forms in shore leads, primarily in Norton Sound and off the coast of St. Lawrence Island. Northerly winds then move the ice southward, where it melts in the warmer waters near the edge of the Bering Shelf. This pattern of ice formation, migration, and melt is driven by a constant stream of low pressure systems that track across the northern North Pacific Basin, producing the Aleutian Low. Day-to-day variability in the ice cover reflects changes in the atmospheric circulation caused by low pressure systems moving through the region, while the inter-annual variability depends on the preferred location of the storm tracks during the season. A high frequency of storms tracking northward along the western margin of the Bering Sea produces southerly winds that push the ice margin northward. Conversely, a storm track that moves low pressure systems across the southern edge of the Bering Sea, or a northward track along the eastern margins of the sea, produces northerly winds that result in extensive ice cover.

Inter-annual variability in the storm tracks is related to larger-scale features of the ocean-atmosphere system in the Pacific, most significantly the position of the Aleutian Low and hemispheric patterns of El Nino–Southern Oscillation (ENSO). Prior to 1978, there is a clear and regular relationship between alternating patterns of El Niño—La Niña events, circulation over the North Pacific, and the lagged responses of sea ice cover in the Bering Sea. In the late 1970s, the circulation switched to an almost continuous El Niño mode, and there has been a shift in the mean location of the Aleutian Low and reduction in Bering Sea ice extent (Niebauer 1998). This switch may be best explained by the influence of the Pacific Decadal Oscillation.

Changes in the extent, distribution, and thickness of Arctic Ocean sea ice have been largely attributed to increases in the Arctic Oscillation Index (Thompson and Wallace 1998; Morison *et al.* 2000). Increases in this index lead to decreases in the strength of the Beaufort High, which ultimately affects Arctic Ocean circulation patterns. Reduced convergence within the Beaufort Gyre leads to more open water, increased ocean heating, and therefore increased ice melting (Morison *et al.* 2000).

PERMAFROST

Permafrost temperatures, extent, and thickness are significant elements of the physical environment that have the potential to significantly influence patterns of whaling, because the permafrost serves as the essential refrigeration of meat stores. If permafrost temperatures rise, the longevity and quality of meat storage become compromised. Atmospheric warming has already been documented and evidence of increasing permafrost temperatures and thawing widely recorded across the Arctic, including extensive areas of Alaska. Lachenbruch and Marshall (1986) reported increased permafrost temperatures from Arctic Alaska of some 2–4°C over the past century. This trend continued to be observed over the past decade and a half (Clow *et al.* 1998). Osterkamp and Romanovsky (1996) working in the continuous permafrost zone of northern Alaska reported cooling of permafrost in the late 1980s, but through the 1990s have recorded extensive permafrost warming and associated thawing. They have also examined permafrost farther south in the discontinuous zone of central and southern Alaska and noted areas of warming and thawing (Osterkamp and Romanovsky 1999). Anisimov and Nelson (1996, 1997) predict century to millennial time scale reductions in permafrost extent in Alaska of 12 to 28%. Consequences of this warming include considerable thickening of the active layer, especially in the continuous permafrost zone, while in the discontinuous zone considerable retreat of permafrost would result in a trend already noted in many areas (Nelson and Hinkel 2002). At smaller temporal and spatial scales, the pattern of permafrost warming and thawing is highly variable. Inter-annual variability in thaw depth is directly related to air temperature. As air temperature rises, thaw depths increase. There is, however, considerable spatial variability in thaw depth depending on such factors as the nature of vegetation cover, substrate characteristics, snow-cover dynamics, and terrain (Nelson and Hinkel 2002).

DISCUSSION AND CONCLUSIONS

Climate change over the last half millennium in northwestern Alaska and the last millennium in southern Alaska are recorded in tree-ring chronologies. These chronologies show great inter-annual variability, but over time spans of 500–1000 years they have reflected decadal-scale periods of warming and cooling. Both of these temporal-scale records permit evaluation of multi-scale climatic forcing mechanisms. An appreciation of the operation of climatic forcing mechanisms operating at different scales is essential to understanding controls on those environmental factors that influence whale availability and accessibility.

These multiscalar climatic forcings, namely positions of the Aleutian Low, ENSO patterns, and Pacific Decadal Oscillation circulation, exert profound influences on regional temperature patterns, sea ice extent, thickness, and concentration, sea surface temperature, storminess, and permafrost distribution and temperature. All of these environmental phenomena strongly influence whale availability, patterns and timing of whale movement, accessibility to migrating whales, and whaling success, as well as the ability to store meat through the winter. Variations in the magnitude of environmental parameters occur at multiple scales in response to the multiscalar forcings. In examining spatial and temporal variability in whale availability and hunting success, it is essential to match appropriate scales of inquiry, realizing that inter-annual and inter-decadal variability is most likely being driven by different forcing mechanisms.

Cultural adaptations to changing environmental conditions in the Western Arctic over the past 20,000 years have been evaluated by Ackerman (1988), who greatly relies on the terrestrial glacial record for the period (Black 1966) and draws heavily on earlier reconstructions of Arctic Ocean sea ice extent (Borisov 1982) and sea level histories

(Hopkins 1973). These human-environment reconstructions are for periods of up to a millennium and as such represent broad generalizations about cultural history and human responses to environmental change.

Calenderal–scale data sets, such as those that have been the focus of this chapter, permit the development of an understanding of historic patterns of whale migration and success or failure of hunting seasons both regionally and locally on inter-annual scales. In addition, examination of decadal- and century-scale patterns of environmental parameters provides an opportunity to predict likely recurrences of environmental conditions that influence patterns of whale migration and either favor successful whale hunting or are likely to be associated with poor hunting. Additionally, calenderal-scale data sets such as tree-ring analyses and annual satellite observations permit evaluation, understanding, and prediction of environmental parameter variability between various locations within the Western Arctic and, thereby, potential impacts on specific whaling communities.

Acknowledgments: This research was supported by NSF grant OPP-9807051 to the University of Arkansas. The comments of Allen McCartney, volume editor, and an external reviewer are greatly appreciated. Any errors in data interpretation, however, remain the responsibility of the author.

REFERENCES

Ackerman, R.E.
    1988   Settlements and Sea Mammal Hunting in the Bering-Chukchi Region. *Arctic Anthropology* 25: 52-79.
Anisimov, O.A. and F. Nelson
    1996   Permafrost Distribution in the Northern Hemisphere under Scenarios of Climatic Change. *Global and Planetary Change* 14: 59-72.
    1997   Permafrost Zonation and Climate Change: Results from Transient Circulation Models. *Climatic Change* 35: 241-258.
Bailey, R.G.
    2002   'Ecoregions,' pp. 235-245 in A.O. Orme, ed., *The Physical Geography of North America.* Oxford: Oxford University Press.
Barclay, D.J., G.C. Wiles, and P.E. Calkin
    1999   A 1119-year Tree-ring Width Chronology from Western Prince William Sound, Southern Alaska. *The Holocene* 9: 79-84.
Bjørgo,E., O.M. Johannessen, O.M., and M. Miles
    1995   Analysis of Merged SMMR-SSMI Time Series of Arctic and Antarctic Sea Ice Parameters 1978-1995. *Geophysical Research Letters* 24: 413-416.
Black, R.F.
    1966   Late Pleistocene to Recent History of Bering Sea-Alaska Coast and Man. *Arctic Anthropology* 3: 7-19.

Borisov, P.M.
    1982   'Reconstruction of the Late– and Post–Pleistocene Arctic Basin
            Ice Sheet,' pp. 53-63 in A.I. Tomachev, ed., *Arctic Ocean and its*
            Coast in the Cenozoic Era. New Delhi: Amerind Publishing.

Calkin, P.E.
    1988   Holocene Glaciation of Alaska (and Adjoining Yukon Territory,
            Canada). *Quaternary Science Reviews* 7: 159-184.

Cavalieri, D.J., C.L. Parkinson, P. Gloersen, J.C. Comiso, and H.J. Zwally
    1999   Deriving Long-term Time Series of Sea Ice Cover from Satellite
            Passive-Microwave Multisensor Data Sets. *Journal of*
            Geophysical Research 104 (C7): 15803-15814.

Clow, G.D., R.W. Saltus, A.H. Lachenbruch, and M.C. Brewer
    1998   Arctic Alaska Climate Change Estimated from Borehole
            Temperature: Past, Present, Future. *Eos, Transactions of the*
            American Geophysical Union 79: F883.

Coachman, L.K., K. Aagaard, and R.B. Tripp
    1975   *Bering Strait: the Regional Oceanography*. Seattle: University of
            Washington Press.

D'Arrigo, R.D. and G.C. Jacoby
    1993   Secular Trends in High Northern Latitude Temperature
            Reconstructions Based on Tree Rings. *Climate Change* 25: 163-
            177.

D'Arrigo, R., G. Wiles, G.C. Jacoby, and R. Villalba
    1999   North Pacific Sea Surface Temperatures: Past Variations
            Inferred from Tree Rings. *Geophysical Research Letters* 26:
            2757-2760.

Denton, G.H. and W. Karlen
    1977   Holocene Glacial and Tree Line Variations in the White River
            Valley and Skolai Pass, Alaska and Yukon Territory.
            Quaternary Research 7: 63-111.

Ellis, J.M. and P.E. Calkin
    1984   Chronology of Holocene Glaciation, Central Brooks Range,
            Alaska. *Bulletin, Geological Society of America* 95: 897-912.

Gallant, A.L., E.F. Binnian, J.M. Omernik, and M.B. Shasby
    1995   Ecoregions of Alaska. *U.S. Geological Survey Professional Paper*
            No. 1567.

Hamilton, T.D.
    1981   'Episodic Alluviation in the Central Brooks Range—Chronology,
            Correlations and Climatic Implications,' pp. 21-24 in *United*
            States Geological Survey in Alaska Accomplishments during
            1979. U.S. Geological Survey Circular No. 823B.

Hamilton, T.D., T.A. Ager, and S.W. Robinson
    1983   Late Holocene Ice Wedges near Fairbanks, Alaska, U.S.A.:
            Environmental Setting and History of Growth. *Arctic and Alpine*
            Research 15: 157-168.

Harritt, R.
    1995   The Development and Spread of the Whale Hunting Complex in
            Bering Strait: Retrospective and Prospects, pp. 33-55 in A.P.
            McCartney, ed., *Hunting the Largest Animals: Native Whaling in*
            the Western Arctic and Subarctic. Edmonton: Canadian
            Circumpolar Institute Press, University of Alberta, Occasional
            Publication No. 36, Studies in Whaling No. 4.

Hopkins, D.M.
    1973   Sea Level History in Beringia During the Last 250,000 Years.
           Quaternary Research 3: 520-540.

Hussey, K.M. and R.W. Michelson
    1966   Tundra Relief Features near Point Barrow, Alaska. *Arctic* 19:
           162-184.

Jacoby, G.C.
    2001   'Tree Rings and Whales Tales,' pp. 36-38 in R. Harritt, ed., *In
           Pursuit of Agviq.* Anchorage: Environment and Natural
           Resources Institute. University of Alaska.

Jacoby, G.C. and R. D'Arrigo
    1989   Reconstructed Northern Hemisphere Annual Temperature
           Since 1671 Based on High–Latitude Tree-ring Data from North
           America. *Climate Change* 14: 39-59.
    1995   Tree-Ring Width and Density Evidence of Climatic and
           Potential Forest Change in Alaska. *Global Biogeochemical
           Cycles* 9: 227-234.

Jacoby, G.C., G. Wiles, and R. D'Arrigo
    1996   'Alaskan Dendroclimatic Variations for the Past 300 Years
           along a North-South Transect,' pp. 235-248 in J.S. Dean, D.M.
           Meko, and T.W. Swetnam, eds., *Tree Rings, Environment and
           Humanity, Radiocarbon.* Tucson: University of Arizona Press.

Johannessen, O.M., M. Miles, and E. Bjørgo
    1995   The Arctic's Shrinking Sea Ice. *Nature* 376: 126-127.

Jordan, J.W. and O.K. Mason
    1999   A 5000 Year Record of Intertidal Peat Stratigraphy and Sea
           Level Change from Northwest Alaska. *Quaternary International*
           60: 37-47.

Koerner, R.M.
    1977   Devon Island Ice Cap: Core Stratigraphy and Paleoclimate.
           Science 196: 15-18.

Latif, M. and T.P. Barnett
    1994   Causes of Decadal Climate Variability over the North Pacific
           and North America. *Science* 266: 634-637.

Lachenbruch, A.H. and B.V. Marshall
    1986   Changing Climate: Geothermal Evidence from Permafrost in
           the Alaskan Arctic. *Science* 234: 689-696.

Lea, P.D. and C.F. Waythomas
    1990   Late Pleistocene Eolian Sand Sheets in Alaska. *Quaternary
           Research* 34: 269-281.

Mann, M.E., R.S. Bradley, and M.K. Hughs
    1999   Northern Hemisphere Temperatures During the Last
           Millennium: Influences, Uncertainties and Limitations.
           Geophysical Research Letters 26: 759-762.

Mason, O.K. and J. Beget
    1991   Late Holocene Flood History of the Tanana River, Alaska. *Arctic
           and Alpine Research* 23 : 392-403.

Mason, O.K. and J.W. Jordan
    1993   Heightened North Pacific Storminess During Synchronous Late
            Holocene Erosion of Northwest Alaska Beach Ridges.
            Quaternary Research 40: 55-69.
    1994   Eustatic Seal-Level Changes in Northwest Alaska During the
            Last 3000 Years. Abstracts with Programs, Annual Meeting,
            Geological Society of America, p. A309.
Mason, O.K., J.W. Jordan, and L. Plug
    1995   'Late Holocene Storm and Sea Level History in the Chukchi
            Sea,' in C.W. Finkl, ed, Holocene Cycles: Climate, Sea Levels
            and Sedimentation. *Journal of Coastal Research* Special
            Publication 17: 173-180.
Mason, O.K., D.K. Salmon, and S. Ludwig
    1996   The Periodicity of Storm Surges in the Bering Sea from 1898 to
            1993, Based on Newspaper Accounts. *Climatic Change* 34:
            109-123.
Miller, A.J., D.R. Cayan, T.P. Barnett, N.E. Graham, and J.M. Oberhuber
    1994   Interdecadal Variability of the Pacific Ocean: Model Response
            to Observed Heat Flux and Wind Stress Anomalies. *Climate*
            Dynamics 9: 287-302.
Moore, S.E. and R.R. Reeves
    1993   'Distribution and Movement,' pp. 313-386 in J.J. Burns, J.J.
            Montague, and C.J. Cowles, eds., *The Bowhead Whale.*
            Lawrence, KS: Allen Press, Society for Marine Mammalogy,
            Special Publication No. 2.
Morison, J., K. Aagaard, and M. Steele
    2000   Recent Environmental Changes in the Arctic: A Review. *Arctic*
            53: 359-371.
Mysak, L.A.
    1986   El Nino, Interannual Variability and Fisheries in the Northeast
            Pacific Ocean. *Canadian Journal of Fisheries and Aquatic*
            Science 43: 464-497.
Nelson, F.E. and K.M. Hinkel
    2002   'The Far North: A Geographic Perspective on Permafrost
            Environments,' pp. 249-269 in A. Orme, ed., *The Physical*
            Geography of North America. Oxford: Oxford University Press.
Nelson, F.E., K.M. Hinkel, and S.I, Outcult
    1992   'Palsa-scale Frost Mounds,' pp. 305-325 in J.C. Dixon, and
            A.D. Abrahams, eds., *Periglacial Geomorphology.* Chichester:
            John Wiley and Sons.
Niebauer, H.J.
    1998   Variability in Bering Sea Ice Cover as Affected by a Regime
            Shift in the North Pacific in the Period 1947–1996. *Journal of*
            Geophysical Research 103 (C12): 27717-27737.
Niebauer, H.J. and D.M. Schell
    1993   'Physical Environment of the Bering Sea Population,' pp. 23-43
            in J.J. Burns, J.J. Montague, and C.J. Cowles, eds., *The*
            Bowhead Whale. Lawrence, KS: Allen Press, Society for Marine
            Mammalogy , Special Publication No. 2.

Oliver, J.E.
　　2002　'Climatic Regionalization,' pp. 112-129 in A.O. Orme, ed., *The Physical Geography of North America*. Oxford: Oxford University Press.

Orme, A.R.
　　2002　'Ocean Coasts and Continental Margins,' pp. 425-455 in A.O. Orme, ed., *The Physical Geography of North America*. Oxford: Oxford University Press.

Osterkamp, T.E. and V.E. Romanovsky
　　1996　Characteristics of Changing Permafrost Temperatures in the Alaskan Arctic, U.S.A. *Arctic and Alpine Research* 28: 267-273.
　　1999　Evidence for Warming and Thawing of Discontinuous Permafrost in Alaska. *Permafrost and Periglacial Processes* 10: 17-37.

Parkinson, C.L.
　　1992　Spatial Patterns of Increases and Decreases in the Length of the Ice Season in the North Polar Region, 1979-1986. *Journal of Geophysical Research* 97 (C9): 14377-14388.
　　2000a　Variability of Arctic Sea Ice: The View from Space, An 18-year Record. *Arctic* 53: 341-358.
　　2000b　Recent Trend Reversals in Arctic Sea Ice Extents: Possible Connections to the North Atlantic Oscillation. *Polar Geography* 24: 1-12.

Patterson, W.S.B, R.M. Koerner, D. Fisher, S.J. Johnsen, H.B. Clausen, W. Dansgaard, P. Butcher, and H. Oeschger
　　1977　An Oxygen Isotope Climatic Record from the Devon Island Ice Cap, Arctic Canada. *Nature* 266: 508-511.

Schell, D.M., B.A. Barnett, and K.A. Vinette
　　1998　Carbon and Nitrogen Isotope Ratios in Zooplankton of the Bering, Chukchi and Beaufort Seas. *Marine Ecology Progress Series* 162: 11-23.

Stone R.S.
　　1997　Variations in Western Arctic Temperatures in Response to Cloud Radiative and Synoptic Scale Influences. *Journal of Geophysical Research* 102 (D18): 21769-21776.

Stringer, W.J. and J.E. Grove
　　1991　Location and Areal Extent of Polynyas in the Bering and Chukchi Sea. *Arctic* 44: 164-171.

Thompson, W.J. and J.W. Wallace
　　1998　The Arctic Oscillation Signature in the Wintertime Geopotential Height and Temperature Fields. *Geophysical Research Letters* 25: 1297-1300.

Trenberth, K.E.
　　1990　Recent Observed Inter-decadal Climate Changes in the Northern Hemisphere. *Bulletin of the American Meteorological Society* 71: 993-998.

Vinnikov, K.Y., A. Robock, R.J. Stouffer, J.E. Walsh, C.L. Parkinson, D.J. Cavalieri, J.F.B. Mitchell, D. Garett, and V.F. Zakharov.
　　1999　Global Warming and Northern Hemisphere Ice Extent. *Science* 286: 1934-1936.

Wahrhaftig, C.
　　1965　Physiographic Divisions of Alaska. *U.S. Geological Survey Professional Paper* No. 482.

Walker, H.J.

    1985   'Alaska,' pp. 1-10 in E.C. Bird and M.L. Schwartz, eds., *The World's Coastline*. New York: Van Nostrand Reinhold.

Wiles, G.

    1997   North Pacific Atmosphere-Ocean Variability over the Past Millennium Inferred from Coastal Glaciers and Tree Rings. *Preprints 8th Symposium on Global Change Studies*, 218-220. *Boston:* American Meteorological Society.

Wiles, G., D. Barclay, and P.E. Calkin

    1999   Tree-Ring Dated Little Ice Age Histories of Maritime Glaciers from Western Prince William Sound, Alaska. *The Holocene* 9: 163-195.

Wilson, J.G. and J.E. Overland

    1987   'Meteorology,' pp. 31-56 in D.W. Hood and S.T. Zimmerman, eds., *The Gulf of Alaska: Physical Environment and Biological Resources*. National Oceanographic and Atmospheric Administration. Minerals Management Study, OC 86-0095.

Re-Examining Wales' Role in Bering Strait Prehistory: Some Preliminary Results of Recent Work

Roger K. Harritt

Abstract. Archaeological sites at the village of Wales, located in Bering Strait on the western tip of the Seward Peninsula, are the focus of an ongoing study. During contact and early historic times, the Wales settlement was a cultural center with approximately 500 Iñupiat inhabitants, with a strong subsistence focus on marine mammals including the bowhead whale. A review of Euroamerican contact during the early historic era reveals that Wales maintained its large population up to the time the 1918 Spanish influenza pandemic decimated the population by approximately one half. A primary focus of the project is to illuminate the processes whereby prehistoric whaling methods were transferred from their center of development on the Chukotka coast and Bering Sea islands to mainland northwest Alaska, as viewed from Wales. A fine-grained comparison will also be made of components from the three main Wales sites including the Hillside site (TEL-025), the Beach site (TEL-026), and Kurigitavik Mound (TEL-079) to demonstrate possible differences between the inhabitants, or alternatively, the homogeneity of a shared culture.

Since 1996 a series of new radiocarbon dates have been obtained. For Kurigitavik, the new dates are approximately one-half the age of the dates obtained previously for the same levels; for the Hillside site, the new dates are younger than were obtained previously, but are not considered conclusive; the six new dates for the Beach site are the first to be obtained for this site. Presently, a total of 17,000 faunal elements between 1998–1999 collections from the three sites have been analyzed; faunal remains from the 2000–2001 field seasons are currently undergoing analysis. From a series of tests conducted on the Hillside site, intact cultural deposits were revealed in only four tests. Excavation of a 4 x 6 m block in a southeastern, partially disturbed portion of the Beach site from 1998–2001 revealed a series of features, including an upper level cache, a portion of a semisubterranean house, and a deep level cache. The major effort at Kurigitavik Mound is focused on exposing intact house remains on the northern end of the site. A 5 x 6 m block, excavated from 1998–2001, has exposed an estimated 70% of the house remains, including a kitchen or storage area and portions of a main room.

INTRODUCTION

An archaeological project presently being conducted at the village of Wales, located on the Seward Peninsula coast, focuses on issues related to the prehistoric development of whaling villages in northwest Alaska, as defined in a recent article (Harritt 1995). Current knowledge of how prehistoric whaling developed in northwestern and northern Alaska continues to reflect Collins' initial interpretation (e.g., Collins 1964), in which whaling technology and techniques initially developed on Siberian shores and on the Bering Sea islands then spread to the east and north onto the mainland Alaska coast. Wales is an ideal location for an investigation of this process because of its geographic position and because of the prehistoric cultures that have previously been documented at the locality (Fig. 1).

The modern village of Wales is situated on the northwesternmost point of mainland North America, nestled under the northwest side of Cape Mountain where the southeastern terminus of Cape Prince of Wales abuts the mountain slope. Approximately 152 residents presently inhabit Wales. Of this number, Iñupiat people constitute approximately 90% of the population, and the remaining 10% are non-Natives (ADCED 2002).

SOME CONSIDERATIONS ABOUT WALES' HISTORY

Historically, Wales has been a mainland Alaska contact point for Siberian Yupik peoples, and it is generally accepted that the relations across Bering Strait extend back into an unknown depth of prehistoric time. Contacts between Alaskans and Siberians were historically comprised of trading relations, with trade fairs held in the vicinity of the Seward Peninsula—including one at Wales, specifically—but the relations also included warfare between the two sides of Bering Strait (e.g., Morrison 1991:102-105; Ray 1975:87-89). These activities undoubtedly provided impetus for the congregation of a population of at least 500 at Wales proper, the largest known settlement for Bering Strait and northwestern Alaska during the early 1800s (e.g., Ray 1975:110-111).

Wales' traditional mainland territory extended from the village northeast to the mouth of the Nuluk River on the northwestern shore of the Seward Peninsula and southeast to Cape York on the southwest shore; also included were inland areas as far east as the headwaters of the Pinguk River (Ray 1983). According to Ray (1983:208), the greater Wales territory also included both Diomede Islands and contained as many as 850 persons overall. The concentration of individuals at Wales itself therefore supports a perception that the village was a headquarters or capital for the area (Ray 1975:149).

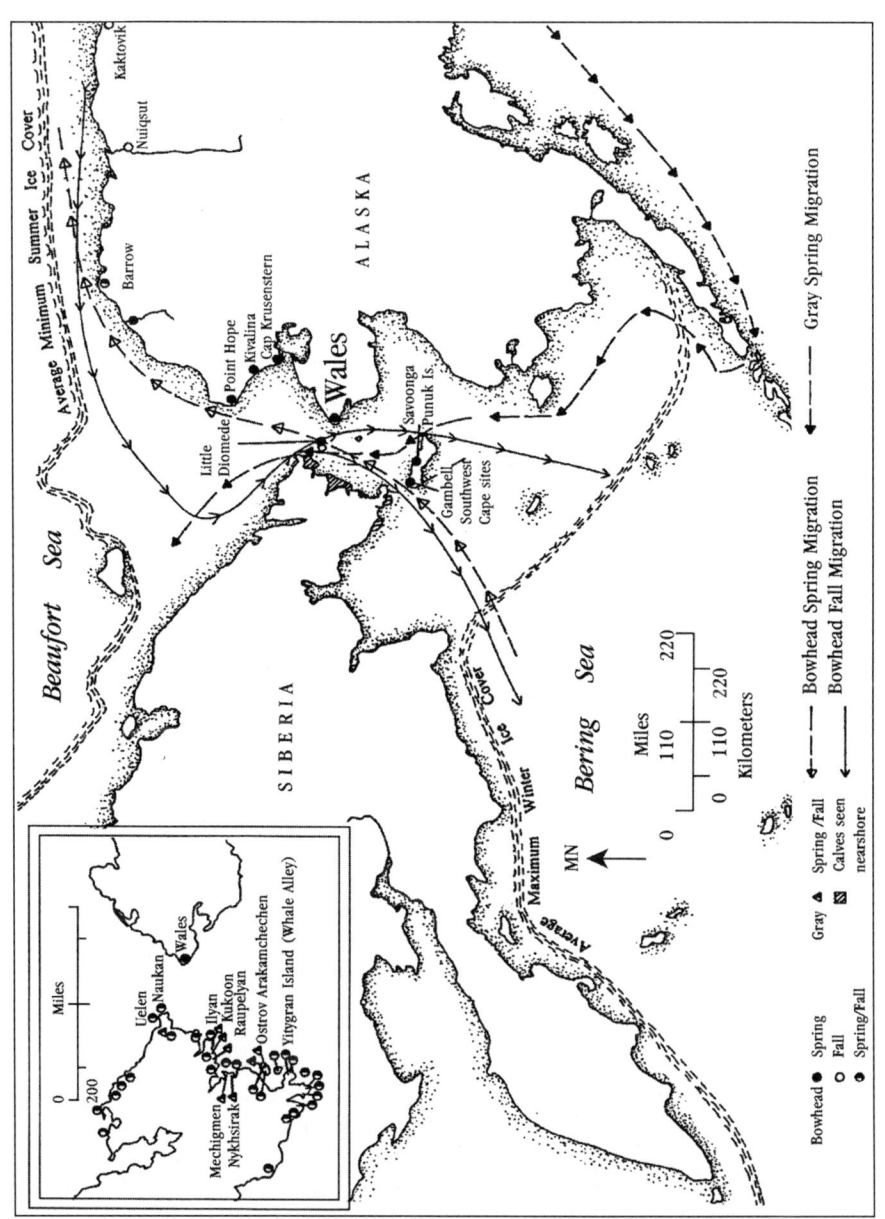

Figure 1. *Location of Wales and some of the major sites in the region (Harritt 1995:Fig. 1)*

THE CONTACT PERIOD

Documentation developed by Ray (1975:21-26; compare with Bockstoce 1977:2-9) indicates that Wales also has historical importance as the first mainland Native village to be observed by the Russian explorers Gvozdev and Fedorov in 1732, nine years prior to the better–publicized discovery by Bering. During their voyage, the Russians actually landed on American shores near Wales and although they did not enter the village itself, they nevertheless made general observations from some distance offshore. The earliest firsthand description of the settlements at Cape Prince of Wales are provided by Ivan Kobelev, who actually entered the settlement in 1791 (Ray 1975:53). Kobelev observed 50 dwellings, all of which appeared to be deserted.

British activities in the region during the middle of the 19th century were focused on the lost Franklin Expedition, with a primary objective of searching for its members initially in northwest Alaska, then farther eastward along the shore of the Beaufort Sea. As part of this effort, a few observations of Wales, or Kingigan were recorded in a log kept by Richard Collinson, Captain of *H.M.S. Enterprise* during the period 1851–1854 (Collinson 1889). Collinson never entered the village, but he recorded brief encounters with Kingigan residents offshore, including one in 1851 in which he observed several *umiaks* making trips between the village and the ice edge, engaged in 'walrus fishing' (Collinson 1889:128; also Ray 1975:147). In 1854, Collinson (1889:330) notes meeting Captain Henry Trollope, of the HMS *Rattlesnake*, another searcher for the lost Franklin Expedition. Trollope entered the settlement in January of 1854, only the second time Kingigan was visited by Europeans. Trollope's is the first recorded account of the inhabitants. He notes that there were two villages—an 'upper' village and a 'lower' or 'spit' village, that both communities had approximately 200 to 250 residents apiece, and that there were four *qargit* in the collective settlement (Trollope 1854; Ray 1975:149-151). Russian trade goods were also observed by Trollope (e.g., Ray 1975:150), indicating Wales' early participation in this activity.[1]

---

[1] Russian activities in Alaska were initially concentrated in the southern areas, but in the later stages of operations, some expansion northward did occur. Ray (1966:8-22; 1975:121-125) points out that even after the fort was established at St. Michael in 1833, trade relations between the Russians and Natives were rather uneven, with some groups receiving considerably more attention from the Russians than others. As examples, she notes (Ray 1975:124-125) that King Islanders, Sledge Islanders, and Kauwerak people were affected most by trade with the Russians at St. Michael. Wales continued in its role as major Alaskan agents in trade with Siberians, and this situation was not changed by the establishment of the Russian post, only 240 mi away. Russian traders paid little attention to

The next visit by non-Natives occurred 12 years later, in 1866, by members of the Western Union Telegraph Expedition who traveled to Kingigan for a four-day stay. As they note in their publication, *The Esquimaux*, they were hosted in a different dance house each of the three nights they were there, and they observed at least three *qargi* in the village. Their estimate of a population of 900 Natives inhabiting 90 houses considerably exceeds the numbers estimated in other early accounts (Ray 1975:165).

The commercial Yankee Pacific bowhead whale fishery began in 1848 in the waters of northwestern and northern Alaska and persisted in the Bering Strait area through the first decade of the 20th century. The presence of Yankee whalers had a major impact in most of the coastal villages in northwestern and northern Alaska for a relatively brief period of 50 years (e.g., Bockstoce 1986; Cassell, this volume). In the early years of pelagic whaling, the basic pattern of interactions between Yankee whalers and Eskimo coastal inhabitants can be characterized as that of occasional visits by ships' crews to refill their water barrels and to trade for fresh meat and other items such as furs and ivory (e.g., Bockstoce 1986). As the numbers of whales decreased, the industry began to shift around 1860, during which commercial shore-based whaling stations were established in a few villages and other locations. Trading activities increased and a focus on harvesting walrus for oil and ivory developed (Bockstoce 1986:130-136; Cassell 2000; Ray 1975:199). Specific accounts of interactions between the commercial whalers and Wales residents are sparse, however, and it is apparent that Wales' large population and its reputation for aggressiveness convinced many whaling ships' captains to forego stopping at the village (e.g., Bockstoce 1986:188-189). There were also reports that the residents sometimes took possession of ships and thereafter, whatever items they wanted (Nelson 1983:299).

Under these circumstances, the use of alcohol as a trade item contributed to the aggression and led rather predictably to violence. In 1877, the infamous Gilley affair took place aboard a ship anchored some distance offshore from Wales (Bockstoce 1986:189-191). A number of *umiat* had gone out to the brig *William H. Allen* (of which George Gilley was captain) to trade. While onboard there was an altercation between the Natives and the small crew, with the result

---

Wales after St. Michael Redoubt was established in 1833 (Ray 1975:124-125). Trade between Wales people and Russians and Europeans was therefore limited, but trade goods were also probably transferred to Wales through middlemen prior to the time that the Yankee whaling fleet and trading vessels began routine travel in the area, beginning in 1848. Evidence of this activity is found in the Russian trade goods that were observed at Wales by Trollope (1854), which is a clear indication that a network was in operation in the area over the preceding two decades.

being the death of 13 Wales residents and 1 of Gilley's crew. As a result of the Gilley incident, '... for many years vessels gave Cape Prince of Wales a wide berth, and if they stopped there at all, they did not anchor and, instead, merely backed their sails while they allowed only a few Eskimos aboard at a time' (Bockstoce 1986:191).

Figure 2. *The 'North Village' of Wales (also the area of the Beach site and the modern village), ca. 1923–1930. This scene is no doubt virtually the same as those experienced by Thornton and Lopp during their work at Wales from 1890–1902. The photograph was taken by C.L. Andrews during the time he was employed by the Education and Reindeer Service under Thomas Lopp, the former Wales school teacher (Bromberg 1976:70, 73; Smith 1995; C.L. Andrews Collection, Division of Special Collections and University Archives, University of Oregon Library System).*

Under these circumstances, it seems very improbable that the type of commercial shore-based whaling operations that were established at Cape Smythe and Point Hope could have been established at Wales.[2] As mentioned above, by the 1860–1870s Wales

---

[2] Stoker and Krupnik (1993:602; see also van Veenen 1977) incorrectly state that shore-based whaling operations were located at Wales and Port Clarence, and drew large populations to those villages in the late 19th century. By other accounts, shore-based whaling operations were confined to the vicinities of Point Barrow and Point Hope, and the Wales population in particular remained relatively stable from the contact period of the late 18th century until the outset of epidemic disease around 1900 (*cf.* Bockstoce 1986:231-254; Ellanna 1983; Ray 1975:198-201; 1983:178-181).

had developed a reputation for aggressiveness with regard to its treatment of ships' crews, a pattern of coastal villages that was directly related to the substantially greater number of residents than numbers of crew men employed on each of the ships (e.g., Bockstoce 1986:184-185, 188-189).

Edward Nelson visited Wales in 1881 (Nelson 1983:98). As did Trollope, he describes two distinct villages—the 'hill' and 'spit'—and states that they were separated by about 75 yards. Nelson also mentions an ongoing feud between the two settlements. Athough he collected a total of approximately 142 artifacts from Cape Prince of Wales, he gives no population estimate for the village, nor did he mention the presence of the *qargit* that Trollope observed in 1854 (Fitzhugh 1983:105-106; Nelson 1983:257). However, a census of Wales ('*Kingigamute*') inhabitants conducted in 1880 lists the population of the village as 400 (Fitzhugh 1983:98).

Near the end of the 19th century, Sheldon Jackson, U.S. agent of education, was the principal force behind establishing three pioneering mission schools in the northwestern and northern parts of Alaska. As a result of this project, an Episcopalian church was established at Point Hope, a Presbyterian church at Barrow, and a Congregational mission school at Wales (Ray 1975:214-215; Fig. 2). The first teachers assigned to Wales were William Lopp and Harrison Thornton. Construction of the wood frame school, the first frame building constructed in Wales, began in the summer of 1890 (e.g., Ray 1975; Thornton 1931). The experiences of the missionaries, including the eventual murder of Thornton, have been described in great detail elsewhere and will not be revisited here (e.g., *The American Missionary* 1890-1901; Education Report 1894; Montgomery 1963a, 1963b; Ray 1975:214-222; Smith and Smith 2001; Thornton 1931).

In addition to the Congregational mission school, Jackson was also responsible for establishing a reindeer herding station at the village (e.g., Ray 1975:233). The impetus for developing this industry was a general perception that the Natives were facing starvation due to decimation of whales and walrus to an extreme degree by commercial whaling operations by the end of the 19th century (e.g., Ray 1975:226-240). Caribou were also sparse during this time (Stern *et al.* 1980). During the start-up phase, 171 reindeer were transported to Port Clarence in 1892 and the Teller Reindeer Station was established, the first of its kind in the area. Two years later, Thomas Lopp had been enlisted as an agent in the new reindeer service in addition to continuing as principal of the village mission school, and he transported the first herd of 118 animals to Wales (Ray 1975). This initiated the start of the herding industry in the village that has continued through varying levels of success to the present day (*cf.* Stern *et al.* 1980).

WALES IN POST-CONTACT TIMES

The effects of Euroamerican contact on the Natives in the 19th century were relatively mild compared with those that occurred with the beginning of the 20th century. Small-scale epidemics of respiratory disease accounted for the deaths of 44 residents in 1890 and in 1898, 18 deaths occurred in a summer population of 300 (Smith and Smith 2001:196-198). (A population estimate apparently made at a different time of the year during part of the same period—1890–1893—places the number of residents at 539 [Thornton 1976:21]). Subsequently the 'Great Sickness' of 1900, believed to have been an outbreak of scarlet fever (Smith and Smith 2001:249), followed quickly in the same year by episodes of measles, which in turn was followed by either pneumonia or influenza (Fortuine 1989; Smith and Smith 2001:285). By the missionaries' estimates, the population was reduced by one fifth—107 individuals—of a population of 539, leaving approximately 432 survivors.

Although smaller outbreaks occurred after 1900, by far the greatest impact occurred when the 1918 Spanish influenza pandemic reached Wales in November of that year. It is believed that mail carriers brought mail and the sickness from Teller by way of Palazruk, several miles south and east of Wales, arriving on November 9th (Nagozruk 1919; see also Greist 1952:Chap. III, p 1). In short order, an alarming number of residents became sick and died before the first relief party arrived on December 4. All told, the population of the village was reduced by nearly half within two months' time, with as many as 200 deaths occurring during the winter of 1918–1919 (Ellanna 1983:285). At the height of the sickness, the situation deteriorated to epic proportions, with the sick gathering at the school seeking care from a single nurse who attempted to provide some measure of medical attention (Greist 1952:Chap. III). Others, too weak to get themselves to the school, died in their houses (*loc cit.*). By 1920, the population had reached its all time recorded low of 136 individuals, a number that also included non-Natives and teachers (Ellanna 1983:Table 63 & text). Some of those who survived the epidemic moved to population centers such as Nome to be closer to health care and also to find paying jobs (Greist 1952:Chap. III).

By all accounts, the several hundred individuals inhabiting Wales during the contact period and into the second decade of the 20th century were a vigorous and robust group that, for the most part, continued to pursue the same whale, walrus, seal, and caribou hunting traditions that they had practiced since precontact times. Euroamerican influences that were in evidence, as mentioned previously, included Russian trade goods that were found in the village as early as 1854, as well as a growing trade in guns, implements such as axes, and trade beads (Bockstoce 1977, 1986).

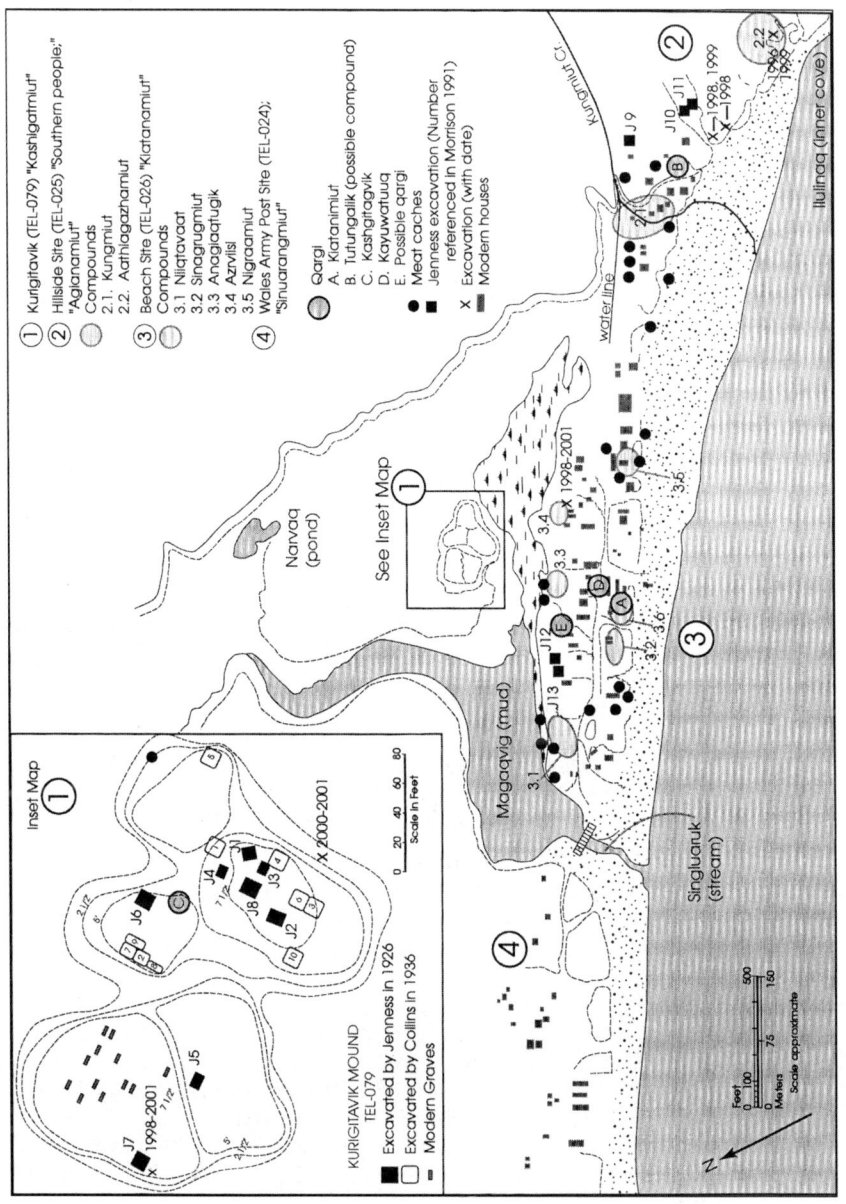

Figure 3. *The Wales locality sites and some of the buildings of the modern village*

The increasingly destructive effects of alcohol were reflected perhaps most clearly in the Gilley incident of 1877. Although the depredations of the Yankee whaling fleet on the whale and walrus populations in the area presented perhaps a more fundamental impact on the culture through the drastic reduction of important sources of food, Wales hunters nevertheless continued to have surprising success in the numbers of sea mammals harvested around the turn of the century (see the Subsistence section, below). Wales maintained the major portion of its population to as late as 1918, when the major impacts of epidemic disease substantially reduced the Native population (e.g., Ellanna 1983; Ray 1975). The persistence and vitality of the traditional culture until 1918 is apparent in the consistency of the observations made initially by Trollope (1854) and later by Thornton and Lopp (e.g., Thornton 1931). In this regard, the sheer numbers and aggressiveness of the Wales inhabitants (Bockstoce 1986:188-191) no doubt aided in preserving the culture for a brief time, in the face of influences carried by relatively small groups of whalers and traders aboard wooden vessels powered by wind and steam. The observations made by Greist (1952) in 1920 of the enormous impacts of the 1918 epidemic are, therefore, poignant in their reflections on the fate of a once populous and vital center in Bering Strait.

THE SITES

Historically, Wales has sometimes been perceived as a single settlement, known as *Kingigan* in the most widely used rendering of the designation (e.g., Braund and Moorehead 1995; Ellanna 1983; Ray 1983:208, 210-211, 262; Stoker 1984).[3] The modern day Wales locality is a National Historic Landmark that was established in 1962 (the designation for the landmark in the Alaska Heritage Resource Survey files, Anchorage is TEL-010, which subsumes all of the Wales locality sites). Social and spatial subdivisions are also known that correspond with the archaeological sites in the locality. They are

---

[3] A variety of different names that were recorded for Wales during the contact period can be found in the literature. Among these are Lantis' (1947) rendition of *Kinugmiut* for the Wales collective settlement appended with the meaning 'high place of Cape Prince of Wales,' a name perpetuated indiscriminately by Montgomery (1963a:47) who also uses *Kingigan* in reference to the same settlement. Montgomery's (1963a:46-51) discussion of the two historic Wales districts (the Hillside and Beach) is, nevertheless, consistent with descriptions by others. A number of similar, early orthographic renditions listed by Ray (1975:293) for the Wales settlement include *Kingigmen, Kigin Eliat, King-a-ghe, Kingaghee, Kingigmute, Kymgovei, Kyngovei, Kyng-Myn, Kynyntsy*. But the preferred name for the Wales collective that she uses last (Ray 1983:262) is *Kingigan*.

designated as *Agianamiut* for the Hillside site, or TEL-025; *Kiatanamiut* for the Beach site, or TEL-026; and *Kazhgitavik* for Kurigitavik Mound, or TEL-079 (Anungazuk and Bodenhorn 1996; Ray 1983; Fig. 3). The Birnirk Midden (TEL-031), located approximately 2.5 km northwest of the modern village, was also a part of the landmark, but it is now completely destroyed as a result of intensive artifact mining (This site is not to be confused with the Birnirk–type site near Barrow, Alaska). The social divisions, with their corresponding habitation zones, within the collective Wales settlement can be conceptualized as 'districts,' however, a parallel Iñupiat term, *nunaqthliq*, designates similar types of divisions (H. Anungazuk, pers. comm. 2000). In the present discussion, these sociospatial divisions with their corresponding habitation zones within the Wales, or *Kingigan*, collective will be referred to as 'districts.'

The social divisions, with their corresponding habitation zones, within the collective Wales settlement can be conceptualized as 'districts,' however, a parallel Iñupiat term, *nunaqthliq*, designates similar types of divisions (H. Anungazuk, pers. comm. 2000). In the present discussion, these sociospatial divisions with their corresponding habitation zones within the Wales, or *Kingigan*, collective will be referred to as 'districts.' *nunaqthliq*, designates similar types of divisions (H. Anungazuk, pers. comm. 2000). In the present discussion, these sociospatial divisions with their corresponding habitation zones within the Wales, or *Kingigan*, collective will be referred to as 'districts.'

It is now clear that smaller segments of the Wales districts also once existed and that they reflect some aspect of the 'family compound' grouping described by Burch (1980:266). An example of a segment within a district are the *Kungmiut*, a group named after the main stream that drains the northern end of the Hillside site, in an area claimed by the *Agianamiut* (Ray 1975:215; 1983:262). This group occupied a mound at the mouth of the stream, in the vicinity of the mission school building (H. Anungazuk, pers. comm. 1999). A second example of a segment are the *Aathlagazhamiut*,[4] a group that occupied Mound A on the southern end of the Hillside site, also in the *Agianamiut* district, in the vicinity of the original location of the Lopp and Thornton house (H. Anungazuk, pers. comm. 1999). It is reasonable to assume that each of the other mounds at the Hillside and the Beach sites, and the mounds that constitute the Kurigitavik

---

[4] *Aathlagazhamiut* is also considered to be an authentic distinction for a compound within the *Agianamiut* district, but for slightly different reasons as the rationale given in noted for *Kungmiut*. *Aathlagazhamiut*, translates into English roughly as 'a place that is separate or apart from other places' (H. Anungazuk, pers. comm. 2000) and, therefore, it distinguishes its inhabitants from those of other places.

site reflect the same types of groupings, provided correspondences between named segments and specific mounds can eventually be demonstrated. It is important to point out that in spite of the existence of the subdivisions within each district, there nevertheless were strong ties between each segment, and inhabitants of the districts possessed a Kiatanamiut or Agianamiut identity, as is reflected in Nelson's (1983:98) description of a 'feud' going on between the upper and lower villages that he observed during his visit.

It is important to note that no mention is made in Trollope's first account of Wales of the presence of a 'village' on Kurigitavik Mound in 1854 (Ray 1975). Therefore, because Kurigitavik was uninhabited at the time of contact, a study of the ethnicity of the mound's inhabitants must therefore rely most heavily on archaeological reconstruction.

A fourth site, located on the northwestern side of the 'Wales River,' is recorded in the Alaska Heritage Resource Survey (AHRS) files and on the National Register Nomination forms as the historic 'Wales Army Post' (TEL-024), a facility constructed during World War II that was nearly completely razed as part of the Air Force Clean Sweep project in 1999. Ray (1983:245) documents an Iñupiat name for the area where the former Army post was located, *Singluraruk*. This designation is the name of the main stream in the village area, and is sometimes now referred to as 'Village Creek.' At present, although the name is not known to define a specific area on the northern side of the stream, recent oral history corroborates that of Ray with statements that there was also once a fourth 'village,' *Singuarangmiut*, located in the area of TEL-024 (Walter and Winton Weyapuk pers. comm., 2001). It is therefore reasonable to consider that the *Singuarangmiut* district was a distinct segment of Kingigan and on this basis, it should be distinguished from the Wales Army Post that was constructed in the 1940s. Although further interviews with Wales elders may reveal additional information about this part of Kingigan, based on an inspection in 2001 by Harritt, it was apparent that the installation of the Army Post and subsequent activities have removed most surface traces of a precontact period settlement district from this area.

CURRENT CONDITIONS

Hillside Site

The Hillside site consists of a series of five middens arranged roughly along a north-south axis along the edge of a bluff, estimated to be 5–10 m above the southeastern end of the Cape Prince of Wales beach ridge system. Small granitic outcrops and scree occur on the northwest slope of Cape Mountain above the Hillside site. In precontact times and occasionally in the historic era, the slope area above the Hillside middens has been a graveyard for deceased Wales residents. The bluff lies at the base of Cape Mountain and was formed during a time of higher sea levels as a wave-cut scarp of the mountain slope. The bluff edge is riven by seasonal springs runoff from upslope that produce both gully erosion and small landslides on the bluff slope down to the high water line on the modern shore. In the warmer months the midden mounds appear as green lobes between the gullies, and springs and are covered by growths of *Artemesia*, which also hide artifact mining pits. Many active mining pits are present on the Hillside, and with the exception of Mound E, most of the site has been subjected to this process as well as the ongoing erosion. Considerable numbers of large whale and walrus bones are scattered across the broken surfaces of the mounds, and many can be seen partially exposed in the mining pits.

Figure 4. *A view of the Wales locality, looking north, taken by Thomas Lopp in 1895. The location of the Lopp and Thorton mission house that was built on the Hillside in 1892 is visible in the foreground, and various meat racks and boat racks appear behind it on the Hillside and on the shore of the Beach site in the background (Smith and Smith 2001:Fig. 21; Thomas Lopp Collection, No. 94-132-83N, Alaska and Polar Regions Department, University of Alaska Fairbanks).*

As part of the mission school effort, a wood frame house was constructed on the southern end of the Hillside site in 1892, in an area described as 'the outskirts of 'A-gen-amete,' (var. *Agianamiut*) or south village on the hill' (Ray 1975:218). A sketch map drawn by Diamond Jenness in 1926 shows this structure on the southern end of the site, in a location on the bluff edge that roughly corresponds to the area where the beach narrows considerably and the wave-cut scarp of the bluff edge meets the steep slope of the mountain (H. Anungazuk, pers. comm. 1999). A photograph taken around 1895 (Smith 1995; Fig. 4) corroborates Jenness' sketch map; nevertheless, neither provides any degree of precision in locating the original building site. As seen by Greist in 1920, the house was at the 'very edge' of the advancing erosion of the scarp and in imminent danger of destruction (Greist 1952:Chap. I). Due to the Greists' need of lodging for their one-year stay in Wales, they dismantled the structure and re-erected it on the northeastern side of the modern village on the Beach site, near its midpoint.

The other sites in the locality are found at the southeastern end of the Cape Prince of Wales beach ridge system in the area where the Cape abuts the base of Cape Mountain. Mason's (Mason n.d.; Harritt and Mason 1997; Mason and Jordan 2002) recent analysis of the Cape has revealed correspondences between accretional and erosional episodes with those reflected in similar coastal features, such as Cape Espenberg.

Cape Prince of Wales

Mason (n.d.) suggests that the bulk of the sand deposit that forms the Cape was transported from the Yukon Delta by ocean currents around 2500 years ago. The shapes and alignments of the depositional units that compose the Cape reflect variations in the processes that built the landform and molded the modern shape and characteristics. In this respect, the cycles of accretion and erosion that are reflected in the modern beach ridges also represent shifts in climatic regimes over time (e.g., Mason and Jordan 2002).

Kurigitavik Mound is situated on a low beach ridge that is a portion of depostional Unit II which Mason (Mason n.d.) suggests was formed sometime before AD 400. The beach ridge comprised the shoreline in the area for several centuries, perhaps as long as 1000 years. The Unit II ridge was superceded in the horizontal stratigraphy by the formation of Units III and IV sometime after 1,000 years ago. Mason (n.d.) suggests that the Unit IV dune, upon which the Beach site and the modern village are situated, was formed no earlier than 1,000 years ago; it was probably stabilized by AD 1400, or roughly 500 years ago.

The Beach Site

The central district of present-day Wales is superimposed over the Beach site, and some portion of the northern end of the *Agianamiut* district that was centered on the Hillside. The Beach site and village are situated on a beach dune only a few meters above the present-day average high-water line on the beach. In times of higher sea levels, the beach dune may have comprised a barrier bar in front of a lagoon, such as those that can now be seen along the Seward Peninsula coast to the north and east of Cape Prince of Wales. Local lore includes stories about the sudden abandonment of Wales in precontact times for varying reasons. In one case that appears to relate directly to the Beach site, raids by Siberian Eskimos was given as the cause for the sudden desertion, with the evacuees fleeing to Barrow (Jenness 1928a:73).

The Lopp and Thornton schoolhouse, which was built in the summer of 1890, still stands in the location of its original construction (H. Anungazuk, pers. comm. 1999), on the eastern end of the Beach site. In modern times the building is used as a church and appears to be in reasonably good condition, despite its 112-year tenure in Wales' climate and shoreside setting. This is a testament to the skills of the whaling ships' carpenters who built the structure (Ray 1975:215) and to those who have maintained the building over the years. A log house was also constructed in the vicinity of the school in 1896 by Thomas Lopp and his wife (Smith 1995:40; Smith and Smith 2001:Fig. 51 and text). This building served as the teachers' residence until they departed Wales in 1902 (*op cit*). This is undoubtedly the 'mission house' seen by Greist approximately 18 years later (Greist 1952:Chap. I) and located less than 60 ft from the 'chapel.' By 1920, this structure was in dilapidated condition, with most of the roof missing, large holes in the floor, and all the windows and doors either missing or broken.

Many of the frame houses in the central portion of the village are built over sod houses that were still inhabited in the early part of the 20th century. In many cases, the modern occupants live in the same locations as their ancestors, and in some cases, structural members of the sod houses may be seen directly beneath the wood frame structures of the present day.

Kurigitavik Mound

Kurigitavik Mound is situated approximately 60 m inland from the Beach site on a low beach berm that is only slightly elevated above the terrain (Fig. 3). The land around the mound is now boggy and prone to flooding due to extensive buried ice lenses that prevent precipitation and spring meltwater from sinking into the ground during spring runoff and times of heavy summer rainfall. The mound consists of a roughly oval midden, with a maximum elevation approximately 5 m above the surrounding muskeg. A description by Jenness (1928a:72), written well over 70 years ago, still provides an accurate description of Kurigitavik:

> The mound was roughly oval in shape, 75 yards long by 50 yards wide, with a modern graveyard at its west end. Its surface was pitted with rectangular depressions partly filled with water, and here and there the rib of a whale protruded above the ground. Evidently these were old house sites, whale ribs being still employed by the Eskimos for the walls and roofs of their dwellings. Skirting the mound was a creek that expanded lower down into a small lagoon.

The designation for the well-known midden site is a reference used by the *Agianamiut* (H. Anungazuk, pers. comm.), and it refers to a ceremonial house believed to have once been located in the area shown in Figure 3 (inset). The names the Kazhgitagmiut had for themselves and for their settlement were undoubtedly different from the Agianamiut designations. A prominent, linear, slightly curved depression bisects the mound, running roughly east-west at its approximate midpoint, a feature that was illustrated on Jenness' 1926 map. The linear depression feature indicates that Kurigitavik is grossly comprised of two adjacent, substantial middens, rather than being a single midden deposit. A careful examination of the contours illustrated on the Jenness map also reveals three distinctive lobes in the contours of the southern segment, providing a further indication that this segment was originally composed of three interfacing small mounds. Other, less distinct surface contours on the northern segment of Kurigitavik suggest that it is actually composed of two interfacing middens. Based on the surface contours of the site, it appears that as a whole, the Kurigitavik site bears strong resemblances to the *Agianamiut* district, in actually being composed of a group of middens.

The mound was not occupied by the time historic documentation of the Wales locality began to be developed but archaeological data, including the available radiocarbon dates, presented below, indicate that the site was inhabited at least until AD 1500. Local lore attributes the abandonment of the mound variously as having been precipitated by raids on Wales by Siberians (Jenness 1928a:73) or by the murder of a *Kiatanamiut* by one of the mound residents. As the second account goes, the incident precipitated a sudden evacuation of the mound residents who fled to Barrow, the same destination as those who abandoned the Beach site, as noted above. The second account seems to be the most popular and is still told by present-day Wales residents. Meat caches and presumably other features were constructed on the mound by residents of the Beach site after the departure of the mound dwellers (H. Anungazuk, pers. comm. 1999). In this regard, Jenness' (1928a) mention of 'a modern graveyard' on the western end of the mound in 1926 indicates a relatively long-term historical use of the site by Wales residents, and it is worth noting that the last burial in this portion of the site took place only some 40 years ago.

The mound has also been affected by artifact mining in the 20th century, with the use of hydraulic excavation methods reported; presumably the more conventional shovel technique was used as well. A hiatus in artifact mining at the mound was imposed by the village corporation sometime in the early 1960s, following its inclusion in the Wales National Register District, in order to preserve the site. As with the Hillside site, Kurigitavik midden deposits are now covered under a lush growth of *Artemesia*. In the 40 years since the prohibition on artifact mining went into effect, the old mining pits now have fully developed vegetation, and there is now no indication of relative ages from one depression to the next.

EARLY RESEARCH AT WALES

The Wales sites have played important roles in reconstructions of prehistoric Eskimo ethnogenesis that were initially advanced over 70 years ago, first by Diamond Jenness in 1926, followed 10 years later by Henry Collins (e.g., Collins 1964; Jenness 1928a; Morrison 1991). Both Jenness and Collins excavated features at the Hillside site and Kurigitavik Mound. Jenness' excavations cannot be definitely located because of inaccuracies in his sketch map and because of intense artifact mining carried out at the site in the 75 years since his effort. Although the locations of Collins' excavations at the Hillside were never documented by Collins, an old photograph he took during the 1936 work appears to show, albeit without certainty, that some of his excavations took place near the northern end of the site, near the location of a frame house that was razed sometime after 1926 (Collins' 1936 photograph, in files of D. Dumond). The frame house appears in the right side of a photograph taken by Jenness in 1926 (Morrison 1991:Fig. 2). Knowledge of the specific locations of Collins' Hillside excavations was lost with Collins' death.

Diamond Jenness

Jenness' 1926 excavations at Wales were his first excavations in the Alaskan Arctic. This work was oriented toward investigating Eskimo origins as an 'ethnological problem' and toward finding evidence of human migrations from Asia to America (Jenness 1928a; Collins 1964:91-92; Morrison 1991:1). A considerable volume of early writings on northern prehistory was dedicated to the problem of Eskimo origins, and the Wales, Diomede, St. Lawrence Island, and Point Barrow sites provided some of the primary information for the first formulations of Eskimo prehistory (e.g., Collins 1940). Jenness' Wales investigation included the excavations at Kurigitavik Mound, the Hillside site, and the Beach site, as described above (Morrison 1991:3-4; Fig. 2). A total of 1820 artifacts were excavated and purchased at Wales by Jenness in 1926 (Jenness 1928a:72-73); he describes excavating a total of 'eight pits' on Kurigitavik Mound. Of the eight, five produced sufficient collections to define the range of artifacts that Jenness believed to be present (Fig. 3). Work at the Hillside site consisted of partial excavation of two superimposed houses and two nearby houses; work at the Beach site consisted of at least partial excavation of two houses in the modern village (Morrison 1991; Fig. 3). None of the houses excavated by Jenness were described with regard to form or construction, either in Jenness' notes or published articles (*cf.* Morrison 1991).

Based on these results, Jenness (1928a:73-74; Morrison 1991:5-6) concluded that three periods of occupation were represented, including a contact period represented by items of European industrial origins such as glass beads and iron items. The two precontact periods consisted of an early presence represented by a deeply buried house at the Hillside site in which harpoon head forms differed from those of less deeply buried forms found in the upper deposits at both the Hillside and Kurigitavik. In Jenness' treatment of the 1926 collections 'from Bering Strait,' the Diomede and Wales collections are presumed to collectively represent a single cultural interaction sphere (e.g., Jenness 1928a:78). In this regard, although Jenness' historic initial encounter with Old Bering Sea elements was restricted geographically to Little Diomede Island, he interpreted the remains as the cultural antecedent to the late prehistoric and historic cultures at Wales. Based on comparisons between Thule artifact traits described by Mathiessen (1927) from the central Canadian Arctic, and his discoveries at Wales and the Diomede Islands, Jenness concluded that Eskimo precursors were present in the Bering Strait area as early as 2000 years ago (Collins 1940:561).

The extent of Jenness' reporting on his Wales excavations consists of two articles (1928a, 1928b), in which comparisons are drawn with Mathiessen's data from the Canadian Arctic. However, Morrison (1991) has developed a published report on Jenness' collection, wherein the meager information for collection proveniences and excavation procedures and the collections themselves are reviewed. Some interpretations are given as well, in light of Collins' subsequent work at Wales and the developments that have taken place in arctic prehistory since 1936. A discussion of Morrison's work is presented below.

Henry Collins

Collins (e.g., 1937b, 1940, 1964) conducted an investigation at Wales in 1936, ten years after Jenness' study. As noted above, Collins' study was carried out as a series of excavations focused at Kurigitavik Mound and as excavations in unspecified locations at the Hillside site and at the so-called Birnirk Mound (TEL-031). Two radiocarbon dates were obtained by Collins for Thule-Punuk deposits at Kurigitavik Mound, and they define a range of 607–727 years AD. Lower deposits at the mound were assigned to Birnirk culture, and a single radiocarbon date obtained by Collins yielded a date of AD 637 (Collins 1964:99; Rainey and Ralph 1959). A fourth radiocarbon determination also produced a date of AD 477 for the Birnirk levels of a location that Collins (1964:99) refers to as the 'nearby beach midden,' actually located near the base of the Hillside site slope somewhere adjacent to the modern shore (Rainey and Ralph 1959). All of the radiocarbon dating done by Collins was based on the early solid carbon method, a technique with significantly large sigma factors, by modern standards, and the reliability of the dates was questioned for other reasons as well (Rainey and Ralph 1959:368).

An important result of Collins' work was the first documentation of Thule culture in Alaska and its stratigraphic and temporal position, a new discovery of the time that was derived from Mathiassen's newly formulated late prehistoric culture. Collins' (e.g., 1940, 1964) interpretations of Wales prehistory were couched in Mathiessen's (1930) formulation of the Thule culture found in the eastern Canadian Arctic, and explanations for Thule manifestations in the Western Arctic therefore proceeded from the perspective of an Eastern Arctic expression of the culture, through comparisons of traits lists (e.g., Collins 1940:560). On this basis, Collins (1940:562) suggested that Birnirk culture represents an intermediate stage of development between Old Bering Sea culture and Thule, and that Thule developed somewhere in coastal Alaska, east of Wales, and subsequently diffused to both the east and west after its initial development (see also Stanford 1976:1-3,112-114). In this scenario, the Punuk culture was centered on the Bering Sea islands and Chukotka coast and the relative strength of its influences on North American Thule culture is viewed as a function of relative distance. Alaskan inhabitants of Bering Strait shores therefore felt influences of the nearby Punuk heartland most strongly. In some alternative interpretations of arctic prehistory that followed Collins (e.g., Dumond 1987:118-119, 128-133, 139-145), the developments leading to Thule culture are portrayed as occurring in the Bering Sea region and spreading eastward into the Canadian Arctic and eventually to Greenland.

Only a few articles concerning his Wales data were published by Collins from 1937-1964. In these writings, Collins refers to the archaeological data for the Wales locality, but he never published a full report on his findings from the 1936 field season. Collins did, however, produce a draft manuscript containing descriptions of the Wales collection, which has been maintained at the Smithsonian Institution. Don Dumond of the University of Oregon has developed the manuscript into a published monograph (Dumond and Collins 2000). Descriptions of house forms and construction features are limited, consisting of structural members, a single complete room and portions of other rooms (e.g. Dumond and Collins 2000:Figs. 2.2, 2.3 and 3.1, and text). Even these limited descriptions were excluded from Collins' published reports (e.g., Collins 1940, 1964).

LATER STUDIES

Although no further field investigations were conducted at the Wales sites until the current effort began in 1996, additional analyses of the Jenness and Collins collections maintained at the musuems in Hull and Washington, D.C., have been carried out by Yama'ura (1979a, 1979b, 1984) and Morrison (1991). In these studies, the attributes of artifact forms and decoration are the primary bases on which the analyses rest and to a large extent, these studies are continuations of the early pre-radiocarbon dating artifact seriation studies done in the region (e.g., Collins 1937a). In the case of the Wales collections, precise contextual information was not recorded, and therefore proveniences of the items are minimal or nonexistent.

Yama'ura's studies (1984) focus on the attributes of sealing harpoon heads recovered from Kurigitavik Mound by Collins in 1936, and he utilizes considerable additional data developed from sites in the region where work was carried out subsequent to Collins' work. With only a handful of minor deviations from Collins' (1940) interpretations of the distributions of attributes, such as the similarity of some Kurigitavik harpoon head forms (e.g., certain Sicco, Natchuk, and Nunagiak forms) with examples found at the Nunagiak site near Barrow, Yama'ura's analysis otherwise confirms Collins' assessments (e.g., Collins 1940). Yama'ura's (1984:221-223) argument for a distinctive 'Kurigitavik culture' follows Collins' prior reasoning as well, insofar as the derivations of decorative styles represented in the Kurigitavik Mound collection are attributed to strong influences from Punuk cultural centers in the Bering Strait area.

Morrison's (1991) report on the 1926 Jenness collection from Wales and Little Diomede includes a review and analysis of late prehistoric artifact trait distributions in the region that includes archaeological investigations up to and including Yama'ura's study. Morrison acknowledges the lack of documentation of contexts for the Wales and Little Diomede collections (Morrison 1991:2-3, 32). His interpretation of the collection and, concomitantly, of cultural ascriptions of the house features from which they came, varies somewhat from Jenness' (1928a) interpretation, largely due to the benefit of investigations carried out in the Bering Strait area since 1926. In this respect, Morrison (1991:5-6) notes that 'Igloo 12' at the Hillside site appears to actually be late prehistoric rather than historic, based on the discovery by Giddings (1952) that fishnets and associated implements were prehistoric technology rather than one introduced by Russian explorers or traders. In a somewhat different vein, Morrison (1991:5-6, 29-30) suggests that 'many of the Kurigitavik houses ... are better considered Late Prehistoric than Thule' based on technological elements that suggest some combination of Punuk and Thule influences. Although this interpretation is a fine distinction by comparison with definitions of late prehistoric entities in other areas of western Alaska, it is nevertheless consistent with Collins' (1964:99) prior suggestion for the later occupations of Kurigitavik Mound, and it does not seem to differ to any significant degree from Yama'ura's (1984) formulation of 'Kurigitavik Culture.' Morrison's important contributions are in providing thorough descriptions of the collection and of Jenness' work at Wales and Diomede. Morrison therefore presents an analysis of the items in the collection based on their attributes, and places the collections in a modern regional perspective.

SUBSISTENCE

Studies of traditional Wales subsistence that have been carried out are based solely on ethnographic and historic information developed over the past 40 years, without the benefit of a background and context provided by an analysis of prehistoric subsistence. This circumstance is due mainly to the fact that faunal remains were neither completely collected or fully analyzed by Jenness or Collins during their work at the Wales sites (*cf.* Morrison 1991:2-3). Changes that have occurred in subsistence patterns following European contact in the mid-19th century are, therefore, not well understood because of this limitation (*cf.* Braund and Moorehead 1995:264-265; Ellanna 1983; Ray 1975:111-120; Stoker and Krupnik 1993:599-622). And, it has generally been assumed that the use of local resources and trade goods during prehistoric times approximated that of historic times.

The early historic data that are available are both less reliable and less substantial than is necessary to establish a baseline model of a late prehistoric Wales subsistence. The existing data are limited to those of sea mammal catches recorded by Thornton and Lopp during their tenure at Wales from 1890-1902 (*cf.* Ray 1975; Ellanna 1983). More recent studies by Ellanna (1983), Braund and Moorehead (1995), and Stoker (e.g., Stoker and Krupnik 1993) concentrate on early historic to present-day patterns.

The presumption of the primacy of whaling at Wales (e.g., Braund and Moorehead 1995; Stoker and Krupnik 1993) is routinely accompanied by a perception that bowhead whales provided the substantial subsistence base for supporting the large population of the village. The limited records of Lopp and Thornton for contact-period whaling indicate that one whale was taken in 1888, one in 1889, none in 1890, one in 1891, and none in 1895. The overall catch for eight years was therefore three whales, but this record is for the period when the Bering Sea bowhead whale stock had been tremendously decimated by the predations of the Yankee whaling fleet (e.g., Bockstoce 1986). It is therefore not certain if the figures for the number of whales taken during the eight-year period is a reflection of substantially diminished numbers of animals available, or if it is representative of the number of whales taken in prehistoric times, over a similar period. Therefore, a glaring question concerning traditional subsistence at Wales is whether whaling was more effective in prehistoric times than that described historically, and it is equally important to establish the roles of the other prey species in the prehistoric Wales economy. Also of import is the question of how the radical reduction in the numbers of whales by the commercial whaling operations affected the 19th century residents of Wales, who were able to continue congregating in extremely large numbers through the time of the Yankee whaling fleet, up to 1900 (*cf.* Bockstoce 1986; Ellanna 1983:281-285). Regardless of what the numbers of whales caught in the late 19th century may indicate about the impact of commercial whaling, it nevertheless demonstrates the prowess of the Wales hunters under documented conditions of a diminished whale population.

As noted previously, there is no indication that the number of Wales residents declined substantially before the onset of the 1918 Spanish influenza pandemic (*cf.* Ellanna 1983; Ray 1975:109-111) and in this regard, the catch of three whales over eight years does not appear sufficient to sustain the 500 individuals concentrated at the village. On the other hand, it is clear that walrus were a major resource at 19th century Wales, King Island, and Little Diomede Island whose residents Ray (1975:111) describes as 'walrus hunters, sometimes whalers' (Fig. 5).

Early records of walrus use at Wales indicate that the documented catch in one early historic year, 1890, consisted of 322 animals (Ray 1975:112; *cf.* Ellanna 1983:458-466; see also Smith and Smith 2001:113-114). By any reckoning, 322 is a considerable number of these enormous animals and based on this documented walrus catch, the ratio for Wales is 0.64 animals per person for 500 individuals. By comparison, an annual catch of 50 animals was sufficient to support a population of 175 King Island residents in 1933 (Ray 1975:112, n.13), a catch that translates to a ratio of slightly less than 0.30 animals per person for King Island residents in

1933. Wales residents indicated that prior to the substantial reduction of the walrus population by the Yankee whaling fleet in the latter half of the 19th century, 600–700 of the animals were taken each year (Ray 1975:112), a harvest level that would undoubtedly support large numbers of human hunters.

It is important as well to consider the effects of climate changes on hunting strategies. The conditions under which whales are hunted have not been advantageous during historic times. An example of this circumstance found in the ethnographic data indicates that bowhead migration routes were some distance from the village during the late 19th century, with at least one description of hunts taking place 'in the middle of the strait' rather than immediately offshore from the village (Ray 1975:113). The effects of changes in sea ice, prevailing winds, and gross climatic changes have had impacts on present-day whale and walrus hunting strategies, and no doubt similar processes occurred in prehistoric times (e.g., Harritt 2001:70). As a case in point, modern Wales hunters note that the later formation of sea ice in recent times makes it necessary to go farther out from shore to hunt walrus.

Figure 5. *A semisubterranean house decorated with walrus skulls at Wales, ca. 1890-1902, photographed by Thomas Lopp (Smith 1995:42; Smith and Smith 2001:Fig. 19). The use of the skulls in this manner suggests that this was the residence of a successful walrus hunter (Thomas Lopp Collection, No. 94-132-18N, Alaska and Polar Regions Department, University of Alaska Fairbanks).*

Therefore, although historic information supports the perception of the importance of bowhead whales to the inhabitants of the locality, it also suggests that other prey species and other resources played major roles as well. 'Common seals' (presumably the ringed seal, *Phoca hispida*; *cf.* Ellanna 1983:Fig. 47 and text), were also taken in large numbers, with between 4000 and 5000 (or around nine animals per person, for subsistence and/or trade) reported as annual catches for the decade of the 1890s (Ray 1975:114). Among other game, the Seward Peninsula caribou herds were well-known for their abundance during the mid-19th century, but a population decline that occurred by the end of the century has been attributed variously to the introduction of firearms among the Natives and a possible natural shift in migration routes (e.g., Ray 1975:174,177). In any case, the decline in numbers of sea mammals and the relative scarcity of caribou, in concert with episodic epidemic diseases, created a crisis for the inhabitants of Wales in the first decades of the 20th century. In a fundamental way, this resulted in a break from the old lifeways to one more closely connected to the American economy and society of the early 20th century.

Wales residents ceased whaling in the late 1930s and then commenced again in 1969 (Anungazuk 1995; Braund and Moorehead 1995; Ellanna 1983:450-473). The landing of a bowhead in 1969 was followed in 1970 by another successful hunt, but Wales' success has been less frequent than that of some of the other Alaskan whaling villages from 1969 to the present day (Braund and Moorehead 1995:264-265, Table 2; Ellanna 1983:Table 93).

The most frequent reason given for the nearly 40-year cessation of whaling by modern day residents is that it was caused by the 'epidemics' of the early 1900s (Anungazuk and Bodenhorn interviews 1996). During a portion of the whaling hiatus in the 1960s, it is said that some families specialized in seal hunting. Modern Wales residents continue to rely heavily on subsistence harvests: bowheads, bearded seal, walrus, beluga, shellfish, fish, ducks, greens, berries, and a limited number of musk-ox (Braund and Moorehead 1995; Ellanna 1983).

THE PRESENT PROJECT

Phase I, 1996
In the start-up phase of the current project, a brief, two-week effort was conducted at Wales that combined sociocultural fieldwork and archaeological and geomorphological field evaluations. Objectives in 1996 included a field evaluation of sites and a preliminary analysis of modern practices related to whaling. Interviews were conducted with four whaling captains, two crew members, two whaling captains' wives, and three elders who had been involved in whaling in the mid-20th century. A geomorphological survey of the Cape Prince of Wales beach ridge system was conducted, and the conditions of the archaeological sites were inspected.

Results of the 1996 effort were threefold: First, the locations of four *qargi* or ceremonial structures and traditional meat caches were established on the basis of interviews. Second, it was established that artifact mining at the sites has destroyed substantial portions of them, but undisturbed stratigraphy was found. Charcoal and peat samples were collected, and one archaeological sample from the Hillside site was assayed. Radiocarbon samples collected in 1996 from selected Wales beach ridges also provided index ages for the Cape Prince of Wales beach ridge sequence, which enabled comparisons with more extensively studied beach ridge complexes in the region.

Finally, a small collection of artifacts was made from the surface at the Hillside site, from looting disturbances, in order to develop an interpretive exhibit for the new Wales school. A supplement to the 1997 sociocultural component grant was awarded for the development of this exhibit. Interviews were conducted with five Wales elders in order to help identify the functions of some of the pieces. The interviews have been transcribed, and the information was incorporated into text descriptions for the artifacts in the display and the names of most of the objects. The displays were completed in July 1999.

Phase II, 1998–2002

The second phase of the project focused on testing and excavation efforts at the three main sites in the locality. The 1998 effort served two main purposes. First, new samples of upper-to mid-level cultural deposits at all three sites were obtained, including new artifact collections and radiocarbon samples. Second, the locations of intact, subsurface prehistoric features at Kurigitavik Mound and the Beach site were established. Some of the results of the 1998 through 2001 field seasons are discussed below.

Hillside Site. At the Hillside site pits excavated by artifact miners were examined in 1998 on each of the six midden mounds for potential exposure of deep, intact deposits. For reference purposes each mound was assigned an alphabetical designation, with Mound A designating the southernmost mound in the site; the next mound northward is designated Mound B, and so forth. Following this assessment, locations that were judged likely to yield information about intact cultural deposits were chosen. A side of the selected mining pit was then made into a vertical wall by troweling to expose a clear profile of the deposits, down to the bottom of the pit. A 1 x 1 m test unit was then excavated adjacent to the base of the profile through disturbed deposit, then into undisturbed deposits. The deepest of the 1998 tests was located in Mound C and the excavated test proceeded to contact with sterile deposits; the overall depth of the midden deposit revealed by this test was approximately 3.8 m below the modern surface. Two substantial house floors were exposed in the faced-off portion of Pit C-1 and a third compacted habitation surface was observed as well, along with a considerable deposit of faunal remains, implements, and other cultural debris.

In 1999, the specific objective for the single test excavated at the Hillside site in Mound A was to obtain additional radiocarbon samples to confirm the 1270 BP age assay of the 1996 sample. This is the oldest date obtained by the present project so far, and in conjunction with Collins' solid carbon date of 1480 years (AD 477) for the Hillside, it provides further evidence for the maximum age of the initial occupation of the Wales locality (Table 1). The 1999 test in Mound A was located approximately 3.0 m upslope from the 1996 test, on the same midden and in an area that appeared to have had approximately 1.0 m of overburden removed as a result of mining activities. This test was excavated to sterile deposits at a depth slightly in excess of 1.0 m below the bottom of the mining pit. Only a handful of artifacts, such as animal bones and pottery fragments, were found in a relatively thin cultural deposit in this test excavation. A radiocarbon sample was collected from the lowest extent of the deposit, which may provide additional evidence for the age of the human presence at Mound A. This radiocarbon sample has not yet been dated.

At Mound C, a single 1 x 1 m test was excavated to corroborate the depth of the deposit encountered in this midden in 1998; a radiocarbon date obtained from the deepest cultural deposit in the 1998 test yielded an age of approximately 950 years, and this appears to be the maximum age for Mound C occupations. A second important objective for the 1999 test was to obtain radiocarbon samples for undisturbed middle-level deposits in this midden. The 1999 test was located approximately 3.50 m downslope from the 1998 test. The eastern, upslope side of the 1999 test consisted of the faced-off wall of a mining pit that had been excavated to a depth of approximately 1.20 m below the modern surface. The southern, adjoining wall of the test consisted of the faced-off wall of the same mining pit; the southern wall in this test provided a considerable amount of new information concerning the profile of the slope deposits at Mound C, which in conjunction with the profiles of the 1998 test, revealed much about the development of the midden over the past 1,000 years. In the process of facing off the two adjoining walls of the mining pit in preparation for excavating the test, an upper-level occupation surface was found at a depth of approximately 0.80 m below the modern surface; this surface had been dug through as a result of the artifact mining process. Excavation of the test into the bottom of the mining pit eventually revealed the presence of two deeper occupation deposits. In the mid-level deposits in Mound C, 1.40–1.90 m below the modern surface, a substantial, 0.50 m thick deposit of wood chips and fragments with interlayered, compacted segments of surfaces was found. This deposit also contained substantial numbers of wood shaft fragments and modified wood sections, all of which collectively suggest that the area had been used as a workshop. Because the deposit followed the natural contour of the slope, it is apparent that the workshop was an outdoor activity area, used during the warmer months of the year. The Mound C test was excavated to a depth of approximately 0.80 m below the bottom of the mining pit. Work was halted at this point due to the rapid accumulation of meltwater and runoff, following heavy rains.

Kurigitavik Mound

North Excavation Block. In 1998, a 1 x 5 m trench was excavated on the northern side of Kurigitavik Mound, oriented north-south and placed so that the midpoint of the trench intersected one corner of a surface depression believed to be a corner of Jenness' 'Igloo 7' that was excavated 73 years earlier. Intact, undisturbed deposits were encountered on the western side of the trench, and in the deepest levels of 1998, a log and other wood members of a constructed feature were found extending eastward beyond the trench. In 1999, an expansion on the eastern side of the 1998 trench at Kurigitavik Mound revealed additional portions of a structure, and collectively they represent most of an entry tunnel and a front portion of a room of a buried house, all lying at depths slightly in excess of 1.0 m below the modern surface. It was at that point clear that the intact house found in the 1998–1999 excavation is <u>not</u> Jenness' Igloo 7, inasmuch as Jenness' activities in the area reportedly comprised complete excavations of eight houses, including Igloo 7 (Morrison 1991:3). Excavations in 2000 focused on exposing additional areas of the house entry tunnel and the portion of the room that had been exposed in 1998–1999 in the northern excavation block at depths slightly in excess of 1.0 m (Fig. 6). Efforts were focused on exposing the buried portions of the house that lay to the east and south, in 1 m expansions along those sides of the block. In the process of exposing the house members and floor, a distinctive walrus skull ring feature was encountered above the house floor in the southwestern quadrant of the excavation block. This ring appears to be approximately 2.5 m in diameter and was comprised of male walrus skulls with the tusks broken out of their sockets.

The skulls were placed nose to occiput, and in some segments of the ring they were three skulls deep. A second upper level concentration of walrus bones comprised of two walrus skulls, mandibles, and several long bones was found in a cluster approximately 80 cm across; this grouping extended into the southern wall of the excavation. A third grouping of walrus skulls was found near the southeastern corner of the excavation block. Although the distribution of these skulls appeared to be unorganized, the overall pattern suggests that it may have been a second skull ring feature. Intact and disturbed rings of walrus skulls such as these have been found at many sites on the Chukotka coast (Orekhov 1999).

Figure 6. *An oblique view of a part of the north excavation block at Kurigitavik Mound at the end of the 2000 field season. Shown here are elements of the partially excavated house and the walrus skull ring (upper right-hand corner).*

In 2000, excavation of the eastern and southern expansions of the northern block were carried out in thicker levels that those used in the initial 1998 excavation and the expansion of 1999, based on enlightened understanding of the deposits provided by two year's experience excavating at the site. In this respect, the excavation proceeded in 20 cm arbitrary levels and real levels down to contact with the house members. At this point an effort was made to delineate the outline of the house by establishing where floor deposits corresponded with horizontal and uprights of the house's log construction. In many cases the relationships were clear cut, but in some instances, such as in the southern end of the block the edge of the house floor deposit appeared to overlay some of the wall members, rendering the relationships less certain than one would hope. Floor deposits were collected by unit quadrant for screening. Screening was done by saturating the sample in the screen with water from the nearby lagoon; the fraction of cultural material remaining in the screen was then collected as a unit quadrant collection.

In 2001, the block was expanded 2 m to the south, bringing the overall length of the block to 6 m north–south, and 1 m to the west, bringing the overall width of the block to 5 m east–west. Although the primary objective was further uncovering the house that was partially excavated in 1998–2000, another important goal in 2001 was to establish a terminal date for the abandonment of the Mound by Kurigitavik people. The legend of the abandonment of TEL-079 is still well known to Wales residents, but the actual time of the departure of Kurigitavik people is vague in the oral history, and the event has never been established by researchers in modern times. In the initial stages of the 2001 excavation, a section of a small log was selected from a layer that lay beneath the sod, from a context that appeared to have escaped the historic artifact mining that was carried out at the site. It is also important to note that the deposit from which the log section was recovered did not contain any historic materials or objects. The result of the radiometric assay is 450±50, or calendar date of approximately AD 1500 (Table 1, below).

Excavation of the new expansions of 2001 extended to the same levels as the 2000 excavation, and exposed additional portions of a house structure that extended from approximately the mid-point of the excavation block to the south, southeast, and southwest.

Figure 7. *Crew working in the north excavation block at Kurigitavik Mound (TEL-079) in 2001. Wooden structural members of partially-excavated, apparently intact house remains are visible.*

The general configuration of the house remains now include a relatively small compartment or room (excavated in 1999 and 2000) located in the north-central portion of the excavation; to the south, a north-south oriented wall segment composed of a series of short uprights, each approximately 40 cm in length, was uncovered in 2001. The short wall joins with more substantial, broken-off posts curving approximately 150 cm to the east and slightly south to a crescent-shape grouping of uprights. The grouping of uprights is tentatively considered to be a possible interior corner of a room, insofar as this type of arrangement would presumably function to reinforce a juncture of horizontal wall logs. Deposits found within the house area in 2000 and 2001 consist of sediments containing a large percentage of decayed organic materials, bones and bone fragments, and a relatively large number of artifacts and debitage. Samples of seeds collected from four different unit quadrants of floor deposits exposed in 2001 have been identified as cloudberry (*Rubus chamaemorus;* G. Davies, pers. comm), thereby providing clear-cut evidence of the use of this plant resource by the inhabitants, and probable habitation of the house during summer months.

A distinct and substantial lens of hearth material was found in 2001 on the southern side of the excavation near the mid-point, and extending into the excavation wall. Horizontal timbers discovered in 2001 included two horizontally split sections of logs, one which bore clear-cut signs of wear along one edge, such as may be expected to occur on a step or on the edge of an entry as result of rubbing by human feet. Additional walrus skulls were uncovered in 2001, however aside from a moderate degree of proximity to the skull ring uncovered in 1999, no type of organized distribution could be discerned (Fig. 7).

A combination of new information became available during the 2001 field season that resulted in an alteration of the prior interpretation of the structure. New data on the Kurigatavik site was published in the spring of 2000 (Dumond and Collins 2000). An illustration of a part of 'House 2' (Dumond and Collins 2000:Fig. 2.3) excavated in 1936 by Collins is presented in this new release, and it resembles the TEL-079 house currently being excavated, with regard to the general construction methods reflected in the walls and in the configuration and relative size of a compartment or small room that was completely uncovered by Collins. The illustration of a 'blubber soaked' deposit in a small compartment of the 1936 house is also consistent with the decayed sea mammal tissue found in the 1999–2000 compartment and, in general, the two examples suggest that the small compartments functioned as storage rooms, caches, or possibly as small kitchens.

At the present stage of progress, the house components now uncovered are minimally equivalent to the extent of the 1936 House 2 excavated by Collins.

Southern Trench. A new trench was also begun at TEL-079 in 2000 on the southern side of the mound in the area where Collins discovered a Birnirk harpoon head in 1936 at a depth of approximately 2.10 m below the surface. The trench area includes two substantial surface depressions that presumably are products of the artifact mining done at the site prior to its listing on the National

Register in 1962. As with the tests that have been excavated at the Hillside site (TEL-025) during the project, the trench actually takes advantage of the mining pits, as a means of reaching deep, intact desposits without spending the substantial amounts of time that would be required to excavate the upper level deposits down to the deepest cultural deposits at the mound. In this approach, the deep level where Collins recovered a Birnirk harpoon head should be exposed and new samples for radiocarbon dating will be obtained, to be used in acquiring new, more accurate dates for the deposits. In 2001, excavations proceeded in the five northern units of the trench to a depth of approximately 1.5 m. Although it had been presumed that the northern end of the excavation had penetrated into intact, undisturbed deposits, a layer containing a dozen tin cans was uncovered in this portion of the excavation. The presence of the recent items at this depth demonstrated that the deposits excavated through 2001 have been disturbed in recent historic times. Excavation of the deeper deposits is planned for the future.

Beach Site

Southeastern Excavation Block. In 1998, a 1 x 4 m trench was excavated at the Beach site on the landward, eastern edge of the beach berm on which present-day Wales is situated. The trench was oriented parallel to the slope of the berm. An abandoned frame house built in the 1940s is located approximately 5 m west of the western end of the trench, and the location of the trench has many modern trash pits and trash both in the pits and around them. The 1998 trench was placed between two pits, but it partially intersected one of these on its southern side. Deposits encountered in this excavation included back dirt from excavation of the modern trash pits; these deposits contained both modern trash and prehistoric artifacts that are presumed to be displaced from their original contexts during the trash pit excavations. Underlying the upper, disturbed deposits was an intact, undisturbed deposit containing prehistoric artifacts and a corner of a constructed log feature. Excavations in this trench extended to 1.20 m below the modern surface in 1998. In 1999, excavation of the area on the northern side of the 1998 trench at the Beach site revealed structural elements related to those of the 1998 excavation, but the function of the structure was still not clear by the end of the 1999 field season. The exposed portion of the feature consisted of a more or less horizontal arrangement of whale ribs and timbers associated with several posts. Although collectively they suggested a house structure, a firm identification of the function was not established. A 1 x 2 m area in the bottom of the 1998 trench was also further excavated to a deep deposit of what appeared to be undisturbed beach sand deposits. A handful of artifacts were recovered from the deep deposits, none of which is diagnostic beyond a general consistency with late prehistoric Thule technology. A thick, lens-like deposit of sea mammal oil-cemented sand was also found that oozed liquid oil as thawing progressed. A post mold was found in the north wall of the excavation in deep deposits below the 1998 levels, associated with a disjuncture in deposits that collectively represent a deeper and older feature than that of the mid-level feature exposed in the 1998–1999 excavations. A 1999 radiocarbon sample produced the first date to be obtained for the Beach site, which is 460± 60 BP for the mid-level remains.

The excavation block was expanded 1 m to the north and 1 m to the west at TEL-026 in 2000. Excavation of the expansion area at the Beach site surprisingly revealed the presence of an undisturbed upper-level bone scatter and a hearth. Both of the upper-level features lay above the structural members that were encountered in the 1998–1999 excavations. In addition, a large toggling whaling harpoon head was found in association with the bone scatter. The upper-level hearth lay beneath most of the bone scatter near the southwest corner of the excavation block. The hearth proper was approximately 50 cm in diameter, delineated by a lens of wood ash and oxidized deposits. A considerable number of small ground slate chips and powder was found in the vicinity of the hearth as well as a relatively large number of implements of various types, including many well preserved types made of bone, antler, ivory, and wood. Neither of the upper level features contained historic materials or objects, and, therefore, it is clear that they are precontact remains. Based on their relative stratigraphic position above the mid-level structural remains of 1998, that are radiocarbon dated to 400–500 years in age, the age of the upper level features is placed at 400–150 years.

In 2000, the mid-level cultural remains were sufficiently exposed to demonstrate that the wood uprights and horizontal timbers, wood sections, and whale-rib sections exposed in 1998 and 1999 were in fact not members of a house. Instead, the elements of the feature, which had been excavated over three field seasons can now collectively be regarded as a cache, on a provisional basis, pending final analysis. The structural members were distributed across an area of approximately 2 x 4 m and the arrangement appears to have been for the purpose of supporting cached material. Excavation below the cache remains revealed a lower-level feature, stratigraphically separated from the cache. In 2001 this feature was completely excavated through the middle levels and it was revealed that it was, in fact, the entry tunnel of a house; an estimated 70% of the main room of the house was uncovered in 2001, and was defined by a floor deposit containing a high percentage of organic material and artifacts. This feature lay beneath the cache excavated in 1999–2000. Among the artifacts found in the floor deposit were a section of a woven baleen mat, basket, or net and a large labret.

Table 1. Radiocarbon Dates for the Wales Locality (to 3/2002).

Lab No.	Rainey and Ralph[1]	Collins[2]	Current Project 1996, 1998, 1999, 2001.[3]	Dumond Calibs.[5] 510yrs	720yrs
			Kurigitavik Mound		
P-65	1320±230	AD637			
P-67	1230±240	AD727			
P-69	1350±360	AD607			
B123364			640±60 (AD1310)		
B123365			580±60 (AD1370)		
B138746			680±60 (AD1270)		
B138747			500±60 (AD1450)		
B139113[4]			1020±40 (AD 730)	AD1240	AD1450
B164466			450±50 (AD1500)		
			Hillside Site		
P-63	1480±240	AD477			
B098939			840 ±60(AD1220)		
B098938			1270±70 (AD 770)		
B123366			950±60 (AD1170)		
			Beach Site		
B 129590			460±50 (AD1405)		
B134829[4]			1030±50(AD1170)	AD1430	AD 1640
B138745			410±60(AD1540)		
B164464			520±50 (AD1430)		
B164465			320±50 (AD1630)		

[1] *Rainey and Ralph (1959:368) note that Collins' radiometric dates ' were among the last ones dated by the solid carbon method. (and) the runs were made during a period of fall-out and were not as reliable as desired.'*

[2] *Collins (1964:99) refers only to calendar dates for his Wales dates. An offset of 7 years is included in the dates in the 1964 article, making each 7 years younger than the value resulting from simply subtracting the age assay from AD 1950; prior to 1960, radiometric values were subtracted from the current year, which in the case of Collins' 1964 article was 1957.*

[3] *Ages for the present project 1996-2001 are conventional radiocarbon ages, adjusted for carbon 13c/12c isotope fractionation.*

[4] *AMS determinations provided by D.E. Dumond; B139113 is from a seal bone associated with a grass sample, B138747, and B134829 is from a seal bone from the same stratum as the B129590 date from a peat sample; calendar dates are provided by BETA.*

[5] *Values developed by Dumond (1998:114-118) for marine reservoir effect corrections for samples with marine origins. The ages shown for B139113 and B134829 are values based on subtracting each of Dumond's corrections from the assayed age, the resulting calendar ages are simple subtractions of the resulting corrected ages from 1950.*

Below the house floor, a layer of clean beach sand up to 20 cm thick was encountered across the extent of the excavated block and it was believed, as in 1999, that it defined the lowest extent of cultural deposits in this area of the site. However, in order to ensure that this was the case, a series of shovel probes were excavated in the middle of each of the 1 x 1m units of the block. As a result, a deep feature was discovered at depths between 140–160 cm below the modern surface. This feature was a circular cache measuring approximately 150 cm in diameter; a floor was made of sea mammal oil soaked logs, planks, and two walrus skulls, collectively arranged to cover the circular feature outline. A radiometric age determination obtained from a sample of peat collected from the cache floor surface indicates an age of 520 years (conventional radiocarbon date); an age determination from a section of wood from the cache floor indicates an age of 320 years (Table 1). Under the circumstances, it is apparent that an age of 520 years is consistent with the dates from samples collected above the cache floor, and it is considered the most accurate of the two determinations. In addition, it is important to point out that the 520 year age places the cache at the approximate time of a brief period of a '...sand pulse of eolian or marine origin (that) co-occurred with northerly Little Ice Age storms...' (Mason n.d.; Mason and Jordan 2002:Fig. 3 caption). This circumstance provides a provisional, preliminary explanation for the layer of clean beach sand that overlays the deep cache feature, and is also generally consistent with the age of the initial formation of the Cape Prince of Wales Unit IV lithostratigraphic unit, sometime immediately preceding AD 1396–1520, that TEL-026 occupies (Mason n.d.; Mason and Jordan 2002:Fig. 3 caption). Following excavation of the deep cache feature, additional shovel probes excavated in the area were negative and, therefore, excavations in the southeast block were concluded.

Northeastern Test Trench. A new test trench was begun on the northeastern end of TEL-026 in 2001. The trench is oriented roughly east-west and is placed on the eastern slope of the beach ridge, extending down to approximately 2 m above the high water mark of the lagoon behind the TEL-026 Beach site. The new trench was only excavated to a depth of approximately 30 cm below the modern surface during the 2001 field season, but complete excavation is planned for the future.

New Radiocarbon Dates
A total of 16 new radiocarbon dates for the Wales sites obtained between 1996–early 2002 are summarized on Table 1, below. This total includes dates from two seal bones provided to Don Dumond of

the University of Oregon for developing a marine correction for radiocarbon dating samples from the Bering Strait area.

One of the conventional determinations from a 1996 sample associated with deep cultural deposits at the Hillside falls within the 1200–1300 year range, a period during which whale hunting is presumed to have begun in Alaska (*cf.* Stanford 1976) and it establishes an early age for the site's occupations and suggests that it may predate the other two major sites in the locality. This is presently the oldest of the recent assays for cultural deposits in the Wales locality, but it is approximately 210 years younger than the 1480 age Collins obtained as a result of a solid carbon assay from an unknown location at the same site (*cf.* Collins 1964; Rainey and Ralph 1959). A second date from a 1996 geological sample of peat exposed by gully erosion produced an age of 840 years, and represents a time at which the southern portion of the Hillside site was stabilized. A third radiocarbon assay obtained from a 1998 sample from a midden deposit at approximately 3.25 m below the modern surface is 950 years, and along with the 1996 date and in conjunction with the other dates for cultural deposits, it suggests differential ages for occupations from one end of the Hillside site to the other. Although none of the new determinations demonstrate consistency with Collins' prior determinations, it is quite possible that Collins' samples were from deeper deposits than those sampled during the current project.

Radiometric assays of samples collected from 1998–2001 at Kurigitavik Mound have yielded age determinations of 450 and 680 years (1500–1270 AD). They indicate a considerably more recent age than Collins' (1964) prior radiometric age determinations of 1223, 1320 and 1343 years (AD 607, 637 and 727) for the upper to mid-level deposits. And, they suggest that a later age range may be more representative for the Thule–Punuk and late Birnirk periods than was indicated by Collins' determinations.

The seal bones provided to Dumond for marine calibration purposes were selected on the basis of their close associations with organic samples assayed with standard radiocarbon techniques. The samples assayed with standard techniques included (a) a grass sample (B138747) from the house floor in the northern excavation block at TEL-079 and (b) a sample of peat (B138745) from the TEL-026 excavation. Dumond's AMS assays of the seal bones are listed below as B139113 and B134829 (see Table 1, Footnotes 4 and 5). The new corrections for marine samples have recently been published (Dumond and Griffin 2002). The new determinants are among the first radiocarbon dates to be obtained for the Beach site and they provide a crucial, initial basis for comparing the artifact assemblage from this site with those from the other Wales sites. The two radiocarbon assays of 2001 samples from the deepest cultural deposit in the southeast block in the Beach site yielded conventional ages of 520±50 (B164464) and 320±50 (B164465; see Table 1). As discussed above, the 520–year age is considered to be the most accurate for the deep cache feature, and is consistent with ages obtained from samples from stratigraphic positions above the deep cache.

Completion of Phase II of the project will consist of completing the field investigations that are now focused at the Beach site and Kurigitavik mound. The fieldwork will concentrate on completing the excavations begun in 1998 down to sterile deposits and conducting smaller tests in additional locations at each site, as described above. Primary objectives are to develop complete documentation of the cultural deposits at the three sites with respect to their ages and to characterize the cultural deposits represented in each.

CONCLUSION

At this stage of the current project, it is clear that the inhabitants of the different Wales districts used a wide variety of local resources and the faunal analysis completed so far reveals, not surprisingly, that a large volume of marine products including shellfish and a variety of birds were utilized. Although a final evaluation of the role of marine mammals in the Wales economy will not be completed until all of the data are obtained, it can be noted at this point that whales do not appear to be the crucial resource for sustaining the settlement's large population at the time of contact. Instead, smaller sea mammals, especially walrus and seals as well as shellfish, appear to have provided major and important contributions throughout the year. The role of terrestrial fauna including caribou appears to have been surprisingly small, based on the numbers of remains recovered. The implication of these preliminary assessments for the major goals of the project is that the settlement was strategically positioned to access a rich variety of marine and terrestrial resources including whales, when conditions permitted. As viewed from this perspective, the acquisition of late prehistoric Siberian whaling techniques by Wales residents undoubtedly occurred under favorable conditions in which the large numbers of people needed for manning whale-hunting crews and butchering the huge animals could congregate with an assurance of having adequate food supplies, regardless of the outcome of the whale hunt.

In the present study, it is important to consider that at the time that Jenness' and Collins' investigations were conducted at Wales, a number of cultures in the region had not yet been identified that, as a group, have increased the complexity of the late prehistory of northwest Alaska shores by a considerable degree. In Alaska these included the Denbigh Flint complex and the Old Whaling, Choris, Norton, and Ipiutak cultures, all of which occupied locations along northwestern Alaskan shores during at least one season of the year. Of these cultures, it is generally conceded in some measure that Denbigh, Choris, Norton, and Ipiutak represent an 'Arctic Small tool tradition' lineage in which seal hunting figures prominently. This lineage is nevertheless separated from the Northern Maritime tradition lineage by the absence of open-water whaling methods practiced in the region from contact in the first half of the 19th-century into early historic times (e.g., Giddings and Anderson 1986:313-314, 316-317). It is important to consider as well that the Birnirk culture, a crucial link in Collins' (1964) Northern Maritime tradition sequence, had been identified on the basis of remains found at the Birnirk site at Point Barrow, but the distribution of Birnirk sites on Alaskan shores had not yet been established in the first three decades of the 20th century (e.g., Collins 1933; 1940:556-565; 1964; J. A. Mason 1930).

The Old Whaling culture has retained some aura of mystery since its discovery in 1958, primarily due to its extremely brief appearance, dating to nearly 3,200 years ago, and its single Alaskan occurrence that has been documented so far at Cape Krusenstern (Giddings and Anderson 1986:32, 231-267). In spite of the romantic images elicited by its designation, and aside from the large implements that appear to be most suitable for hunting and butchering large whales, direct evidence of subsistence pursuits of 'Old Whaling' people—that is, the actual bones of the hunted animals—indicates the taking of walrus, seal, and polar bear but no whales (Ackerman 1998:252-255). And although the presence of a related culture on Wrangel Island during the same period as the Cape Krusenstern occupation has illuminated some aspects of Old Whaling, important questions remain, such as how sea mammals were hunted. In the case of Wrangel Island at least, the use of watercraft has been suggested, based on the circumstantial evidence of the insular location of the *Chertov Ovrag* site. The role of Old Whaling in the development of the Western Arctic whaling villages during the first millennium AD, therefore, is not yet established.

Among the data that has accumulated over the past several decades are a large number of radiocarbon dates that relate to prehistoric Eskimo ethnogenesis. As noted, Collins' radiocarbon dates from the mound and Hillside sites were among the last to be obtained by the solid carbon assaying technique (Rainey and Ralph 1959). After 1957, the time when Collins' samples were assayed, the gas method soon became the standard, and radiometric assaying has become increasingly refined and sophisticated, both in the areas of developing corrections for radiocarbon enrichment and leaching that affect sample assay values and in the development of calibrations for raw ages that seek a high level of precision in translating radiocarbon years into calendar years. Advancements in the radiocarbon technique and an increasing interest among some prehistorians in examining chronological relationships on the basis of radiocarbon assays have essentially made radiometry a special focus in arctic Alaska over the past ten years (e.g., Dumond 1998:111-119; Dumond and Griffen 2002; Gerlach and Mason 1992; O. Mason 1998). In this regard, it remains to be seen whether increasingly sophisticated techniques of calibrating and correcting raw assay values—especially those derived from marine samples—will engender an unprecedented precision and faith in the method, or if the persistence of 'tweaking' for the sake of unquestionable accuracy will lead to a general distrust among archaeologists. For the time being, the decision of whether to accept a given assay as being accurate continues to reside in the realm of 'best fit' with associated data and—in the absence of accompanying data—with a certain faith in the technology.

In spite of the data that have been amassed since Jenness and Collins carried out their Wales work, the major issues have remained the same with continued emphasis on the relationships, both temporal and cultural, between the prehistoric cultures of the region. In this regard, analyses of the Alaskan Denbigh Flint complex/ Norton/ Birnirk/ Ipiutak relations as well as relations between these Alaskan entities and those of the Asian Eskimo area continue (e.g., Dumond 1998; O. Mason 1998). And, they bear striking similarities to the issues that concerned Diamond Jenness and Henry Collins.

Acknowledgments. Support for the project was provided by National Science Foundation as awards in 1996, 1998, 1999 and 2000. One of these, 'Western Arctic Whale Hunting Societies: Whale Biometrics Component,' (Allen McCartney, James Savelle, and John Dixon, PIs, University of Arkansas), partially supported the 1996 archaeology effort at Wales, which was carried out in conjunction with the 1996 ENRI-UAA sociocultural work at the village. Subsequently the Wales efforts were supported by 1998, 1999, and 2000 awards to ENRI-UAA, Roger Harritt, P.I. This support has enabled the archaeology-geomorphology studies at Wales, as well as a substantial effort in sociocultural analysis of whaling, which has been carried out by Barbara Bodenhorn (Cambridge University), Herbert Anungazuk (National Park Service, Anchorage), and Carol Jolles (Indiana University at Indianapolis). The Alaska Eskimo Whaling Commission has sponsored our efforts in project planning and activities since 1994, and their support has also been important to the success of the project. The administration of NSF funding has been done by the Environment and Natural Resources Institute, University of Alaska Anchorage. None of the Wales studies would have been possible without the support of the Wales Native Corporation, City Council, and the IRA Council, in their approval of our efforts and their support of the goals of the work at Wales.

Use of the C.L. Andrews photograph (Fig. 2) is by permission of the Division of Special Collections and University Archives, University of Oregon Library System; further reproduction or citing requires permission from the Division of Collections and University Archives. Use of the Thomas Lopp photographs (Figs. 4 and 5) is by permission of the Archives, Alaska and Polar Regions Department, University of Alaska Fairbanks.

REFERENCES

Ackerman, Robert E.
    1998   Early Maritime Traditions in the Bering, Chukchi, and East Siberian Seas. *Arctic Anthropology* 35(1): 247-262.

ADCED—Alaska Department of Community and Economic Development
    2002   Alaska Community Database, Wales. Community Information Summary. Webpage: www.comregaf.state.ak.us/CF_CIS.cfm

American Missioner
    1890-1901 Vols. 44-55, with the exception of Vol. 52 (1898). The American Missionary Association.

Anungazuk, Herbert
    1995   'Whaling: A Ritual of Life,' pp. 339-345 in A.P. McCartney, ed., *Hunting the Largest Animals: Native Whaling in the Western Arctic and Subarctic.* Edmonton: Canadian Circumpolar Institute (CCI) Press, University of Alberta, Occasional Publication No. 36; Studies in Whaling No. 3.

Anungazuk, H., and B. Bodenhorn
    1996   Wales August 1996 field notes and map. In files of Anungazuk and Bodenhorn.

Bockstoce, J.R.
    1977   *Eskimos of Northwest Alaska in the Early 19th Century.* T.K. Penniman, ed. Oxford: University of Oxford Pitt Rivers Museum Monograph Series, No. 1.
    1986   *Whales, Ice and Men: The History of Whaling in the Western Arctic.* Seattle: University of Washington Press.

Braund, S.R. and E.L. Moorehead
    1995   'Contemporary Alaska Eskimo Bowhead Whaling Villages,' pp. 253-279 in A.P. McCartney, ed, *Hunting the Largest Animals: Native Whaling in the Western Arctic and Subarctic.* Edmonton: Canadian Circumpolar Institute (CCI) Press, University of Alberta, Occasional Publication No. 36; Studies in Whaling No. 3.

Bromberg, N.A.
    1976   Clarence Leroy Andrews and Alaska. *The Alaska Journal* 6(2): 66-77.

Burch, E.S. Jr.
    1980   'Traditional Eskimo Society in Northwest Alaska,' pp. 253-304 in Y. Kotani and W.B. Workman, eds., *Alaskan Native Culture and History.* Osaka, Japan: Senri Ethnological Studies No. 4.

Cassell, Mark
    2000   Feeding Industry: The Creation of an Industrial Labor Force in Late 19th Century North Alaska. *Anthropological Papers of the University of Alaska* 25(1): 3-15.

Collins, H.B., Jr.

1933 'Archeological Investigations at Point Barrow, Alaska,' pp. 45-48 in W.P. True, ed., *Explorations and Fieldwork of the Smithsonian Institution in 1932*. Washington, D.C.: Smithsonian Institution.

1937a Archeology of St. Lawrence Island, Alaska. Washington: Smithsonian Miscellaneous Collections 96(1).

1937b *Archaeological Excavations at Bering Strait. Explorations and Field-Work of the Smithsonian Institution in 1936*. Washington, D.C.: Smithsonian Publication No. 3407, 63-68.

1940 Outline of Eskimo Prehistory. Washington, DC: Smithsonian Miscellaneous Collections 100: 533-592

1964 'The Arctic and Subarctic,' pp. 85-114 in J. Jennings and E. Norbeck, eds., *Prehistoric Man in the New World*. Chicago: University of Chicago Press.

Collinson, R.

1889 *Journal of H.M.S. Enterprise on the Expedition in Search of Sir John Franklin's Ships by Behring Strait, 1850-1855*. Edited by Major-General T.B. Collinson. London: Samson Low, Marston, Searle and Rivington.

Dumond, Don E.

1987 *The Eskimos and Aleuts*. New York: Thames and Hudson.

1998 *The Hillside Site, St. Lawrence Island, Alaska: An Examination of Collections from the 1930's*. Eugene, OR: University of Oregon Anthropological Papers No. 55.

Dumond, Don E. and H.B. Collins

2000 *Henry B. Collins at Wales, Alaska, 1936*. Eugene, OR: University of Oregon Anthropological Papers, No. 56.

Dumond, Don E. and D.G. Griffin

2002 Measurements of the Marine Reservoir Effect on Radiocarbon Ages in the Eastern Bering Sea. *Arctic* 55(1): 77-86.

Education Report

1894 Report on Education in Alaska, pp. 923-960 in RCE for 1890-91 2 (whole number 208), and pp. 873-892 in RCA for 1896-92 2 (whole number 212). Washington DC: US Government Printing Office.

Ellanna, L.J.

1983 *Bering Strait Insular Eskimo: A Diachronic Study of Economy and Population Structure*. Anchorage: Alaska Department of Fish and Game Division of Subsistence Technical Paper No. 77.

Fitzhugh, W.

1983 'Introduction to the 1983 Edition,' pp. 5-106 in E.W. Nelson, *The Eskimo About Bering Strait*. Washington, D.C.: Smithsonian Institution Press.

Fortuine, R.

1989 *Chills and Fever: Health and Disease in the Early History of Alaska*. Fairbanks: University of Alaska Press.

Gerlach, C. and D.K. Mason

1992 Calibrated Radiocarbon Dates and Cultural Interaction in the Western Arctic. *Arctic Anthropology* 29 (1): 54-81.

Giddings, J.L.
1952   *The Arctic Woodland Culture of the Kobuk River.* Philadelphia: University of Pennsylvania, University Monograph 8.
Giddings, J.L. and D.D. Anderson
1986   Beach Ridge Archaeology of Cape Krusenstern. *National Park Service Publications in Archaeology* 20. U.S. Department of the Interior, Washington, D.C. Greist, H.W.
1952   *Seventeen Years With the Eskimo.* Draft ms. Dartmouth College Library, Special Collections.
Harritt, Roger K.
1994   *Eskimo Prehistory on the Seward Peninsula.* Anchorage: U.S. National Park Service, Alaska Region, Resources Report NPS/ARO/RCR/ CRR-93/21.
1995   The Development and Spread of the Whale Hunting Complex in Bering Strait: Retrospective and Prospects, pp. 33-50 in A.P. McCartney, ed., *Hunting the Largest Animals: Native Whaling in the Western Arctic and Subarctic.* Edmonton: Canadian Circumpolar Institute (CCI) Press, University of Alberta, Occasional Publication No. 36, Studies in Whaling No. 3.
2001   [R.K. Harritt, Compiler]. *In Pursuit of Agvik: Some Results of the Western Whaling Societies Regional Integration Project.* Anchorage: Environment and Natural Resources Institute, University of Alaska.
Harritt, Roger K. and Owen K. Mason
1997   The Western Arctic Whale Hunting Societies Regional Integration Project: Recent Archaeology and Geomorphology at Wales. Paper presented at the *24th Annual Meeting of the Alaska Anthropological Association,* Whitehorse, Yukon, Canada.
Jenness, D.
1928a   'Archaeological Investigations in Bering Strait, 1926,' in *National Museum of Canada Annual Report for 1926.* Ottawa: Canada Department of Mines, Bulletin No. 50.
1928b   'Ethnological Problems of Arctic America,' pp. 167-175 in W.L.G. Joerg, ed., *Problems of Polar Research: A Series of Papers by Thirty-One Authors.* New York: American Geographical Society Special Bulletin No. 7.
Lantis, M.
19457   Alaskan Eskimo Ceremonialism. American Ecological Society Monograph 11. Seattle: University of Washington Press.
Mason, J.A.
1930   Excavations of Eskimo Thule Sites at Point Barrow, Alaska. *Proceedings of the 23rd International Congress of Americanists,* 383-394 New York, 1928.
1998   The Contest Between Ipiutak, Old Bering Sea and Birnirk Polities and the Origin of Whaling During the First Millenium AD Along Bering Strait. *Journal of Anthropological Archaeology* 17: 240-325.
Mason, Owen K.
n.d.   Report on 1999 Radiocarbon Results, Wales Beach Ridge Complex. Unpublished ms. in R. Harritt files.

Mason, Owen K. and J.W. Jordan
    2002   Minimal Late Holocene Sea Level Rise in the Chukchi Sea: Arctic Insensitivity to Global Change? *Global and Planetary Changes* 32: 13-23.

Mathiessen, T.
    1927   Archaeology of the Central Eskimo, 2 vols. Copenhagen: Report of the Fifth Thule Expedition 1921-24, Vol. 4 (1-2)
    1930   Archaeological Collections from the Western Eskimos. Copenhagen: Report of the Fifth Thule Expedition 1921-24(10) 1.

Montgomery, M.
    1963a   *An Arctic Murder: A Cultural History of the Congregational Mission at Cape Prince of Wales, Alaska, 1890-1893.* Master's thesis in History. Eugene: University of Oregon.
    1963b   The Murder of Missionary Thornton. *Pacific Northwest Quarterly* 54(4): 167-174.

Morrison, D.
    1991   *The Diamond Jenness Collections from Bering Strait.* Ottawa: Canadian Museum of Civilization, Mercury Series Paper No. 144.

Nagozruk, A.
    1919   *Annual Report, 1918-1919.* United States Public School, Wales, Alaska. Seattle: U.S. Bureau of Education, Alaska Division.

Nelson, E.W.
    1983   (1899). *The Eskimo About Bering Strait.* Washington, D.C.: Smithsonian Institution Press.

Orekhov, A.A.
    1999   An Early Culture of the Northwest Bering Sea. R.L. Bland, translator. (Original published in Russian under the same title by Nauka Moscow, 1987). Anchorage: National Park Service.

Rainey, F.G.
    1947   The Whale Hunters of Tigara. *Anthropological Papers of the American Museum of Natural History* 41(2): 231-483.

Rainey, F.G. and E. Ralph.
    1959   Radiocarbon Dating in the Arctic. *American Antiquity* 24(4): 365-374.

Ray, D.J.
    1966   'Introduction,' in D.J. Ray, ed., The Eskimo of St. Michael and Vicinity, as Related by H.M.W. Edmonds. *Anthropological Papers of the University of Alaska* 13(2): 7-22.
    1975   *The Eskimos of Bering Strait, 1650-1898.* Seattle: University of Washington Press.
    1983   *Ethnohistory in the Arctic: The Bering Strait Eskimo.* Kingston, ON: The Limestone Press.

Smith, K. L.
    1995   Tom and Ellen Lopp and the Natives of Wales, 1890-1902. *Alaska History* 10(2): 36-46.

Smith, K.L. and V. Smith.
    2001   *Ice Window, Letters From a Bering Strait Village: 1892-1902.* Fairbanks: University of Alaska Press.

Stanford, D.J.
    1976   The Walakpa Site, Alaska: Its Place in the Birnirk and Thule Cultures. *Smithsonian Contributions to Anthropology* 20.

Stern, R.O., E.L. Arobio, L.L. Naylor, and W.C. Thomas.
    1980   *Eskimos, Reindeer and Land.* Fairbanks: School of Agriculture and Land Resources Management Bulletin 59. Agricultural Experiment Station, University of Alaska Fairbanks.

Stoker, S.W.
    1984   'Subsistence Harvest Estimates and Faunal Resource Potential at Whaling Villages in Northwestern Alaska,' Appendix A in Stephen R. Braund and Associates, *Subsistence Study of Alaska Eskimo Whaling Villages.* Contract Report for the U.S. Department of Interior.

Stoker S. and Igor Krupnik.
    1993   'Subsistence Whaling,' pp. 579-629 in J.J. Burns, J.J. Montague and C.J. Cowles, eds., *The Bowhead Whale.* Lawrence, KS: Allen Press, Society for Marine Mammalogy Special Publication No. 2

Thornton, H.R.
    1931   *Among the Eskimos of Wales, Alaska, 1890-1893.* Edited and annotated by N.S. Thornton and W.M. Thornton. Baltimore: Jr. Johns Hopkins University Press.

Thornton, H.R.
    1931   Among the Eskimos of Wales, Alaska, 1890-1893. Edited and annotated by N.S. Thornton and W.M. Thonton. Baltimore: J.R. Johns Hopkins University Press.

Trollope, H.
    1854   Great Britain. Parliament, House of Commons 1854-1855. Sessional Papers, Accounts and Papers. Papers relative to the Recent Arctic Expedition in Search of Sir John Franklin. Vol. 35: Journal kept by Commander Henry Trollope during a trip from H.M. Sloop *Rattlesnake* in Port Clarence to King-a-ghee, a village four or five miles round Cape Prince of Wales. January 9, 1854—January 27, 1854.

van Veenen, E.
    1977   *Subsistence Whaling in Alaska.* Unpublished dissertation in Polar Studies. Cambridge, UK: Scott Polar Research Institute

Yama'ura, K.
    1979a   'On the Origins of Thule Culture as Seen from the Typological Studies of Toggle Harpoon Heads,' pp. 474-484 in A.P. McCartney, ed., *Thule Eskimo Culture: An Anthropological Retrospective.* Ottawa: Canadian Museum of Civilization, Mercury Series, Paper No. 88
    1979b   On the Development Process of the Toggle Harpoon Heads Around Bering Strait. *Journal of the Archaeological Society of Nippon* 64(4). In Japanese with English summary.
    1984   Toggle Harpoon Heads from Kurigitavik, Alaska. *Bulletin of the Department of Anthropology, Tokyo* 3: 213-262.

A Paleo-Geographic Preface to the Origins of Whaling: Cold is Better

Owen K. Mason and Valerie Barber

Abstract. *Climatic deterministic models often over-simplify cultural processes; nonetheless, successful whale hunting is constrained by climate–forcing both in its influence on whale demography and the development of ice leads that determine access to whales. Geologic proxy records allow the reconstruction of past climatic changes in northwest Alaska. Five major episodes of northwest Alaska storm erosion are recognized: (a) 1600–1200 B.C., (b) 800–200 B.C., (c) A.D. 750–950, (d) A.D. 1030–1200 and (e) A.D. 1450–1800. The second episode, 800–200 B.C includes the expansion of whaling at East Cape and Point Hope, with the third, A.D. 750–950 co-occurs with the development of the Thule culture and the Ipiutak 'collapse' which may be causally linked to adverse climatic conditions.*

INTRODUCTION

The enervation associated with 'ameliorating' warmer weather remains the accepted explanation for the dispersal of Thule whalers to the High arctic during an early Medieval 'optimum,' placed during A.D. 900–1200 (Bockstoce 1976; McGhee 1969/70, 1972, 1981; Anderson 1983; Morrison 1999). The favored mechanisms implicate improved hunting returns (i.e., surpluses) that led to increased populations, stressed local resources, and eventually forced entire communities to migrate (Stanford 1973, 1976; Krupnik 1993:204ff; Stoker and Krupnik 1993:584). Warmer climes, to Taylor (1963:461), Stoker and Krupnik (1993:584ff), fueled a positive amplifying feedback between an increased need for open water mobility and the development for dog transport dependent on the ice platform. Although many archaeologists assume warming was a beneficial circumstance for northern peoples, ethnographic accounts often indicate the reverse.

A variety of climatic parameters, e.g., winds at hemispheric, mesoscale and local levels, influence ice cover and constrain aboriginal hunting [*cf.* Anderson (1983:76ff), reviews in Mason and Gerlach (1995a) and Mason (1998)]. In brief, ice serves as a platform for many marine mammals and for hunters, its absence renders hunting more difficult, unfamiliar and even suggests cosmic distress (Rasmussen 1931:136ff). For ice-obligate and open-lead seal and whale hunters, more reliable ice cover is better. Among the Netsilik,

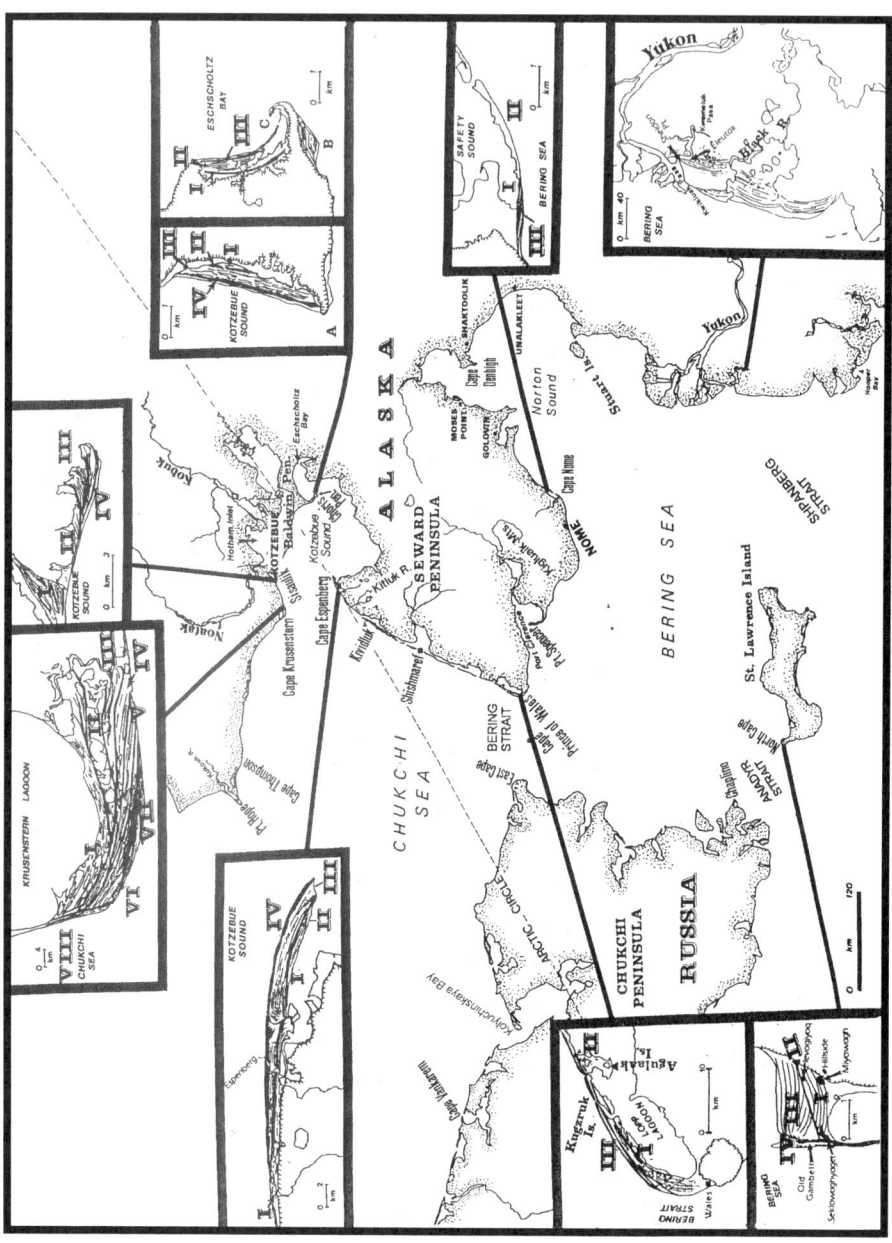

Figure 1. *Principal sites occupied in Alaska and adjacent Chukotka during the late 1st millennium AD.*

dire consequences followed the year without a winter; without the critical ice platform for hunting seals, people were unable to improvise or innovate open water fishing techniques (Rasmussen 1931:136ff). Doubtless, very cold, catastrophic summers, e.g., in A.D. 1783, also produced famine and led to migration (Jacoby *et al.* 1999). By contrast, some areas with a long history of open water whaling, for example, St. Lawrence Island (*cf.* Rainey 1941) and off the coast of Chukotka (Dinesman *et al.* 1999), were pre-adapted to warmer, less icy conditions and may have benefited from 'climatic amelioration.'

Despite the predilection of middle latitude residents for the perceived benefits of warmer climes, ecological data show that the intensified upwelling associated with colder conditions increases subarctic and arctic productivity (Fraker 1989; Schell *et al.* 1998; Finney *et al.* 2002). Proxy evidence for 'paleo-productivity' in the North Pacific derives from lake cores into the accumulated decayed organic material from spawned and subsequently decayed salmon (Finney *et al.* 2002). In the North Pacific, salmon abundance tracked cold peaks, with two periods of interest: (a) the last centuries B.C. and the (b) the Little Ice Age, following A.D. 1200—similar trajectories in abundance likely prevailed in the Bering and Chukchi Seas.

Dumond (1972:314ff) and Lutz (1982:146), proposed that sustained colder weather, 'deteriorating climates,' *ca.* 500 B.C. produced a resource collapse for the Arctic Small Tool tradition, leading, paradoxically, to a retrenchment in sedentary communities and technological elaboration (ceramics, slate, oil lamps) during the Norton culture and its descendants along the Bering Sea. This paradigm still guides most researchers (*cf.* Mason and Gerlach 1995a). While most cultural transitions co-occur with cold climates, notable exceptions occur in the Brooks Range during the Little Ice Age after the elaboration of dog traction (Hall 1978; Gerlach and Hall 1988).

Contrasting paradigms are used to explain the Thule intensification of whaling and migration toward the high arctic: (a) entire impoverished communities relocated *en masse* following the effects of local extinctions of ringed seal (Stanford 1973, 1976); (b) the necessity of finding whales led people out of a warm water devastated biota in Alaska and into newly emergent ice edge environments in Coronation Gulf (McGhee 1972, 2000; Morrison 1999) or (c) the efforts of lower status individuals or groups seeking new avenues for higher status (Whitridge 1999a, 1999b:95ff). An implicit assumption in nearly all formulations is that locally population was sufficiently high to act as a stress on resources. North Pacific bowhead whales penetrated the High arctic during the early Holocene with no human predators on their tails (Dyke and Savelle 2001); likewise, the spread of Arctic Small Tool/Sarqaaq occurred in the absence of bowheads in the western arctic (Grunnøw and Pind 1996; Dyke and Savelle 2001:378).

FRAMING THE QUESTION OF CLIMATIC CONSTRAINTS ON WHALING

To discuss the role of climate in fostering whaling, we will first review the limited data for the earliest whaling in Bering Strait, subsequently discuss the data sets for constraining past climates, then close with archaeological data that document correspondences between 500 B.C. to A.D. 900 to 1200, focusing on the latter period.

Two hypotheses structure the argument:

(a) Whaling arose during the last centuries B.C. on the south shore of the Chukotsk Peninsula, from Ekven to Sirenki, during a cooler interval.

(b) Whaling intensified in terms of success rates and geographic distributions *ca.* A.D. 800–1200, the Early Medieval glacial expansion, and during the Little Ice Age. Conversely, whaling lessened or declined in importance during intervening periods.

ENVIRONMENTAL CONSTRAINTS IN THE BERING STRAIT REGION

Sea mammal and fish populations in Bering Strait (Fig. 1) track food availability, i.e., phyto- and zooplankton (Fraker 1989; Moore and Reeves 1993; Mason and Gerlach 1995a), reflecting the contrast in geostrophic currents that upwell nutrients and water temperature onto the western Chukotkan shore away from the eastern Alaskan shore (Walsh *et al.* 1989:293; Coachman and Hansell 1993; Springer and McRoy 1993. Schell *et al.* 1998). In addition, large freshwater sources on the Alaskan coast, i.e., the Yukon River, inhibit productivity (Walsh *et al.* 1989). Salmon numbers fluctuate in tandem with sea surface temperature variability that is driven by atmospheric patterns (Mantua *et al.* 1997).

Inter-annual variability occurs in upwelling to the eastward to the Alaska coasts by anomalous winds and currents through the Strait. Reversals of flow are more common during fall than during spring (Coachman 1993:502). During the late 20th century, water transport through Bering Strait was nearly 33% stronger from the late 1940s to 1970 than in the 1980s (*cf.* Walsh *et al.* 1989:348). In addition, shifts in surface winds that influence ice cover are mediated by hemispheric patterns such as El Niño and the Pacific Decadal oscillation (Niebauer 1988, 1998).

Primary productivity varies between decades and determines the growth of sea mammal populations. Phytoplankton productivity declined significantly between 1976–1995 in comparison to the previous 35 years due to lessened ice during the spring bloom that builds on the ice and/or increased summer cloudiness that decreased solar input (Schell *et al.* 1998:22). The decline in productivity co-occurs with a decline in storm erosion in the southern Chukchi Sea associated with fewer fall storms from the North Pacific (Mason and Jordan 1993). By contrast, isotopic signatures from baleen from the 19th century and the early 1st millennia A.D. shows that Bering Sea productivity was considerably higher than at present (Schell, pers. comm. 1999).

Access to sea mammals, especially whales during the spring migration, depends on the extent of open water in the ice-covered Bering and Chukchi seas (Fraker 1989). Offshore winds foster the development of leads that are useful in tracking whales, walrus and bearded seal. Bowhead movements are also directly controlled by ice thickness which can delay movement north by weeks (Moore and Reeves 1993:317).

A particular constellation of favorable winds for one coastal reach might produce poor ice or lead conditions for a neighboring, often very close, community—circumstances which may have structured or induced alliances and/or rivalries between communities. Along Mys Dezhneva (East Cape) in Chukotka, adverse ice conditions could be buffered by migrating or re-locating only tens of km from Uelen, on the north, to Ekven, on the south. At south-facing Wales in Alaska, the benefits of a north-facing coast, near Mitletavik, are apparent only 20–30 km distant across Lopp Lagoon. A similar, reciprocal relationship existed between Tikigaq at Point Hope and 60 km distant Uivvaq, 2 km east of Cape Lisburne. Whaling at Uivvaq required southerly winds to open leads while northerly winds were critical at Tikigaq. As Burch (1981:53) reported, 19th century Uivvaq residents closely monitored the ice and made plans to join their brethren at Tikigaq, if ice conditions proved unfavorable.

Several archaeological consequences are expected from the biogeographic constraints for the Bering Strait, as detailed by Mason and Gerlach (1995a): (a) greater population densities along the Chukotsk Peninsula shore; (b) higher resource reliability at East Cape (on the Chukotsk Peninsula), due to access to the interception of whale migrations in both spring and fall; and (c) surplus generation was more reliable along the Siberian shore, with the converse on the Alaskan shores: more frequent scarcity. Short-term variations may be introduced by climatic changes that shifted hot spots closer to the Alaskan shore (*cf.* Coachman 1993).

THE ANTIQUITY OF WHALING ALONG BERING STRAIT

Whaling emerged fully elaborated in eastern Chukotka very early in the first millennium A.D. or in the last centuries B.C. (Dinesman *et al.* 1999); nonetheless, its prior development remains obscure despite a long history of seaside residence and the use of marine resources since the first appearance of humans in the Americas, as early as 11,000 B.P. in California (Erlandson 2002), in southeast Alaska 9000 B.P. (Dixon 2001) and in the Aleutians 8500 B.P. (Dumond and Knecht 2001). The co-occurrence of several houses with whalebones led Giddings (1967) to infer whaling at Cape Krusenstern by 1500-1000 B.C. However, the zooarchaeological data are insufficient to rule out the scavenging of stranded carcasses (Mason and Gerlach 1995b).

Artifacts, e.g., ivory drag float pieces and a single 'whaling' harpoon head, suggest that Okvik people 'hunted whales in the open sea' (Rainey 1941:465) off the southeast coast of St. Lawrence Island. On the west coast of St. Lawrence Island, an Old Bering Sea/Okvik occupation at the Hillside site occurred as early as A.D. 200 (Mason 1998; Dumond 1998); however, firm evidence of whaling is lacking. If both Okvik sites date to A.D. 200 or earlier, then whaling co-occurs with a final phase of the Neoglacial cooling.

'Casual' whale hunting (Harritt 1995:36) was possibly practiced during the late Norton culture during the first centuries A.D., if one accepts the artifactual evidence of Norton (Near Ipiutak) at Point Hope (Larsen and Rainey 1948; Larsen 1982) and the dating interpretation of Mason (1998). At Cape Krusenstern, Anderson (1986b:153-154) suggested that a harpoon head with Ipiutak motifs reflected whaling during the House 17 settlement between A.D. 540–670, the average of two ages (*cf.* Giddings and Anderson 1986:30, calibrated in Mason 1998:252), while a few whale bones were present in the archaeofauna of House 30, based on charcoal from A.D. 429–659 (Mason 1998:252).

The earliest firm evidence of whaling in the Bering Strait can be inferred at Chukotka employing demographic reconstructions of archaeofaunas from East Cape middens (Dinesman *et al.* 1999) that indicate a mortality profile due to human predation, following McCartney (1995). Several sites, some with >100 whale skulls, indicate that gray whales were dispatched at Ekven and Uelen between 180 B.C.–A.D. 560 [2683±149 B.P. (Iemezh-829)], concurrent with the first interments in the Ekven cemetery (Dinesman *et al.* 1999:111ff,136). The bowhead whale was possibly taken as early, or during the century or so afterward, during A.D. 160-665 [2166±68 B.P. (Iemezh-860), Dinesman *et al.* 1999:138]. While the ages are unsatisfying in terms of uncertainty, >500 to 750 years, clearly, whaling in Chukotka developed preceding A.D. 250, concurrent with and causally linked to sedentism, the use of cemeteries and marked social stratification and inter-community forms of organization, including trade, alliance and conflict (Mason 1998). Unfortunately, evidence for pre-whaling community structure is lacking in the immediate zone of the Chukotkan coast. What is its relationship with climatic change?

Taphonomic Constraints on Finding Sites related to the Origin of Whaling

Sea Level Stabilized ca. 500 B.C. Lacking data from the relevant time periods, archaeologists remain handicapped in understanding the prehistory of whaling in Bering Strait. Because maritime hunters require the water's edge, sites adjacent to the sea are especially subject to the erosive effects of rising sea levels that erode and redeposit mobile sand and gravel deposits in the littoral (Mason 1993). The appearance of whaling with stable sea levels may more properly be an artifact of this sea level stabilization. When did sea level stabilize to near modern levels allowing site preservation?

Precise radiometric dating of *in situ* terrestrial peat demonstrated that the flooding of the Chukchi platform and Bering Strait occurred after 11,000 B.P. (Elias *et al.* 1992). Additionally, the appearance of Pacific mollusks on the Arctic Ocean coast reflects the flooding of the Bering Strait as early as 10,500 B.P. (Dyke *et al.* 1996:150).

Sea level continued its rapid rise throughout the early and middle Holocene, transgressing to *ca.* 50 m below present *ca.* 10,000 B.P. until *within* 2 m of modern by *ca.* 5000 B.P.–3000 B.C. (all calendar ages are calibrated using Stuiver *et al.* 1998) Despite a postulated higher eustatic sea level stand in Chukotka during the early Holocene (Mason 2002), eustatic sea level on the Alaska coast remained 5-10 m below present until the late Holocene (Jordan and Mason 1999; Mason and Jordan 1997, 2002). The late Holocene witnessed an exponentially slower rate of sea level rise. Radiocarbon ages (n=>30) from deltaic and barrier marsh peat along Seward Peninsula lagoon margins from Wales to Point Barrow indicate a slow rate of 0.03 mm yr[-1] eustatic sea level rise from *ca.* 1.5 m below modern at 2000 B.C. to 1 m below modern by A.D. 1 (Jordan and Mason 1999; Mason and Jordan 2002).

Nonetheless, much of the early and middle Holocene record on Alaska shores, at least, will remain irretrievable—no higher shoreline deposits are likely and storm erosion has either transgressed or re-deposited the relevant landforms. For that reason, only the appearance of maritime adaptations along Bering Strait appears sudden (Dinesman *et al.* 1999), as sea level stabilized in the last centuries B.C. (Mason and Jordan 2002). If additional Chukotkan evidence for a higher relative sea level in the early and middle Holocene can be found, then possibly the antiquity of whaling could be further elucidated (Mason 2002).

Paleoclimatic Records from Beach Ridges

Due to the combination of a micro-tidal range (*ca.* <50 cm), the dominance of wave energy, and abundant sediments (both terrestrial and shelf), the Alaskan coasts are sensitive proxies for atmospheric processes, if the rate of sea level rise is slow (Mason and Jordan 1993). The fullest records of past storm-mediated sedimentation are located at the termini of long shore drift systems such as beach-ridge forelands that extend over several thousand years, with barrier islands providing partial records up to 2000 years in length. Beach–ridge plains occur at nearly all promontories located at critical shifts in coastal orientation that act as depositional sinks along the eastern coasts of the Bering and Chukchi seas from the Yukon Delta to Point Barrow (Mason 1990; Mason and Jordan 1993; Mason 1999). Calibrated ^{14}C ages (n=>280) constrain the Holocene record, derived from archaeological sites, storm and dune facies within beach ridges and salt-marsh peat graded to sea level (Jordan and Mason 1999; Mason 1999; Mason and Jordan 1997, 2002). The methods employed in mapping and dating beach ridges are described elsewhere (Mason 1990, 1993; Mason and Jordan 1993; Mason *et al.* 1997).

HOLOCENE STORM CHRONOLOGY ALONG BERING STRAIT

During the last two thousand years, progradation predominated along most of the Chukchi Sea coast (Mason 1990; Mason and Jordan 1993); even spits at Point Hope and Cape Nome expanded, especially during the early first millennium A.D. Quite a different situation prevailed prior to A.D. 1. During the Neoglacial, 1600–200 B.C., much of the coast endured intensely erosive storms that precluded the preservation of older landforms. Several factors are involved: (1) intense storms may have directed eroded sediment offshore rather than into the long shore transport system; (2) storm recurrence intervals were too brief for full-recovery and progradation; and (3) rates of sediment delivery had to exceed the rate of sea-level rise. We will briefly summarize the pattern of storm history from the areas with the best dated sequences; fuller presentations may be found in Mason *et al.* (1997), Mason (1999), Mason and Jordan (2002), Mason and Ludwig (1990), Jordan and Mason (1999), and Jordan (1988, 1990).

Although barrier islands front considerable stretches of the Chukchi Sea, especially its northern shore from Icy Cape to Point Franklin, only partial proxy records of landform history are available. Comparatively few archaeological sites are recorded on the barriers and none predates the Little Ice Age (Schaaf 1988). Geological research is limited to the southern Chukchi Sea, e.g., Shishmaref barrier island complex along northern Seward Peninsula (Jordan

1990; Jordan and Mason 1999; Mason and Jordan 1997). The chances of locating archaeological sites dating prior to A.D. 1 or to some extent from A.D. 700–1200, are extremely poor (Jordan 1988; Schaaf 1988), a circumstance which precludes substantiating that the purported less icy conditions of the Medieval 'optimum' led to nomadic whaling (Sheehan 1995).

Intense, closely-spaced storms produced substantial erosion at Cape Espenberg following an Ipiutak occupation (KTZ-157) during A.D. 600–775 (Harritt 1994:111-122, 304). The erosional period is dated from geologic contexts between A.D. 750–1200 within high dunes that contain a basal series of beach berms, succeeded by cross-bedded eolian and interbedded shell-laden storm beds (Mason *et al.* 1997, Mason and Jordan 1997). The Espenberg composite dune represents four principal constructional events, (a) to (d): First, (a) low dunes built atop the beach berms from *ca.* A.D. 775 to the late 10th century A.D., (b) followed during A.D. 960–1020 [two 14C ages, average 1010±47 B.P., A.D. 961–1070, (Mason and Jordan 1997: Appendix D)] by a surface stabilization on two western Espenberg dunes at Cape Espenberg. Subsequently (c), large storms emplaced another ridge, from A.D. 1030–1200: bracketed by an age on basal plant detritus within beach facies and an upper age from shell in one of several storm beds. Surface stabilization (d) occurred A.D. 1220–1302 associated with KTZ-087, an archaeological site atop the dune (Schaaf 1988; Harritt 1994:299-300, as averaged in Mason and Jordan 1993). After A.D. 1400, dunes again built, indicating high winds and storms, alternated with swale formation—progradation under fair-weather wave climates—the pause represented by the swales records that storms were less frequent at Cape Espenberg during the Little Ice Age than during the early Medieval period. Seven dunes of variable lengths added during eight intervals from A.D. 1450–1900. Onshore sand supply apparently remained high because storms were oblique to coast; hence, waves favored alongshore sand transport.

Within sheltered Kotzebue Sound, only a composite ridge occurs on the Deering spit. A substantial Ipiutak settlement, replete with a large *qarigi* (Larsen 2001), a house (Bowers *et al.* 2000) and several burials (Reanier *et al.* 1998). The Ipiutak settlement is associated with an extensive buried organic horizon (Mason 1999, field notes), indicative of non-stormy surface stabilization between A.D. 600-800 (Gerlach and Mason 1992; Larsen 2001). The Ipiutak surface is succeeded by thicker and more gravelly storm beds (Mason 1999, field notes) emplaced during a Thule occupation dated *ca.* A.D. 1000 (Bowers *et al.* 2000).

The most recent erosional truncation at Cape Krusenstern, followed ridge 10 (*ca.* A.D. 700–1200) and was less pronounced, while the last six centuries of Little Ice Age storms co-occurred with a south-

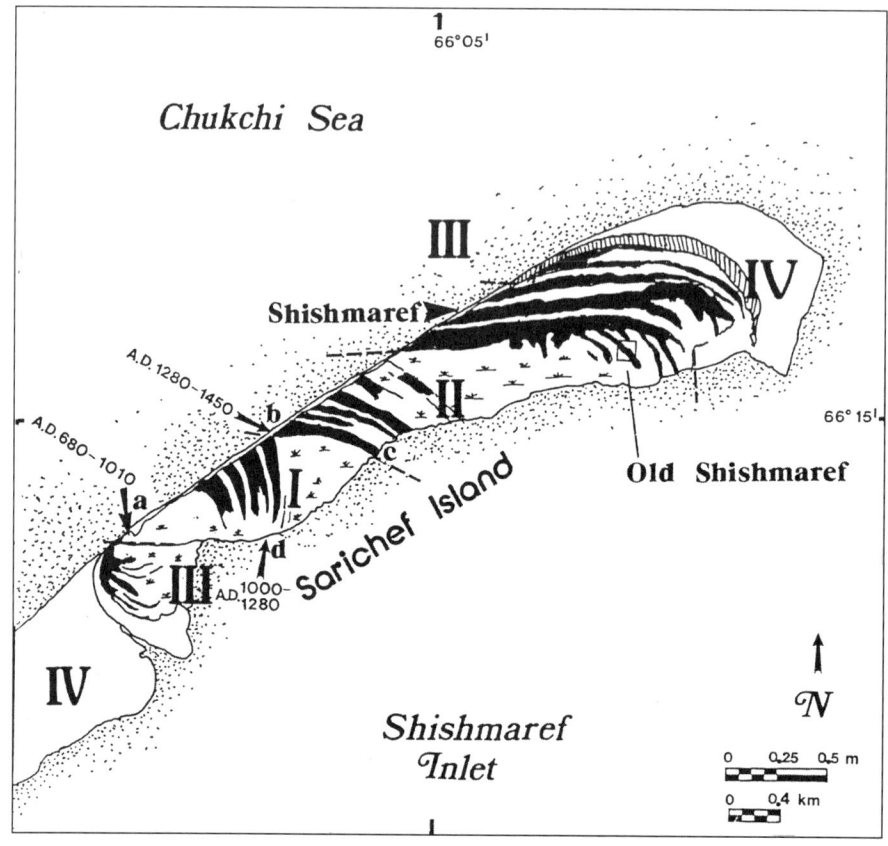

Figure 2. *Enlarged aerial photograph of the northwest portion of the Cape Krusenstern beach ridge complex. Note the north to south composite ridge that formed in the late prehistoric and a truncation of an older northwest to southeast ridge that contains several Birnirk houses dated to the late first millennium A.D. Apparently, storms truncated this ridge.*

eastern addition (Mason and Ludwig 1990). Thule occupations generally follow this erosional disconformity, and are located on the re-deposited series of ridges to the southeast (Giddings and Anderson 1986). However, on Krusenstern's northern margin, several high intensity northwesterly storms constructed composite ridges during the Little Ice Age. The timing and character of the composite ridge may be estimated from the position of several Birnirk houses occupied A.D. 764–1023 (average of two ages, 1136±75 B.P., K-851, K-816). The Birnirk houses are on a ridge perpendicular to the more recent late prehistoric houses, indicating that the erosional truncation occurred after A.D. 1030 (*cf.* map in Giddings and Anderson 1986:41).

Table 1. Unequivocal Radiocarbon ages relevant to the Birnirk, Punuk and early Thule period in northwest Alaska.

Site	^{14}C age Yrs B.P.	Calibrated Age B.C./A.D. 2 σ range	Laboratory Number	Diagnostic Artifacts	Culture	Source for ^{14}C date	Material
Dovelavik	960 ±85	A.D. 900-1230	B-2436	Burial, Slate spear, knife blades, Whale skull, Ivory rods	Termed Punuk ?	Hofmann-Wyss 1987:13	Wood
Kitnepaluk	1130 ±30	A.D. 890-970	B-6476	Sicco harpoon head	Punuk	Blumer 1997	Wood
Kitnepaluk	1050 ±30	A.D. 900-1026	B-4414	Burial K-45, Open socket Punuk IIIa (x) harp. Hd, "Sicco", red chert blade	Punuk	Hofmann-Wyss 1987:30	Wood
Ayveghyaaq	910 ± 145	A.D. 871-1391	P-92	Numerous Punuk objects fr. Village site	Punuk	Hofmann-Wyss 1987:30	Wood
Ayveghyaaq	1070 ±210	A.D. 600-1373	P-69	Numerous Punuk objects fr. Village site	Punuk	Collins 1937	Wood
Mayughaaq	1000 ±70	A.D. 984-1155	B-889	Punuk designs, slate	Punuk	Collins 1937	Wood
Mayughaaq	740 ±80	A.D. 1221-1376	B-888	Punuk designs, slate	Punuk	Bandi 1972	Charred material
Mayughaaq	710 ±50	A.D. 1276-1376	B-891	Punuk designs, slate	Punuk	Bandi 1972	Wood
Kukulik	1680 ±40	A.D. 1071-1331 alternatively, A.D. 1304-1438	Beta-144990	Collins type V(a)x or y Punuk designs	Punuk	This paper	Ivory
Kukulik	1110 ±40	AD 870-1010	Beta-144991	Collins III-a-x. Open socket, lashing slots	Punuk?	This paper	Antler
Wales Hillside	1270 ±70	A.D. 644-943	Beta-98938	N.A.	N.A.	Harritt 2001	Wood
Kirgitavik	680 ±60	A.D. 1243-1404	Beta-13874	N.A.	N.A.	Harritt 2001	Wood
Ekven	1655 ±50	A.D. 530-680	AAR-2777	Naulock, Birnirk harpoon heads	Birnirk	Blumer 1997:63	Human Hair
Ekven	1645 ±45	A.D. 670-890	AAR-2776	Naulock, Birnirk harpoon heads	Birnirk	Blumer 1997:64	Human Hair
Cape Krusenstern	1180 ±110	A.D. 651-1034	K-851	Birnirk harpoon heads House 32	Birnirk	Giddings and Anderson 1986:30	Charcoal, bone
Cape Krusenstern	1100 ±100	A.D. 686-1160 Probability option / ***A.D. 691-1067*** Probability option	K-816	House 33	Birnirk	Giddings and Anderson 1986:30	Charcoal
Cape Krusenstern	1070 ±100	A.D. 694-1206 Probability option / ***A.D. 767-1169*** Probability option	K-817	House 6	Thule	Giddings and Anderson 1986:30	Charcoal
Cape Krusenstern	1000 ±110	A.D. 779-1263 Probability option / ***A.D. 811-1256*** Probability option	K-851	House 4	Thule	Giddings and Anderson 1986:30	Charcoal
Onion Portage	900 ±50	A.D. 1025-1224	P-1112	Ahteut house	Thule	Anderson 1988:44	Wood
Uivvaq	980 ±40	A.D. 987-1160	-26.3	N13 E1 114 cm bd	Thule	Anderson 1988:44	Salix Charcoal
Uivvaq	1010 ±35	A.D. 980-1153	CURL-5947	-80 cm bs. Punuk trident above at -67 cm	Punuk	Hoffecker et al. 2002; this paper	Beetle exoskeleton
Uivvaq	785 ±65	A.D. 1153-1306	CURL-5767	35 to 40 cm bs	Birnirk	Hoffecker et al. 2002; this paper	Beetle exoskeleton
Kugusuguruk	1146 ±95	A.D. 672-1033	P-97	Tasik, Tuquok, Naulock, Oopik harpoon heads, includes a considerable variety of other burial goods	Birnirk	Morrison 2001	Wood
Walakpa	840 ±90	A.D. 1020-1295	Gak 2297	Substantial inventory	Early Thule	Stanford 1976	Charcoal
Walakpa	970 ±90	A.D. 895-1248	Gak-2298	Substantial inventory	Birnirk	Stanford 1976	Charcoal
Utqiagvik	1230 ±60	A.D. 660-976	Beta-2907	Burial 4, two needle cases with Birnirk designs	Birnirk	Brower 1990:35, 38	Wood
Birnirk	970 ±40	A.D. 996-1162	Beta-133361	Mound A Birnirk	Birnirk material adjacent Birnirk	Mason 2000	Wood—outer rings, structural wood

Low dunes produced by northwest winds cap the four oldest Point Hope ridges. The lower limit of spit construction can be estimated from archaeological remains at prior to 200 B.C. (Mason 1990, 1998). Apparently, intense storms precluded the preservation or formation of a spit prior to that time. Point Hope construction may represent a shift in sediment supply as southerly storms re-worked gravel or, less likely, the influence of the small Kuk River (Mason 1990). Two villages indicate Ipiutak occupation on the 24th to 28th ridges from A.D. 550–900 and at Tikigaq on the 3rd to 6th ridges formed during the Little Ice Age, probably after A.D. 1400. Spacing between the ridges provides a relative measure of storm recurrence variability; storms were more frequent during A.D. 880–1100 and following A.D. 1665 (Mason 1990).

A composite beach ridge has built east of Cape Lisburne under the occasional influence of large northerly storms that direct gravel southwesterly toward the cape. Storm beds within Uivvaq mound 2 are bracketed by Birnirk and Punuk occupations between A.D. 650 and A.D. 1200 (Hoffecker *et al.* 2002; Mason n.d.).

Point Barrow spit includes 15 gravel and sand ridges that formed since 800 B.C. (Mason and Jordan 1993). The highest and broadest ridges co-occur with other Neoglacial ridges, and are contemporaneous with the truncations farther south, based on driftwood incorporated into the ridges. The first cycle of ridges formed before 800 B.C. until the first centuries A.D., with a second episode of high ridge construction A.D. 650–1200 (Mason and Jordan 1993:61). Storms at Barrow were more intense during the Little Ice Age, considering the development of a scarp and reorientation in ridge development and the addition of high gravel ridges seaward of the Birnirk settlement, likely occupied from A.D. 700–1000 (Mason 1998, 2000, *cf.* tree ring age on Mound A, Table I). High gravel storm ridges also formed landward of Nunagiak following a Punuk and early Thule occupation thought to date during the late 13th century A.D. (Ford 1959:243). What is the overall signal from the coasts of the Chukchi Sea?

The Big Picture: Paleo-Storm History in the Chukchi Sea

The sum of local records from the Chukchi Sea provides a consistent picture of the past intensity of storms. In the Chukchi Sea, west or south-facing beach-ridge complexes record that the most intense storms in the last 2000 years occurred A.D. 750–950 and 1020–1200 based on prominent truncations or composite ridge construction. Heightened coastal erosion along northwest Alaska coasts, due to more frequent fall or summer storms and dune building due to high winter winds, occurred between A.D. 800–950 and 1030–1200, 1450–1550, and during the 17th and 19th centuries, the Little Ice Age (*cf.* Mason and Jordan 1997; Mason 1999). Storms were infrequent and had a low magnitude from A.D. 200–750 when dunes stabilized and progradation dominated the coast (Mason *et al.* 1997). At Cape Krusenstern, progradation was associated with onshore, fairweather movement of gravel under the influence of southwesterly swell, while stronger intensity storms were also from the southwest but led to longshore transport to the southeast (Mason and Ludwig 1990; Mason 1999).

The major shift in storm frequency and direction caused substantial erosion along Chukchi Sea coasts (Mason and Jordan 1993). The entire nearshore transport system was pushed into a net erosional condition and sediments were transferred onto the shelf (*cf.* Hequétte and Hill 1993 for the Yukon coast). The signature of the Neoglacial expansion is evident on the ocean floors as well: ice-rafted detritus was deposited in the Arctic Ocean between 3500 and 1500 B.P. (Darby *et al.* 1997) and eroded peat on the Bering Sea shelf (Mason and Jordan 1993:67) closely parallel to the storm cycles evident in terrestrial records (Mason and Jordan 1993). Although beaches were eroded, sand was also transported higher onto the beach, thus available for dune construction (Psuty 1988). Strong fall and late winter winds sent this sand inland; within decades high dunes had built along the northwest Seward Peninsula coast and elsewhere. Inlet channels were cut through the Shishmaref and other barrier islands while storms produced composite gravel ridges on the northwest shores of Cape Krusenstern, Choris Peninsula and at Point Barrow (Mason and Jordan 1993).

Dendroclimatic Records A.D. 300–1200

Alaska tree-ring indices provide another proxy record of precipitation and water balance, although the cause of anomalous wide rings is problematic (Garfinkel and Brubaker 1980). If wider rings translate into increased precipitation and storminess, then three periods of high storms are apparent in Giddings' (1952) tree-ring record: A.D. 1000–1080, 1350–1370 and 1520–1560. After re-analyzing and de-trending Giddings' (1952) data, Graumlich and King (1998) inferred temperature correlations that show colder conditions prevailed from A.D. 1150–1206, 1369–1408, 1633–1687 and 1763–1852. These records are only indirect measures of northwest Alaska climate due to their driftwood origin.

Data from Barrow and Golovnin Bay permit the extension of Giddings's chronology prior to A.D. 1000 reveal lengthy alternations in climate that correspond to cooler conditions during the first millennium (Figs. 2, 3). Adverse weather was associated with the early years of tree growth in the late 8th century A.D., with a nadir felt A.D. 775 to 780. Subsequently, conditions were very favorable for growth throughout most of the 9th century A.D., particularly during the decades from A.D. 840 to 870. However, after A.D. 870 conditions were unfavorable for nearly 100 yrs, but *ca.* A.D. 980 a drastic shift in weather produced rings nearly as wide as during the 9th century. The older range of Giddings' chronology, held in place by a single tree from Ekseavik, shows a similar spike in the early 11th century A.D. (Giddings 1952:109).

For the earlier sequence, three spruce structural timbers in the Qitchauvik (SOL-143) *qarigi* within inner Golovnin Bay (Mason *et al.* 2001) provide a floating chronology for several centuries prior to A.D. 500–530. The Golovin tree ring sequence (Fig. 3) shows that cold conditions prevailed in either the late 3rd or the early 4th century, followed by a century of decline in growth conditions during the late 4th and/or early to late 5th centuries. Finally, a short pulse of 10-20 yr associated with warmer temperatures followed in the late 5th century (*ca.* A.D. 465) or early in the 6th century (*ca.* A.D. 540). A return to cool conditions occurred afterward, in the late 6th century A.D.

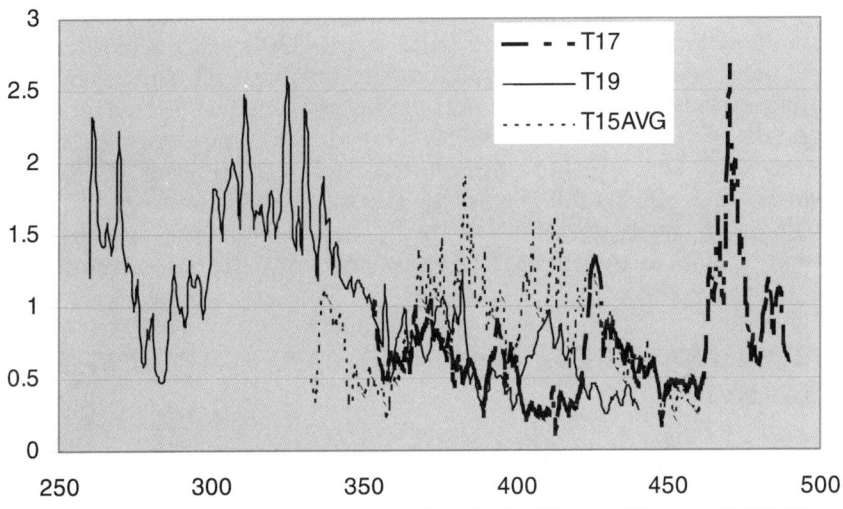

Figure 3. *Tree ring chronology from three structural timbers at Qitchauvik, Golovnin Bay. On left, departures from mean are listed, while on lower axis, calendar years A.D., based on the average of three 14C ages on outer rings (Analyst: Val Barber, Institute of Marine Science, University Alaska Fairbanks). The three structural timbers from the qarigi at Qitchauvik provide an average 14C age of AD 426–561, with a midpoint of AD 533, the age of the outer rings, an uncertainty of several hundred years. For chronological purposes, the age of tree death was placed at AD 500; hence, the sequence may be either older or younger by 100 yrs.*

Comparisons with global records largely confirm the cold induced storm sequence for northwest Alaska. The Greenland ice cores show major storm signals within the 2600–2300, 1450 and 512 year intervals within the 2600–2300, 1450 and 512 yr intervals (Dansgaard *et al.* 1984; O'Brien *et al.* 1995; Mayewski *et al.* 1998). Similar periodicities (2850, 1650, and 600 B.P.), but with a 500 yr error, are linked to the North Atlantic oscillation by Viau *et al.* (2002). Heightened amounts of sea salt in cores indicate the meridional circulation intensified twice during the late Holocene, during 1400–500 B.C. and A.D. 1340–1900 (O'Brien *et al.* 1995). This increase in sea salts delivered northward reflects heightened storminess and correlates with Chukchi Sea beach ridges.

The last 1500 years alternate between decades of warming and cooling, especially in the period from A.D. 600 to 1200. Despite its reputation as the Medieval Warm Period, this interval also records several decades-long temperature declines, some likely as pronounced as the Little Ice Age. Four lengthy cool intervals, 30–80 yr in duration, were defined by $^{18/16}$O concentrations (Dansgaard *et al.* 1975) and include: A.D. 680–710, 825–875, and 1040–1120, with the coolest temperatures from A.D. 1150–1200. The Nile River also serves as a global hydrograph, showing frequent El Niño events and cooling between A.D. 660 and 1000 (Quinn 1992). The global signature of the Medieval Warm Period remains controversial (*cf.* Hughes and Diaz 1994); minimalists define the peak of warming between A.D. 900–1300, with a glacial expansion A.D. 700–900 (Grove and Switsur 1994). The greatest episode of erosional truncation within the Chukchi Sea co-occurs with this early medieval glacial expansion.

COSTS AND BENEFITS OF THE EARLY MEDIEVAL COLD EVENT

Paleoclimatic reconstructions based on beach ridge evolution (Mason and Jordan 1993; Mason *et al.* 1997; Mason 1999) indicate the prevalence of southerly winds during limited decadal to century-length periods, especially from 800 to 200 B.C. Less stormy, but predominantly north and northwest winds prevailed A.D. 300–700 responsible for the addition of beach ridges along eastern Chukchi Sea barrier islands and spits (Mason and Jordan 1993). Between A.D. 750 and 1200, most beach ridge sequences were eroded and truncated because of more frequent and/or ferocious storms due to intense northerly winds, which also co-occurred with colder winters or stable ice conditions (Mason and Jordan 1993).

Advantageous conditions for hunters likely alternated between the north and south facing coasts in relation to climatic conditions. On St. Lawrence Island, the north shore was probably more favorable for hunting between A.D. 300–750 but was possibly less attractive afterward, due to heavier ice and northerly storms, A.D. 750–950 and 1030–1200, and during the Little Ice Age. Conversely, the south shore was more favorable during the colder periods dominated by northerly winds. However, the extant cultural chronology is insufficient to support or refute this perspective.

Several locales may have benefited from both cold and warm intervals, as at Point Hope; unfortunately, conditions for site preservation varied with the prevalence of powerful northerly storms. Cape Lisburne, marked by storm deposits in the late 1st millennium A.D., may have had positive attributes both during southerly winds during spring whaling but with northerly ice-carrying winds during fall whaling. In fact, the impact of intensified ice during fall is a factor little considered by archaeologists. Cape Krusenstern whaling was practiced only during warmer conditions when the leads were displaced south (Anderson 1983:77) apparently associated with northerly and westerly winds that produced erosional truncation and southeastward progradation.

CULTURAL CONSEQUENCES OF THE EARLY MEDIEVAL COLD PERIOD

The Ipiutak Collapse and the Rise of Thule

The enigmatic Ipiutak shamanic cult, or alliance between Old Bering Sea East Cape and Ipiutak Point Hope polities (Larsen and Rainey 1948; Mason 1998), thrived during a less stormy episode during A.D. 550–900 (Gerlach and Mason 1992; Mason and Gerlach 1995a; Mason 2000a). The Ipiutak phenomenon was focused within Kotzebue Sound, especially at Cape Krusenstern, Deering and Point Hope; its influence extended southward to Seward Peninsula (Mason 1998, 2000a) to Golovnin Bay (Mason *et al.* 2003) and across the Brooks Range (Gerlach and Hall 1988, Mason 1998).

By compiling the two sigma ranges of 46 Ipiutak dates, Mason [2000a:233] concluded that no Ipiutak community *definitively* dates after A.D. 900. Several younger ages from the Croxton site (Gerlach and Hall 1988), in the Brooks Range foothills, are an artifact of faulty laboratory procedures and can be rejected (Reuther *et al.* 2002). Ipiutak cult declines as rapidly as it arose; in fact, the disappearance of Ipiutak shamans might have occurred considerably earlier than A.D. 900.

The first appearance of Ipiutak in Kotzebue Sound may have occurred as early as the 2nd or 3rd century A.D. at Cape Krusenstern, and possibly as early at Point Hope (Mason 1998)—although the burial cult is best dated at Deering in the 7th to 8th centuries A.D. (Reanier *et al.* 1998). Although zooarchaeological analyses re-main preliminary, Ipiutak represents an economy dependent on seal, walrus and caribou, and only occasionally, if at all, on whales (Anderson 1986b).

Using recurrence interval estimates from Cape Espenberg, Cape Krusenstern and Pt. Hope, intense fall storms were only an infrequent occurrence during Ipiutak, far fewer than during the previous millennia or during the succeeding centuries, as evident from the surface stabilization during the Ipiutak at Deering (Reanier *et al.* 1998; Mason 1999, field notes). The abundance of driftwood during Ipiutak may also reflect warm spring breakup with upriver ice jam flooding (Mason 1998). The disappearance of the Ipiutak network co-occurs with the profound shift to stormy, cooler climates and with a cultural and technological transformation, the Thule and Punuk cultures.

CLIMATIC CONTROLS OVER WHALING

The Thule Expansion

A persistent belief among arctic archaeologists is that population pressure *and/or* warmer climates A.D. 1000–1200 catalyzed the Thule expansion (Taylor 1963; Bockstoce 1973; McGhee 1969/70, 1972, 1981, 2000; Stoker and Krupnik 1993; Morrison 1999). Typically, the process relationships are un-stated or implicit (*cf.* Anderson 1983:76-77); for example, Dekin (1972:19) simply observes that the period 'A.D. 800–900–1200' was 'considerably warmer than present which gave Thule an advantage.' Apparently, that advantage translated into a greater share of storable protein and ultimately a higher number of Thule people. The erudite William Taylor (1963:461) observed that the two most 'plausible platitudes' are 'population pressure and hunting potential beyond' and postulated a linkage between the climatic amelioration 'in the centuries up to A.D. 1000, and the intensification of dog traction that ultimately lent an ecological advantage to Thule over Dorset. Dog sleds gave greater mobility but required more food; apparently, the need for mobility occasioned the greater investment in dogs' (Taylor 1963:461). Simplistic faunal observations, e.g., percentage calculations, from Cape Krusenstern led Bockstoce (1979:42) to postulate that Birnirk people were sea-mammal specialists who consumed very little caribou and that a broadening of the diet occurred during Thule.

The most refined model for Thule expansion concentrates on an inherent tension, a Marxist dialectic, the opposing and contradictory effects of warming on whales and seals, the position of McGhee (1969/70:176ff) and Stanford (1973, 1976). Warming, apparently, reduced ice cover, adversely affected ice-obligate ringed seals and led to the intensification of open water whaling because 'spring ice-lead whaling became less productive' when seal and walrus were less reliable (McGhee 1969/70:177). More importantly, habitat for whales shifted eastward from Alaska and increased in area, as the fast ice in

Amundsen Gulf and Melville Sound disappeared. Still, the eastward movement of incipient Birnirk/Thule groups did not occur in force until the 13th century A.D. (McGhee 2000), coincident with a warming evident in the stabilization of beach ridges throughout northwest Alaska and the expansion of Thule across Alaska (*cf.* Giddings 1952; Mason 2000b).

Several assumptions about whaling are strongly imbedded in the literature: (a) ringed seal and whale are interchangeable, but possibly mutually exclusive; (b) bowheads can be reliably secured and stored during open water conditions by using tow lines and *umiaq*; (c) small numbers of whalers can secure a surplus for the community, at least under ice-lead whaling (Anderson 1983:68) and (d) whaling required constant meteorological vigilance for weeks on and—apparently—precluded sealing (Bockstoce 1979:43). While the lead system offers a regimented and straightforward method of prey encounter, open-water whaling requires a very different set of perceptions and techniques. Search strategies must have involved covering more area, i.e., increasing mobility and altering foraging patterns to a more collector-based regimen (Binford 1982). The utility of dog teams in achieving this mobility in an ice-lessened northwest Alaska is unclear, although their use on persisting shorefast ice in Canadian locations might have been important.

Birnirk subsistence had a deleterious ecological impact according to Krupnik (1993:237) representing a 'search and destroy' mode, requiring constant movement as the resources in one area are extinguished and require a move onward. This parallel to the Pleistocene over-kill model of Martin (1967) cannot be substantiated with faunal or demographic data in northwest Alaska (*cf.* below).

Most radically, Sheehan (1995) postulates an Early Medieval Optimum at A.D. 1000–1200 associated with 'nomadic' whaling from anywhere on the coast preceding the promontory-based whaling of the Little Ice Age. First, neither site maps nor precise chronological data are available to establish regional population density. Second, as documented above, in Bering Strait proper the period from A.D. 1000 to 1200 was a cooler interval, as reflected in the storm truncation of beach ridge systems across northwest Alaska (*cf.* above) and the tree ring record (D'Arrigo and Jacoby 1992; Gramulich and King (1998), and supplemented by Barber (*cf.* above).

Evidence for whaling intensification can be equivocal; e.g., Birnirk or Punuk people at Ekven increased their yield of 'small' whales, to the exclusion of bowhead bones, as in the occurrence of nine scapulae in one burial, and in the larger toggling harpoons (Krupnik 1987:18). At the Birnirk site, whaling can be substantiated by (a) the widespread use of whale products (structural members, bowls and sleds) (Ford 1959:40ff) and (b) the size of the mandibles and skulls is consistent with human selection of smaller, 7 to 10 m long bowhead whales (McCartney 1995:93ff; Savelle and McCartney

2001). The scale of Birnirk whaling is unknown; the small size of even the largest Birnirk communities (Mason 1998) argues against the elaboration of whaling, although seasonal aggregations might have enabled the formation of sufficiently large crews. Improvements in dog handling by the use of sleds might have proved critical in assembling such crews (Taylor 1963).

Storm Impacts from A.D. 900-1000 across Northwest Alaska

Punuk and its Outward Push. Concrete evidence for whaling in Chukotka is scanty and occurs in mostly undated Punuk communities (*cf.* Rudenko 1961), based on 'whaling' harpoons and, again, the use of whale products for construction and as containers (*cf.* review in Mason 1998). Collins (1937:217) argued that whaling was comparatively recent, Punuk, practice in Seveokok; an idea also espoused by Rudenko (1961:113), and re-introduced by Bandi (1995). Many commentators attribute Punuk success to the introduction of more powerful bows, defensive armor, improved iron supply and social stratification (Mason 1998). Limited chronological data from St. Lawrence Island led Mason (1998) to hypothesize that Punuk prevailed between A.D. 900–1000, coincident with a hiatus in storms and a temporary renewed persistence of the north polynya along St. Lawrence Island. This dating assignment is bolstered by new data (*cf.* Table I) and an independent analysis of Punuk ^{14}C ages from St. Lawrence Island by Blumer (2002) revealed that the central tendency clusters within the century A.D. 900–1000.

While St. Lawrence Island has several sites with Thule objects (Ackerman 1961, 1962), most researchers prefer a Punuk label, following Collins (1937, 1964). A 'pure Thule' round stone-walled house with a long entry 'with only Thule type harpoons,' was excavated at Kitnepaluk, 32 km south of Gambell, by Giddings in 1939 (Rainey 1941:562). The house is not still fully described or dated; a brief inspection (by OKM) of the University of Alaska Fairbanks collection indicates considerably more Punuk design elements and harpoon heads than Thule characteristics.

Although Punuk villages occur on all coasts of St. Lawrence Island, the largest polities were apparently on its eastern shores, and on the south shore of Chukotka (Rudenko 1961; Chard 1955; Ackerman 1961, 1962); unfortunately, most Punuk sites remain undocumented (Ackerman 1961, 1962; Crowell 1985). The circumstance that Punuk occupations overlie half the Old Bering Sea (OBS) occupations fuels the assumption that Punuk occupations were larger in size or duration. Reliable chronological data (n=12) for Punuk include several newly acquired ages from Kukulik and Uivvaq (Table I). These new data from Seveokok and Uivvaq indicate that the Punuk phenomenon was considerably younger than the age of A.D. 500–900 favored by an earlier generation of researchers (Giddings

1960; Collins 1964) including the most extreme contemporaneity proponents (Gerlach and Mason 1992).

Houses and burials in the greater Seveokok region of northwest St. Lawrence Island offer some age control for the Punuk occupations and establish its florescence was coincident with the major cycle of storminess (Table I): (a) Kitnepaluk dated *ca.* A.D. 900–1000 (Blumer 1997a), (b) several burials from A.D. 900–1100, and (c) a Mayughaaq house dated to A.D. 1200 (Bandi 1972).

The Kitnepaluk house had Sicco harpoon heads and dated (Table I) to A.D. 890–970 (B-6476) and A.D. 1035–1165 (B-6475) (Blumer 1997a). Several burials south of Seveokok, at Dovelavik may be early Punuk interments (*cf.* Table I); unfortunately, few of the artifacts are indubitably Punuk. Dovelavik burial M-1 had slate spear and knife blades, a whale skull and or ivory rods (Hofmann-Wyss 1987:13); nothing terribly diagnostic of Punuk outside of its age: A.D. 900–1230 (B-2436) (Table I). Grave K-45 at Kitnepaluk, interred between A.D. 900–1026 (B-4414, Table I] contained a decorated open socket III a(x) harpoon head (Hofmann-Wyss 1987:30-31, Pl. 26:3); its linear motifs termed transitional between Old Bering Sea and Punuk by Hofmann-Wyss (1987:30-31).

Ayveghyaaq, excavated >75 years ago, still provides one of the largest Punuk—albeit also Thule—inventories (Collins 1937:362-364). Two solid carbon ^{14}C assays (P-69 and-92) within A.D. 871–1391 and A.D. 600–1373 (Table I) establish only the broad parameters for its age. Considering its location, landward of the truncation of the beach-ridge plain (Mason and Ludwig 1990), the Punuk occupation should be older than A.D. 1000. This inference is based on the regional coastal reorganization that does not post date A.D. 1150–1200 and whose lower age limit is A.D. 750 (Mason and Jordan 1993; Mason 1999 and elsewhere).

Kukulik mound, one of the largest in Bering Strait, exemplifies the interplay between contemporaneity and descent. Contemporaneity is shown by the interaction—stylistic similitudes—between the various paleo-ethnicities along the Bering Strait (Ackerman 1962; Gerlach and Mason 1992). This interaction is evident in the parallel stratigraphic relationship—Birnirk and Punuk harpoon heads in the same levels within Northeast cut profile at Kukulik (Geist and Rainey 1936:198-200). In addition, the succession of Birnirk/Punuk to Punuk/Thule is evident in the stratigraphy, perhaps separated by decades-long abandonment periods—the enigmatic sod lines whose lateral extent is unclear (Geist and Rainey 1936:45ff). Punuk materials were abundant at Kukulik, often in association with spurred, double bladed 'Birnirk' harpoon heads (Geist and Rainey 1936:227ff).

In a stratigraphic profile 'below the third house' at Kukulik, six Punuk and Birnirk harpoon head types were plotted within a discrete zone 3.35 m below surface; Punuk and Thule types lie above this zone. Apparently, then, a phase of Punuk and Birnirk interactions in the middle levels was supplanted by a 'Punuk/Thule' culture, the phrase coined by Collins (1964, 1973), and employed by Morrison (1991). Only a single chronometric reference point to the age A lower Punuk component at Kukulik is dated from a decorated ivory harpoon head at A.D. 1071–1331 or (b) A.D. 1304–1438, depending on the marine correction used (*cf.* Dumond 1998). The antiquity of this lower Punuk occupation at Kukulik may be younger than expected, and shows contemporaneity with Mayughaaq and Kitnepaluk (Table I).

In sum, an early Punuk occupation at and adjacent to Seveokok occurred during the 10th to 11th centuries A.D., based on ages from Bandi's excavations; nonetheless, several ages may predate this time. Punuk occupation at Kukulik was established possibly at least as early as the 12th or 13th century A.D., although it is likely earlier.

Excluding a poorly–dated bastion at Wales (Collins 1929, 1940; Morrison 1991; Dumond 2000; *cf.* Harritt 2001), Punuk is rare on the Alaskan mainland, especially south of Point Hope. A Punuk–decorated Sicco harpoon head was found in trenching activity during the late 1990's at Deering; its dating and context are uncertain (Mason 1999, field notes). Point Hope is notable for the absence of any authentic Punuk occupation—despite an undated winged object obtained by Knud Rasmussen (Mathiassen 1929:45). To the North, just east of Cape Lisburne, at Uivvaq (XPH-045)) a Punuk 'trident,' atlatl counterweight, one of only a handful known, dates younger than A.D. 980–1153 but is older than A.D. 1153–1306 (Table I, Mason *et al.*, unpublished data). In addition, the case for Birnirk contemporaneity with Punuk (*cf.* Ackerman 1962) is bolstered by three ^{14}C ages from the lower component at Uivvaq that reveals Birnirk occupation between AD 987–1160[980±40 BP, Beta 190816], below the Punuk levels, separated by 50 cm of beach gravel (Mason 2003).

Nunagiak extends along 'an old beach line' >200 m inland beyond several high storm ridges on the sheltered margin of a small lagoon, near Point Belcher, >150 km southwest of Barrow. Nunagiak has 13 mounds, one is large, possibly a *qarigi* (Ford 1959:58), of unknown affiliation. 'Early' Punuk engraved designs occur on goggles, a Sicco harpoon head, and a trident counterweight (Ford 1959:62-64). Unfortunately, Nunagiak is undated; an early Punuk assignment would place it *ca.* A.D. 900–1000, possibly during the period of more southerly winds that might explain both Point Belcher's temporary attractiveness and provide a lower limiting age on the subsequent addition of storm ridges.

The Birnirk to Thule Transition: Chronology and Climate after A.D. 1000

The extant settlement pattern data are insufficient to evaluate regional population changes, both for the period before and after A.D. 1000. Further, adequate chronometric data are lacking for most large, let alone, small settlements on the coasts (Mason 1998). The record from the interior is even less documented (Gerlach and Mason 1992, Morrison 1989). 'Western' Thule occupations in northwest Alaska can be only tentatively dated, despite the initiatives of Morrison (1989, 2001). Rejecting most determinations from the 1950s to 1960s, five consistent AMS radiocarbon ages on diagnostic harpoon heads (Natchuk, Birnirk, Sicco) lead Morrison (2001) to infer that Birnirk was transformed into Thule over a forty-year interval (or less) from A.D. 979–1019, the two sigma average of the ages.

The chronological data on its immediate predecessor, Birnirk, indicate that people lived, if not thrived, in small settlements with only a handful of houses (Mason 1998). The extent of Birnirk on the Chukotkan coast was significant; e.g., one of the largest settlements is at far distant Cape Baranov near the Kolyma River (Okladnikov and Beregovaia 1971). A few ^{14}C dates from north Chukotka (Gerlach and Mason (1992) document an extensive Birnirk presence prior to A.D. 900; Birnirk may even have had its greatest strength on the north Chukotka shore, as Jenness (1940) suggested. The precise nature of any Birnirk occupation in the depths of the Punuk Islands and Kialegek (S.E. Cape) is attested only by a 'few Birnirk heads, all from the lower levels' (Collins 1929:41).

Two locations contain evidence supporting an early age for Birnirk (Table I): (a) Ekven, with ages *ca.* A.D. 700 (Blumer 1996, 1997b; Blumer and Csonka 1998) and (b) Uivvaq with a Natchuk harpoon head within storm beds dated by insect remains between A.D. 591-680 (Table I, Hoffecker *et al.* 2002).

The scale, chronology and duration of the largest Birnirk occupation in Alaska near Point Barrow are poorly known; the twenty mounds at *Piqniq* remain undated but likely were not contemporaneous (Mason 1998). Several burials in Kugok ravine at Utiqiagvik may reflect a Birnirk occupation (Ford 1959:25ff), based on an age of A.D. 660–976 (Beta-2907, *cf.* Table I) associated with purported—but unillustrated—'Birnirk' designs on a needle case (Brower 1990:38), unconvincing to Morrison (1999:77). Two small Birnirk structures (houses?) at Walakpa are possibly younger, within A.D. 937–1248 (Gak-2298) (Stanford 1976; Mason 1998; Morrison 2001:77) noteworthy in their contemporaneity with the 11th–12th century A.D. storms. At some point, possibly only years or decades later, Thule people inhabited Walakpa between A.D. 1020–1295 (Gak-2297 (Stanford 1976:90).

The overlap between the Birnirk and Thule materials at Walakpa may illuminate cultural process or record, as Stanford

(1976:90ff) and Morrison (2001) argue, the precedence of Birnirk *prior* to A.D. 1000 and the elaboration of Thule *after* A.D. 1000. Quite notably, Ford (1959) failed to discern any Thule occupations in stratified contexts at the Birnirk site; very likely the 5–7 m high composite ridges constructed landward in the subsequent centuries after A.D. 1100 (Ford 1959:35) either precluded or contained Thule remains—the ridge was removed for airport fill in the 1960s (Mason *et al.* 1991:12ff). Nonetheless, a gap of several centuries is evident in the Barrow region, virtually no cultural materials are dated between A.D. 1000 and A.D. 1220, the youngest intercept on an AMS age on a Thule 2 harpoon head from the Mound 44 slump block [695±55 B.P., Beta-42391, A.D. 1220–1393 (Mason 1991:65)]. While Walakpa may provide evidence for Birnirk to Thule occupation during the critical stormy centuries A.D. 1000–1200, the Punuk component at Uivvaq (XPH-045) shows the approximately contemporaneous process of transformation from Punuk to Thule (Hoffecker *et al.* 2002).

Birnirk residence at Point Hope is revealed from Burial 88 near Tikigaq and House 2 at Jabbertown, where an irregular, multi-roomed structure (Larsen and Rainey 1948:Fig. 52, p. 170ff) produced both Thule 2 and Natchuk harpoon heads, a diverse slate inventory, curvilinear impress ceramics, and bola weights (Larsen and Rainey 1948:Pl. 95). No radiocarbon data are available for Birnirk at Jabbertown, but Larsen and Rainey (1948:174) argued for contemporaneity with the Kobuk River site of Ahteut, dated by tree-rings to the early 13th Century AD (Giddings 1952b:108). The Ahteut phase was possibly several centuries older, AD 1021–1257, the age of House 13 at Onion Portage (Anderson 1988:48, *cf.* Mason and Gerlach 1995b:106-107). At Nukleet, an Ahteut arrow point also dated AD 1040–1260 [870±40 BP Beta–157237)] Murray *et al.* 2003:Table1). Quite significantly, the lower component at Uivvaq also dated to the 11th to 12th centuries AD (Mason 2003), evidence of a widespread transformation or migration.

This brief review of Thule phenomena in northwest Alaska indicates the continuing insufficiency of the archaeological record, and the overweening constraints of adverse geomorphic forces such as intense storms. Several critical areas lack evidence for early Thule: Safety Sound, Point Hope and Point Barrow; the reasons are unclear but it is likely that intensified erosion has removed the evidence. A major restraint on any analytic posture should avoid characterizing the entire 450 yr period from A.D. 750 to 1200 with the same broad brush of adverse or beneficial. Dansgaard *et al.* (1975) showed that Greenland witnessed drastic decadal long alternations in cold(er) to warm(er) climes. What about Bering Strait?

CLIMATE AND DEMOGRAPHY IN NORTHWEST ALASKA

Were there huddled masses of Thule folks in Bering Strait awaiting the good times associated with a climatic amelioration? If so, the rise in population would be associated with the cooler climates and higher productivity prior to A.D. 1200—while the Thule departure sped considerably after A.D. 1200 (McGhee 2000). However, the task of reconstructing population in Bering Strait is difficult; adequate data are so rare. Notwithstanding the limitations of the database, Mason (1998) provisionally quantified first millennium A.D. population at a few Alaskan sites, and a few Siberian sites. Even the highest estimates lead to the conclusion that northwest Alaska was by no means densely populated during the first millennium A.D. (Krupnik 1983). Mason (1998) employed ethnographic data, both global and regional, to estimate that Ipiutak villages at Deering, Point Hope and Cape Krusenstern had between 100–300 people while the largest Birnirk site, Piqniq, had fewer than 100 inhabitants. Birnirk communities were smaller, except for the several around Wales. Any Wales estimates are fanciful, but in relative terms, a population only half that of Mayughaaq can be extrapolated using the comparative measure (Mason 1998)

Bering Strait population shifts can be examined from two cemeteries—a methodology discussed in Mason (1998) with sizable numbers (>35) of ^{14}C ages. Deaths and, indirectly population, were highest A.D. 1000–1200 with half as many recorded between A.D. 800–1000—contemporaneous with the Ipiutak occupations at Pt. Hope, Krusenstern and Deering. A similar relationship occurs at the Gambell cemetery with more frequent interments between A.D. 1000–1200 and a comparatively lower number during A.D. 600–900. Tentatively, we argue that fewer people lived at Gambell and East Cape relative to Pt. Hope between A.D. 600–900 and subsequently a rapid population increase occurred. As noted, the 11th and 12th centuries were ones of the most storm-prone in the last two thousand years! Northerly winds probably maintained large polynyas off south-facing coasts.

The number of dated burials at Ekven declined by 80% after A.D. 1200, the onset of a warmer period in Bering Strait. Similar declines at Gambell suggest a linkage to a Thule expansion out of Bering Strait; many Alaskan coastal Thule sites date to the 13th century (Mason 2000b). Centuries-long hiatuses, or data gaps, between A.D. 1000–1200 at Point Hope and Point Barrow, including Utqiagvik, present a difficulty for the population stress hypothesis for Thule origins. Were those areas abandoned, or were sites subsequently eroded?

Sparse data from St. Lawrence Island testify that Punuk people entered or innovated A.D. 800–900 (Bandi 1984; Hofmann- 1987;

Blumer 1997a; Mason 1998), during a cold period, apparently during the later phases or after the Old Bering Sea/Ipiutak system collapsed. Unfortunately, the remaining majority of Chukotka and St. Lawrence Island Punuk occupations are undated. Nonetheless, considering that winged objects at Uivvaq and Nunagiak have considerable similarity with Ayveghyaaq on St. Lawrence Island, it is possible to infer that the Punuk movement occurred directly from that locale.

Significantly, the Old Bering Sea/Ipiutak system and cult collapsed *ca.* A.D. 900. As no acceptable ages for Ipiutak are younger than A.D. 900 (Mason 1998, 2000a; Reuther *et al.* 2002), possibly its fate is also inversely related to cold. Ipiutak subsistence, trade and hegemony apparently relied on less stormy conditions. Conditions were not consistently equally adverse across the Chukchi Sea, however; progradation associated with weaker storms occurred at Cape Krusenstern following Ipiutak until the storm events prior and/or during the Thule occupations A.D. 1000–1200. Cape Espenberg also witnessed two episodes of powerful, northwesterly storms that constructed high dunes landward of the Ipiutak house at KTZ-157 (Harritt 1994).

After the collapse of Ipiutak *ca.* A.D. 900, Pt. Hope has no evidence of occupation during A.D. 1000–1200 until Jabbertown was settled after A.D. 1200, based on correlations with the similar inventory from a 13th century Ahteut house in the Kobuk (Giddings 1952:27). Subsequently, a large village and cemetery grew at Tikigaq, contemporaneous with the Little Ice Age following A.D. 1660, if relative beach ridge limiting ages are reliable (Mason 1990). The massive cemetery fence of whale mandibles (n>1000 at Point Hope (Savelle and McCartney 2001) probably dates to the Little Ice Age. The sheer volume of whales preserved at Tikigaq (Point Hope) within its cemeteries is impressive. However, none of these are dated; many may date to the 19th century and the introduction of firearms or, less likely, result from the scavenging of whales not retrieved by European whalers.

Considering that most Kotzebue Sound promontories are marginal for present day settlers and are not used for whaling, Mason and Gerlach (1995a) speculated that when regional population density was high, marginal areas, e.g., Mitletavik or Cape Espenberg, might prove attractive. However, examining the climatic and demographic data more fully (Mason 1998 and above), we suggest that settlement pattern and whaling might be positively correlated inversely with climate: people are more dispersed, even if population was low, during warmer, less advantageous periods—a position advanced by Lutz (1982) who, nonetheless thought the cold periods were disadvantageous. The need to raid the larder during hard times provided the rationale for sheep hunting in the Brooks Range (Campbell 1978).

CONCLUSIONS

What then, is the relationship between climate and whaling? Storm-intensified upwelling during cooler periods and the resulting heightened productivity should foster improved conditions for whaling (Schell *et al.* 1998; Finney *et al.* 2002). While the Old Bering Sea/Ipiutak interaction sphere operated during a warmer climate (Mason 1998), the development of whaling apparently occurred *prior* to that, during a colder interval prior to A.D. 200. Admittedly, the dating of this development remains imprecise. However, the development of the Thule culture in Bering Strait coincides with a stormy interval. Unfortunately, few Alaska sites date to the period A.D. 1050–1200; considerably more occupations are known from the Brooks Range (Gerlach and Hall 1988) or the Kobuk River after A.D. 1200 (Giddings 1952).

The relationship between climate and political/military fortunes witnesses a shift between A.D. 900–1000. Previously, between A.D. 550–900, the success of many Bering Strait and Chukchi Sea polities (Old Bering Sea, Ipiutak) was associated with the persistence of southerly wind conditions during the fall season (Mason 1998), a correlation that suggests warmer climatic conditions prevailed during the late first millennium A.D. The linkage for most sites, especially on St. Lawrence Island, with climate cannot be firmly established because of the lack of ages. The abandonment of Mayughaaq and Ipiutak apparently occurred centuries before A.D. 900, following or during a century of increased storminess with northerly winds pummeling Kotzebue Sound. However, settlement continued at the Northwest Cape (Seveokok), with the Ayveghyaaq site recording a Punuk occupation A.D. 800–1000 and with its cemeteries recording a population increase during this stormy interval, possibly, the consequence of immigration (or war captives?) from disadvantaged localities. More landward Mayughaaq proved attractive to at least one Punuk household as late as the 13th century, if Bandi's (1972) ages are reliable. Northerly wind/storm conditions would have also favored Birnirk settlements on south-facing coasts such as Wales and Krusenstern, and possibly Barrow.

Two centuries are nearly invisible in coastal northwest Alaska: A.D. 1000–1200; few settlements can be securely or solely assigned to this period. Paleo-climatic records indicate that this period witnessed massive northerly storms, apparently associated with colder climates (Mason and Gerlach 1995a). Bering Strait villages at Seveokok and Ekven may have had increased populations, relative to the period before A.D. 800. Archaeological remains from Cape Krusenstern may date to this period; their deep burial by storm beds should serve as a guide for finding evidence from this interval. Unfortunately, storms brought erosion that might have rapidly removed the residential residues of even the most successful of Thule inhabitants. Site preservation inevitably tracks progradation or stasis, conditions which might not mirror successful hunting or subsistence. The 13th century A.D. reveals small Thule settlements at numerous locations from Norton Sound, the middle Kobuk and Wales to Barrow and as far as Thetis Island (Mason 2000b; Mason *et al.* 2003) and on into western Canada (McGhee 2000). People were on the move, stopping briefly for a season or two and not accumulating deep midden deposits. Archaeologists still cannot confidently answer why or assess the direction(s) of the Thule and/or Punuk 'migration(s).'

Acknowledgments. This paper originated in research by O.K. Mason conducted on the Wales beach bridge complex in 1996, as a participant in the NSF Office of Polar Programs grant to examine The Origins of Whaling in the western arctic project, headed by R.K. Harritt and A.P. McCartney. The original version completed by Mason in 1999 benefited from three anonymous reviews and the incorporation of a NSF funded OPP grant for multi-disciplinary research at Uivvaq in 2000 conducted by J.F. Hoffecker, Georgeie Reynolds, Scott Elias and Diane Hanson. A few notable results are incorporated from the 2002 season of research under contract by GeoArch Alaska, which included the NSF researchers, with the addition of Claire Alix. The participation of the Point Hope community has been essential in conducting the Uivvaq research.

REFERENCES

Ackerman, Robert E.
    1961   *Archaeological Investigations into the Prehistory of St. Lawrence Island.* Philadelphia: University of Pennsylvania, Ph.D. dissertation.
    1962   'Culture Contact in the Bering Sea: Birnirk-Punuk Period,' pp. 27-34 in John M. Campbell, ed., *Prehistoric Cultural Relations Between the Arctic and Temperate Zones of North America.* Montreal: Arctic Institute of North America, Technical Paper 11: 27-34.
Anderson, Douglas D.
    1983   'Changing Prehistoric Eskimo Subsistence Patterns: A Working Paper,' pp. 62-83 in H.N. Michael and J.W. VanStone, eds., *Cultures of the Bering Sea Region*, Papers from an International Symposium. Moscow: International Research and Exchanges Board.
    1986b   'The Ipiutak Villagers: Large Populations at Cape Krusenstern,' pp. 117-159 in J.L. Giddings and D.D. Anderson, *Beach Ridge Archaeology of Cape Krusenstern: Eskimo and Pre-Eskimo Settlements around Kotzebue Sound, Alaska.* Washington, D.C.: National Parks Service, Publications in Archeology 20.
    1988   Onion Portage: The Archaeology of a Stratified Site from the Kobuk River, Northwest Alaska. *Anthropological Papers of the University of Alaska* 22(1-2): 1-163.

Bandi, Hans Georg

 1972 Gräber der Punuk-kultur bei Gambell auf der St. Lorenz Insel, Alaska. *Sonderbruck aus dem Jahrbuch des Bernischen Historischen Museums* 51/52: 41-116.

 1984 *St. Lorenz Insel-studien.* Vol. 1. Berner Beitärge zur Archäologischen und Ethnologischen Erforschung des Beringstrassengebeetes. Band I. Allegemeine Einführung und Gräberfunde bei Gambell am Nordwestkap der St. Lorenz Insel, Alaska. Academica Helvetica 5, Bern and Stuttgart. [Partial translation by Stefanie Ludwig, manuscript in possession of the author].

 1995 'Siberian Eskimos as Whalers and Warriors,' pp. 165-184 in A.P. McCartney, ed., *Hunting the Largest Animals: Native Whaling in the Western Arctic and Subarctic.* Edmonton: Canadian Circumpolar Institute (CCI) Press, University of Alberta, Occasional Publication No. 36, Studies in Whaling No. 3.

Binford, L.A.

 1982 The Archaeology of Place. *Journal of Anthropological Archaeology* 1(1): 5-31.

Blumer, Reto

 1996 'Première Expédition Archéologique Internationale en Tchoutotka, Sibérie nord-orientale. Rapport de la Contribution Suisse aux Travaux de l'été 1995,' pp. 110-150 in Jahresberich 1995, Zürich et Vaduz: Fondation Suisse-Leichtenstein pour les Recherches Archéologiques à l'étranger.

 1997a *Le Matériel Archéologique de Kitngipalak, Ile St. Laurent, Alaska,dans le Contexte du Néoeskimo Béringien.* M.S. Archaeologie Prehistoire, Université de Genévé.

 1997b 'Seconde Expédition Archéologique Internationale en Tchoutotka, Sibérie nord-orientale,' pp. 57-78 in *Jahresbericht 1996, Rapport de la Contribution Suisse à la campagne de 1996.* Zürich et Vaduz: Fondation Suisse-Leichtenstein pour les Recherches Archéologiques à l'étranger.

 2002 'Radiochronological assessment of Neo-Eskimo occupations on St. Lawrence Island,' pp. 61-106 in E.E. Dumond and R.L. Bland, eds., *Archaeology in the Bering Strait Region: Research on Two Continents.* Eugene, OR: University of Opregon Anthropological Papers No. 59.

Blumer, R. and Y. Csonka

 1998 'Archaeology of the Asian shore of Bering Strait: Swiss Contributions to the 3rd International Field Season,' pp. 83-130 in *Jahrebericht 1997.* Schweizerisch-Liechtensteinsche Stiftung für Archäologische Forschungen im Ausland.

Bockstoce, John R

 1973 A Prehistoric Population Change in the Bering Strait Region. *Polar Record* 16: 793-803.

 1976 On the development of whaling in the western Thule culture. *Folk* 18: 41-46.

 1979 The Archaeology of Cape Nome, Alaska. Philadelphia: University of Pennsylvania, University Museum Monographs 38.

Brower, Lucinda, Peter M. Bowers, O.K. Mason, R.O. Mills, and Catherine Williams
    1990   'Ceramic Variation at the Utqiagvik site,' pp. 285-298 in E.S.Hall and L. Fullerton, eds., *The Utqiagvik Excavations*, Vol. 1. Barrow, AK: North Slope Borough, Commission on Iñupiat, History, Language and Culture.

Bowers, Peter M., O.K. Mason, R.O. Mills, and Catherine Williams
    2000   *Preliminary Report on the 1999 Excavations in Deering, Alaska. Abstracts, 27th Alaska Anthropological Association Annual Meeting.* Anchorage: Alaska Anthropological Association.

Burch, Ernest S., Jr.
    1981   *The Traditional Eskimo Hunters of Point Hope, Alaska: 1800-1875.* Barrow, AK: North Slope Borough

Campbell, John M.
    1978   'Aboriginal Overkill of Game Populations: Examples from Interior North Alaska,' pp. 179-208 in R.C. Dunnell and E.S. Hall, eds., *Archaeological Essays in Honor of Irving B. Rouse.* The Hague: Mouton Publishers.

Chard, Chester S.
    1955   Eskimo Archaeology in Siberia. *Southwestern Journal of Anthropology* 11: 150-177.
    1957   The Southwestern Frontier of Eskimo Culture. *American Antiquity* 22(3): 304-305.
    1959   The Western roots of Eskimo Culture. *Actas, XXXIII Congreso Internacional de Americanistas*, San José, Costa Rica.

Coachman, L.K.
    1993   On the Flow Field in the Chirikov Basin. *Continental Shelf Research* 13(5/6): 481-508.

Coachman, L.K. and D.A. Hansell, eds.
    1993   ISHTAR: Inner Shelf Transfer and Recycling in the Bering and Chukchi Seas. *Continental Shelf Research* 13(5/6): 473-704

Collins, Henry B., Jr.
    1929   *Prehistoric Art of the Alaskan Eskimo.* Washington: Smithsonian Institution, Miscellaneous Collections 81(14).
    1937   *Archaeology of St. Lawrence Island, Alaska.* Washington: Smithsonian Institution, Miscellaneous Collections 96(1).
    1940   *Outline of Eskimo Prehistory. 100 Essays in the Historical Anthropology of North America Washington:* Smithsonian Institution, Miscellaneous Collections.
    1964   'The Arctic and Subarctic,' pp. 85-114 in J.D. Jennings and E. Norbeck, eds., *Prehistoric Man in the New World.* Chicago: University of Chicago Press.
    1973   'Eskimo Art,' pp. 1-37 in H.B. Collins, F. de Laguna, E. Carpenter, and P. Stone, eds., *The Far North: 2000 Years of American Eskimo and Indian Art.* Washington, D.C.: National Gallery of Art

Crowell, Aron and Henry B. Collins, Jr.
    1985   *Archeological Survey and Site Condition Assessment of Saint Lawrence Island, Alaska.* Report submitted to the Department of Anthropology, Smithsonian Institution, Washington, D.C. and Sivuqaq, Inc., Gambell, AK.

Dansgaard, W., S.J. Johnsen, N. Reeh, N. Gundestrup, H.B. Clausen, and C.U. Hammer
 1975 Climatic Changes, Norsemen and Modern Man. *Nature* 255: 24-28.
Dansgaard, W., S.J. Johnsen, H.B. Clausen, D. Dahl-Jensen, N. Gundestrup, C.U. Hammer, and H. Oeschger
 1984 North Atlantic Climatic Oscillations Revealed by Deep Greenland Ice Cores. *American Geophysical Union Monograph* 29: 288-298.
D'Arrigo, R.D. and G.C. Jacoby
 1992 'Dendroclimatic Evidence from Northern North America,' pp. 296-311 in R.S. Bradley and P.D. Jones, eds., *Climate Since A.D. 1500*. London: Routledge.
Darby, D.A., J.F. Bischof, and G.A. Jones
 1997 Radiocarbon Chronology of Depositional Regimes in the Western Arctic Ocean. *Deep Sea Research* II, 44(8): 1745-1757.
Dekin, A.A.
 1972 Climatic Change and Cultural Change: A Correlative Study From Eastern Arctic Prehistory. *Polar Notes* 12: 11-31.
Dinesman, L.G., N.K. Kiseleva, A.B. Savinetsky, and B.P. Khassanov
 1999 *Secular Dynamics of Coastal Zone Ecosystems of the Northeastern Chukchi Peninsula, Chukotka: Cultural Layers and Natural Depositions from the Last Millennia*. Tübingen: Russian Academy of Sciences, Severtsov Institute of Ecology and Evolution, Mo Vince Verlag.
Dixon, E.J., Jr.
 2001 *Early Maritime Adaptations on the Northwest Coast of North America: Excavations at 49-PET-408*. Abstracts, 28th Alaska Anthropological Association Annual Meeting. Fairbanks: Alaska Anthropological Association.
Dumond, Don E.
 1972 'Prehistoric Population Growth and Subsistence Change in Eskimo Alaska,' pp. 311-328 in B. Spooner, ed., *Population Growth: Anthropological Implications*. Cambridge: MIT Press.
 1998 *Hillside Site, St. Lawrence Island, Alaska: An Examination of Collections from the 1930s*. Eugene: University of Oregon Anthropological Papers 55.
 2000 *Henry B. Collins at Wales, Alaska: A Partial Description of Collections*. Eugene: University of Oregon Anthropological Papers 56.
Dumond, D.E. and R.A. Knecht
 2001 'An Early Blade Site in the Eastern Aleutians,' pp. 9-34 in D.E. Dumond, ed., *Archaeology in the Aleut Zone of Alaska: Some Recent Research*. Eugene: University of Oregon Anthropological Papers 58.
Dyke, A.S., J.E. Dale and R.N. McNeely
 1996 Marine Molluscs as Indicators of Environmental Change in Glaciated North America and Greenland during the 18,000 years. *Géographie physique et Quaternaire* 50: 125-184.
Dyke, A.S. and J.M. Savelle.
 2001 Holocene History of the Bering Sea Bowhead Whale (*Balaena mysticetus*) in its Beaufort Sea Summer Grounds of

Southwestern Victoria Island, Western Canadian Arctic. *Quaternary Research* 55: 371-379.

Elias, S. A., S.K. Short, and R.L. Phillips
    1992   Paleoecology of Late-glacial Peats from the Bering Land Bridge, Chukchi Sea Shelf Region, Northwestern Alaska. *Quaternary Research* 38: 371-378.

Erlandson, Jon
    2002   'Anatomically Modern Humans, Maritime Voyaging and the Pleistocene Colonization of the Americas,' pp. 59-92 in N.G. Jablonski, ed., *The First Americans: The Pleistocene Colonization of the New World.* San Francisco: University of California Press, Memoirs of the California Academy of Sciences No. 27.

Finney, Bruce F., Irene Gregory Eaves, Marianne S.V. Douglas, and John P. Smol
    2002   Fisheries Productivity in the Northeastern Pacific Ocean over the past 2,000 Years. *Nature* 416: 729-733.

Ford, James A.
    1959   Eskimo Prehistory in the Vicinity of Point Barrow, Alaska. *Anthropological Papers of the American Museum of Nat. Hist.* 47(1).

Fraker, M.A.
    1989   'Aspects of the Ecology of the Bowhead Whale (*Balaena mysticetus*) in the Western Arctic,' pp. 252-279 in L. Rey and V. Alexander, eds., *Proceedings of the 6th Conference of the Comité arctique international* 13-15 May 1985. Leiden: E.J. Brill.

Garfinkel, Harriet L. and L.B. Brubaker
    1980   Modern Climate-tree Growth Relationships and Climatic Reconstruction in Sub-Arctic Alaska. *Nature* 286: 872-874.

Geist, Otto and Froelich Rainey
    1936   *Archaeological Excavations at Kukulik, St. Lawrence Island: preliminary Report.* University of Alaska Miscellaneous Publications 2.Washington, D.C.: U.S.Government Printing Office.

Gerlach, S. Craig, and E.S. Hall
    1988   'The Later Prehistory of Northern Alaska: The View from Tukuto Lake,' pp. 107-136 in R.D. Shaw, R.K. Harritt, D.E. Dumond (eds.), *The Late Prehistoric Development of Alaska's Native People.* Fairbanks: Alaska Anthropological Association, Aurora Monograph Series IV.

Gerlach, S. Craig, and Owen K. Mason
    1992   Calibrated Radiocarbon Dates and Cultural Interaction in the Western Arctic. *Arctic Anthropology* 29(1): 54-81.

Giddings, J. Louis
    1952   *The Arctic Woodland Culture of the Kobuk River.* Philadelphia: University of Pennsylvania, University Museum Monographs.
    1960   The Archaeology of Bering Strait. *Current Anthropology* 1(2): 121-138.
    1964   *The Archaeology of Cape Denbigh.* Providence: Brown University Press.
    1967   *Ancient Men of the Arctic.* New York: A.A. Knopf.

Giddings, J. Louis and Douglas D. Anderson
    1986   *Beach Ridge Archaeology of Cape Krusenstern: Eskimo and Pre-Eskimo Settlements around Kotzebue Sound, Alaska.* Washington, D.C.: National Parks Service, U.S. Department of the Interior, Publications in Archeology 20.

Gerlach, S.C. and E.S. Hall
    1988   The later prehistory of northern Alaska: The view from Tukuto Lake, pp. 107-136 in R.D. Shaw, R.K. Harritt, and D.E. Dumond (eds.), in *The Late Prehistoric Development of Alaska's Native People.* Anchorage: Alaska Anthropological Association Monograph Series.

Graumlich, Lisa and John C. King
    1998   'Late Holocene Climatic Variation in Northwestern Alaska as Reconstructed from Tree Rings,' in Grunnøw, Bjarne and J. Pind, eds., *Abstract, 25th Alaska Anthropological Association Annual Meeting.* Alaska Anthropological Association.

Grove, J.M. and R. Switsur
    1994   Glacial Geological Evidence for the Medieval Warm Period. *Climatic Change* 26: 143-169.

Grønnow, B. and J. Pind, eds.
    1996   *The Paleo-Eskimo Cultures of Greenland: New Perspectives in Greenlandic Archaeology.* Copenhagen: Danish Polar Center Publication No. 1.

Hall, Edwin S.
    1978   'Technological Change in Northern Alaska,' pp. 209-230 in R. C. Dunnell and E.S. Hall, eds., *Archaeological Essays in Honor of Irving B. Rouse.* The Hague: Mouton Publishers.

Harritt, Roger K.
    1994   *Eskimo Prehistory on the Seward Peninsula.* Anchorage: National Park Service, Alaska Regional Office, Research Report AR 21.

    1995   'The Development and Spread of the Whale Hunting Complex in Bering Strait: Retrospective and Prospects,' pp. 33-51 in A.P. McCartney, ed., *Hunting the Largest Animals: Native Whaling in the Western Arctic and Subarctic.* Edmonton: Canadian Circumpolar Institute (CCI) Press, University of Alberta, Occasional Publication No. 36, Studies in Whaling No. 3.

    2001   'Re-examining Wales Role in Bering Strait Prehistory: Some Preliminary Results of Recent Work,' pp. 31-35 in R.K. Harritt (compiler), *The Pursuit of Agviq: Some Results of the Western Whaling Societies Regional Integration Project.* Anchorage: Environment and Natural Resources Institute, University of Alaska Anchorage.

Hequétte, A. and P.R. Hill
    1993   Storm-generated Currents and Offshore Sediment Transport on a Sandy Shoreface, Tibjak Beach, Canadian Beaufort Sea. *Marine Geology* 113: 283-304.

Hoffecker, J.F., O.K. Mason, G. Reynold, D.H. Hanson, S. Elias
    2002   *Results from the Uivvaq Project, FY 2000.* Washington, D.C.: National Science Foundation, Final Report to the Office of Polar Programs.

Hofmann-Wyss, Anna Barbara
    1987   *Prähistorische eskimogräber an der Dovelavik Bay und bei Kitnepaluk im westen der St. Lorenz Insel, Alaska.* St. Lorenz Insel-studien. Berner beitärge zur archäologischen und ethnologischen erforschung des Beringstrassengebeites. Band II. Academica Helvetica 5, Bern and Stuttgart.
Hughes, M.K. and H.F. Diaz, eds.
    1994   The Medieval Warm Period. *Climatic Change* Special Issue 26(2-3): 109-325.
Jacoby, Gordon C., Karen W. Workman, and Rosanne D'Arrigo
    1999   Laki Eruption of 1783, Tree Rings and Disaster for Northwest Alaska Inuit. *Quaternary Science Reviews* 18: 1365-1371.
Jenness, Diamond
    1940   Prehistoric Culture Waves from Asia to America. *Journal of the Washington Academy of Sciences* 30(1): 1-15.
Jordan, J.W.
    1988   'Erosion Characteristics and Retreat Rates Along the North Coast of Seward Peninsula,' pp. 322-362 in J. Schaaft, ed., *The Bering Land Bridge National Preserve: An Archaeological Survey.* Anchorage: National Park Service, Research/Resources Management Report 14.
    1990   *Late Holocene Development of Barrier Islands in the Southern Chukchi Sea, Alaska.* Fairbanks: University of Alaska, M.S. thesis, Quaternary Studies.
Jordan, J.W. and O.K. Mason
    1999   A 5000 yr Record of Intertidal Peat Stratigraphy and Sea Level Rise from Northwest Alaska. *Quaternary International* 60: 37-47.
Krupnik, Igor
    1983   'Early Settlements and the Demographic History of Asian Eskimos of Southeastern Chukotka (Including St. Lawrence Island),' pp. 84-111 in H.N. Michael and J.W. VanStone, eds., *Cultures of the Bering Sea Region. Papers from an International Symposium.* Moscow: International Research and Exchanges Board.
    1987   The Bowhead *vs.* the Gray Whale in Chukotkan Aboriginal Whaling. *Arctic* 40(1):16-32.
    1993   *Arctic Adaptations: Native Whalers and Reindeer Herders of Northern Eurasia.* Hanover: University Press of New England.
Larsen, Helge
    1982   An Artifactual Comparison of Finds of Norton and Related Cultures. *Arctic Anthropology* 19(2): 53-58.
    2001   *Deering—A Men's House from Seward Peninsula, Alaska.* Copenhagen: National Museum of Denmark, Publications of the National Museum Ethnographical Series, 19.
Larsen, Helge and Froelich Rainey
    1948   *Ipiutak and the Old Whaling Culture.* American Museum of Natural History, Anthropological Paper No. 42.
Lutz, B.J.
    1982   Population Pressure and Climate as Dynamics within the Arctic Small Tool Tradition, Alaska. *Arctic Anthropology* 19(2): 143-149.

Mantua, N.J., S.R. Hare, Y. Zhang, J.M. Wallace, and R. C. Francis
    1997   A Pacific Interdecadal Climate Oscillation with Impacts on
             Salmon Production. *Bulletin of the American Meteorological*
             Society 78: 1069-1079.

Martin, Paul S.
    1967   'Pleistocene Overkill,' pp. 75-120 in P.S. Martin and H.E.
             Wright, eds., *Pleistocene Extinctions: The Search for a Cause.*
             New Haven: Yale University Press.

Mason, Owen K.
    1990   *Beach Ridge Geomorphology of Kotzebue Sound: Implications for*
             Paleoclimatology and Archaeology. Fairbanks: University of
             Alaska, Ph.D. dissertation, Quaternary Science.

    1991   'Chronological Inferences from Harpoon and Arrowheads from
             Mound 44 slump,' pp. 64-80 in O.K. Mason, S.G. Gerlach and
             S.L. Ludwig, eds., *Coastal Erosion and Salvage Archaeology at*
             Utqiagvik, Alaska: The 1990 Excavation of the Mound 44 Slump
             Block. Final Report to the City of Barrow and the North Slope
             Borough Commission on Inupiat History, Language and
             Culture. Barrow: Alaska Quaternary Center, Occasional Paper
             4.

    1993   The Geoarchaeology of Beach Ridges and Cheniers: Studies of
             Coastal Evolution Using Archaeological Data. *Journal of*
             Coastal Research 9(1): 126-146.

    1998   The Contest between Ipiutak, Old Bering Sea and Birnirk
             Polities and the Origin of Whaling during the First Millennium
             A.D. along Bering Strait. *Journal of Anthropological*
             Archaeology 17: 240-325.

    1999   'At the Tail of Asia: Heightened Storminess during Cold
             Climates in Late Holocene Northwest Alaska,' pp. 9-40 in Y.A.
             Park ed., *Proceedings, First International Symposium on*
             Biodiversity and Geomorphic Change. Korean Journal for
             Quaternary Research in 1999.

    2000a   'Ipiutak/Birnirk Relationships in Northwest Alaska: Master
             and Slave or Partners in Trade,' pp, 229-251 in M. Appelt, J.
             Bergland, and H.C. GullØv, eds., *Identities and Cultural*
             Contacts in the Arctic. Proceedings from a Conference at the
             Danish National Museum. Copenhagen: Danish National
             Museum and Danish Polar Center.

    2000b   Climate Change, Whaling, and the Thule Transformation, in
             Abstracts, 33rd CAA annual meeting. Ottawa: Canadian
             Archaeological Association.

    2002   'Paleoclimatic Records in the Ekven Site: Comparisons with
             Northwest Alaska,' pp. 261-272 in Don E. Dumond and
             Richard L. Bland (eds.), *Archaeology in the Bering Strait Region:*
             Research on Two Continents. Eugene: University of Oregon
             Anthropological Papers No. 29.

    n.d.   *Coastal Geomorphology and Geoarchaeology of the Uivvaq Site.*
             Unpublished manuscript.

    2003   (editor). *Uivvaq Heritage Project. Field season 2002.* Final
             report to Agiaq ConAm, J.V. Geoarch. Alaska, Anchorage.

Mason, O.K., M.L. Ganley, M. Sweeney, C. Alix, and V. Barber
    2003   *An Ipiutak Outlier: A Late 1st millennium AD Qarigi on Golovnin Lagoon*. Final Report on the Qitchauvik Field School, 1998-2000. Final Report to the National Park Service, Shared Beringian Heritage Program.
Mason, Owen K. and S. Craig Gerlach
    1995a   Chukchi Sea Hot Spots, Paleo-polynyas and Caribou Crashes: Climatic and Ecological Constraints on Northern Alaska Prehistory. *Arctic Anthropology* 32(1): 101-130.
    1995b   'The Archaeological Imagination, Zooarchaeological Data, the Origins of Whaling in the western Arctic and 'Old Whaling' and the Choris Cultures,' pp. 1-32 in A.P. McCartney, ed., *Hunting the Largest Animals: Native Whaling in the Western Arctic and Subarctic*. Edmonton: Canadian Circumpolar Institute (CCI) Press, University of Alberta, Occasional Publication No. 36, Studies in Whaling No. 3.
Mason, Owen K., S. Craig Gerlach, and Stefanie L. Ludwig, eds.
    1991   *Coastal Erosion and Salvage Archaeology At Utqiagvik, Alaska: The 1990 Excavation of the Mound 44 Slump Block*. Final Report to the City of Barrow and the North Slope Borough Commission on Inupiat History, Language and Culture. Anchorage: Alaska Quaternary Center Occasional Paper 4.
Mason, Owen K., David M. Hopkins, and Lawrence Plug
    1997   Chronology and Paleoclimate of Storm-induced Erosion and Episodic Dune Growth across Cape Espenberg Spit, Alaska, U.S.A. *Journal of Coastal Research* 13(3): 770-797.
Mason, Owen K. and James W. Jordan
    1993   Heightened North Pacific Storminess and Synchronous late Holocene Erosion of Northwest Alaska Beach Ridge Complexes. *Quaternary Research* 40(1): 55-69.
    1997   *Sea Level and Storm History of Northern Seward Peninsula*. Anchorage: Final Report to the Shared Beringian Heritage Project, Bering Land Bridge National Park and Preserve, Alaska Regional Office.
    2002   Minimal late Holocene Sea Level Change in the Chukchi Sea: Arctic Insensitivity to Global Change? *Global and Planetary Change* 32(1): 13-23.
Mason, Owen K. and Stefanie L. Ludwig
    1990   Resurrecting Beach Ridge Archaeology: Parallel Depositional Records from St. Lawrence Island and Cape Krusenstern. *Geoarchaeology* 5(4): 349-373.
Mathiassen, Therkel
    1929   Some Specimens from the Bering Sea Culture. *Indian Notes* 6(1): 33-56.
    1930   'Archaeological Collections from the Western Eskimos,' in *Report of the Fifth Thule Expedition* 1921-1924, 10(1): 1-98.
Mayewski, P., L.D. Meeker, M.S. Twickler, S. Whitlow, Q. Yang, W.B. Lyons, and M. Prentice
    1998   Major Features and Forcing of High Latitude Northern Hemispheric Atmospheric Circulation Using a 110,000 Year-

long Glaciochemical Series. *Journal of Geophysical Research* 102 (C12): P 26,345-26,366.

McCartney, Allen P.
1995   'Whale Size Selection by Precontact Hunters of the North American Western arctic and Subarctic,' pp. 83-108 in A.P. McCartney, ed., *Hunting the Largest Animals: Native Whaling in the Western Arctic and Subarctic.* Edmonton: Canadian Circumpolar Institute (CCI) Press, University of Alberta, Canadian Circumpolar Institute (CCI) Press, Occasional Publication No. 36, Studies in Whaling No. 3.

McGhee, Robert
1969/70 Speculations on Climatic Change and Thule Culture Development. *Folk* 11-12: 173-184.
1972   'Climatic change and the development of Canadian arctic cultural traditions,' pp. 39-59 in Y. Vasari, H. Hyvarinen and S. Hicks (eds.), Climatic Changes in the Arctic During the Last Ten Thousand Years. Oulu: University of Oulu.
1981   'Archaeological Evidence for Climatic Change During the Last 5000 Years,' pp. 162-179 in T.M.L. Wigley, M.J. Ingram and G. Farmer, eds., *Climate and History: Studies in Past Climates and their Impact on Man.* Cambridge: Cambridge University Press.
2000   'Radiocarbon Dating and the Timing of the Thule Migration,' pp. 181-191 in M. Appelt, J. Berglund, and H.C. Gulløv, eds., *Identities and Cultural Contexts in the Arctic.* Proceedings from a Conference. Copenhagen: Danish National Museum.

Moore, S.E. and R.R. Reeves
1993   Distribution and Movement, pp. 313-386 in J.J. Burns, J.J. Montague, and C.J. Cowles, eds., *The Bowhead Whale.* Lawrence, KS: Allen Press, Society for Marine Mammalogy, Special Publication No. 2.

Morrison, David
1989   Radiocarbon Dating the Thule Culture. *Arctic Anthropology* 26(2): 48-77.
1991   *The Diamond Jenness Collections from Bering Strait.* Ottawa: Archaeological Survey of Canada, Canadian Museum of Civilization, Mercury Series Paper 144.
1999   The Earliest Thule Migration. *Canadian Journal of Archaeology* 22(2): 139-156.
2001   Radiocarbon Dating the Birnirk-Thule Transition. *Anthropological Papers, University of Alaska.* New Series 1(1): 74-86.

Murray, Maribeth S., A.C. Robertson and R. Ferrara
2003   Chronology, culture, and climate: A radiometric re-evaluation of late prehistoric occupations as Cape Denbigh, Alaska. *Arctic Anthropology* 40(1): 87-105.

Niebauer, H.J.
1988   Effects of El Niño Southern Oscillation and North Pacific Weather Patterns on Interannual Variability in the Subarctic Bering Sea. *Journal of Geophysical Research* 93 (C5):5051-5068.

1998   Variability in Bering Sea Ice Cover as Affected by a Regime
        Shift in the North Pacific in the Period 1947-1996. *Journal of
        Geophysical Research* 103 (C12): 27,717-27,737.

O'Brien, S.R., P.A. Mayewski, L.D. Meeker, D.A. Meese, M.S. Twickler, and
S.I. Whitlow
1995   Complexity of Holocene Climate as Reconstructed from a
        Greenland Ice Core. *Science* 270: 1962-1964.

Okladnikov, Aleksei P. and Nina A. Beregovaia
1971   *Drevnie poseleniia Baranova mysa* [Early Settlements of Cape
        Baranov]. Izdatel'stvo 'Nauka,' Sibirskoe Ordelenie,
        Novosibirsk [Partial translation by Richard Bland].

Psuty, N.P.
1988   Sediment Budget and Dune/beach Interaction. *Journal of
        Coastal Research* Special Issue 3: 1-4.

Quinn, W.H.
1992   'A Study of Southern Oscillation-related Climatic Activity for
        A.D. 622-1990 Incorporating Nile River Flood Data,' pp. 119-
        150 in H.F. Diaz and V. Markgraf, eds., *El Niño: Historical and
        Paleoclimatic Aspects of the Southern Oscillation.* Cambridge:
        Cambridge University Press.

Rainey, Froelich G.
1941   Eskimo Prehistory: The Okvik Site on the Punuk Islands.
        *Anthropological Papers of the American Museum of Natural
        History* 37(4).

Rasmussen, Knud
1931   The Netsilik Eskimos: Social and Spiritual Culture. *Report of
        the Fifth Thule Expedition 1921-1924, VIII, Nos. 1-2*

Reanier, R.E., G.W. Sheehan, and A.M. Jensen
1998   *Report of 1997 Field Discoveries City of Deering Village Safe
        Water Cultural Resources Project.* Report to Ukpeagvik Inupiaq
        Corporation Real Estate Science Division, Barrow. Anchorage:
        on file at Alaska State Office of History and Archaeology.

Reuther, Joshua, S. Craig Gerlach, and Carol Gelvin-Reymiller
2002   New Radiocarbon Dates from an Interior ASTt Site in the Arctic
        Foothills of Northern Alaska, in *Abstracts, 35th CAA Annual
        Meeting.* Ottawa: Canadian Archaeological Association.

Rudenko, S.I.
1961   *The Ancient Culture of the Bering Sea and the Eskimo Problem.*
        Arctic Institute of North America, Translations from Russian
        Sources No. 1. Toronto: University of Toronto Press.

Schaaf, Jeanne
1988   *The Bering Land Bridge National Preserve: An Archaeological
        Survey.* 2 vols. Anchorage: National Park Service, Resources
        Management Report 14.

Savelle, James and Allen P. McCartney
2001   'Bowhead Whale and Gray Whale Selection by Prehistoric and
        Early Historic Alaska Whaling Societies,' pp. 44-46 in R.K. Harritt
        (compiler), *The Pursuit of Agviq: Some Results of the Western
        Whaling Societies Regional Integration Project.* Anchorage:
        University of Alaska, Environment and Natural Resources
        Institute.

Schell, D.M., B.A. Barnett, and K.A. Vinette
    1998   Carbon and Isotope Ratios in Zooplankton of the Bering,
            Chukchi and Beaufort Seas. *Marine Ecology Progress* 162: 11-
            23.
Sheehan, Greg W.
    1995   'Whaling Surplus, Trade, War and the Integration of
            Prehistoric Northern and Northwestern Alaskan Economies,
            A.D. 1200-1826,' pp. 185-206 in A.P. McCartney, ed., *Hunting
            the Largest Animals: Native Whaling in the Western Arctic and
            Subarctic.* Edmonton: Canadian Circumpolar Institute (CCI)
            Press, University of Alberta, Occasional Publication O.P. 36,
            Studies in Whaling No. 3.
Springer, A.M. and C.P. McRoy
    1993   The Paradox of Pelagic Food Webs in the Northern Bering
            Sea—III. Patterns of Primary Production. *Continental Shelf
            Research* 13(5/6): 575-599.
Stanford, Dennis J.
    1973   *The Origins of Thule Culture.* Albuquerquy: Unviersity of New
            Mexico, Ph.D. Dissertation, Department of Anthropology.
    1976   *Walakpa: Its Place in the Birnirk and Thule Cultures.*
            Washington, D.C.: Smithsonian Institution, Contributions to
            Anthropology 20.
Stoker, S. and I.I. Krupnik
    1993   'Subsistence Whaling,' pp. 579-630 in J.J. Burns, J.J.
            Montague, and C.J. Cowles, eds., *The Bowhead Whale.*
            Lawrence, KS: Allen Press, Society for Marine Mammalogy
            Special Publication No. 2.
Stuiver, M., P.J. Reimer, E. Bard, J.W. Beck, G.S. Burr, K.A. Hughen, B.
Kromer, G. McCormac, J. Van Der Plicht, and M. Spurk
    1998   IntCal98 Radiocarbon Age Calibration, 24,000–0 cal bp.
            Radiocarbon 40(3): 1041-1084.
Taylor, William E., Jr.
    1963.   Hypotheses on the Origin of the Canadian Thule Culture.
            American Antiquity 28(4): 456-464.
Viau, A.E., K. Gajewski, P. Fines, D.E. Atkinson, and M.C. Sawada
    2002   Widespread Evidence of 1500 yr Climate Variability in North
            America During the past 14,000 yr. *Geology* 30(5): 455-458.
Walsh, J.J., C.P. McRoy, L.K. Coachman, J.J. Goering, J.J. Nihoul, T.E.
Whitledge, T.H. Blackburn, P.L. Parker, C.D. Wirick, P.G. Shuert, J.M.
Grebmeier, A.M. Springer, R.D. Tripp, D.A. Hansell, S. Djenidi, E.
Deleersnijder, K. Henriksen, B.A.Lund, P. Andersen, F.E. Müller-Karger and
K. Dean
    1989   Carbon and Nitrogen Cycling within the Bering/Chukchi Seas:
            Source Regions for Organic Matter Effecting AOU Demands of
            the Arctic Ocean. *Progress in Oceanography* 22: 277-359.
Whitridge, Peter J.
    1999a   The Prehistory of Inuit and Yupik Whale Use. *Revista de
            arqueologia Americana* 16: 99-154.
    1999b   *The Construction of Social Difference in a Prehistoric Inuit
            Whaling Community.* Tempe: Arizona State University, Ph.D.
            dissertation.

Ekven—A Prehistoric Whale Hunters' Settlement on the Asian Shore of Bering Strait

Yvon Csonka

Abstract: *Between 1995 and 1998, excavations were conducted by an international team in the settlement of Ekven, on the Asian shore of Bering Strait. This paper presents some of the preliminary results of this project as they relate to the question of whaling, adding a missing link between the zooarchaeological data (Dinesman and Savinetsky, this volume) and the interpretations derived from the previous excavation of the well-known Ekven cemetery. During the period ca. A.D. 600 to 1600, bearers of the Birnirk, Punuk, and Western Thule cultures settled at Ekven in large dwellings whose superstructures are almost exclusively made of whale bone. Baleen was an abundant raw material used for different types of artifacts. Despite considerable circumstantial evidence of whaling, other species of fauna, especially sea mammals, also formed an important part of the diet. Further analyses might confirm that Ekven was a center of Birnirk culture and a bridgehead for the diffusion of Punuk traits and/or people to the New World.*

INTRODUCTION

Most of what we know about Neoeskimo[1] prehistory on the Asian shore of Bering Strait derives from the excavations of cemeteries conducted by Russian archaeologists since the 1950s. Every summer from 1995 to 1998, an international team has excavated at the settlement of Ekven, situated on the coast about 18 km west of East Cape, the easternmost point of land of the Old World. The ruins lie a few hundred meters from the well-known cemetery of the same name. The purpose of this project is to gather the first detailed settlement data in the Asian Eskimo zone (architecture, subsistence, settlement patterns, cultural affiliations, and successions, etc.), to allow comparison with data collected in the adjoining cemetery and with sites of similar age from nearby Alaska and other regions settled by Neoeskimos.

Data are currently under analysis. The intent here is to present some preliminary results as they pertain to the issue of whale hunting. This paper adds archaeological interpretation to Dinesman

---

[1] Neoeskimo is used here as a close equivalent of what has been termed 'Northern maritime tradition' (Collins 1964), or 'Thule tradition' (Dumond 1987:101-139).

and Savinetsky's results (this volume), derived from the examination of whale bones from prehistoric coastal settlements on northeastern Chukchi Peninsula. It will also be of interest to compare the present data with those from the roughly contemporaneous sites of Wales (Harritt, this volume), located on the other continent less than a 100 km across the strait (Fig. 1).

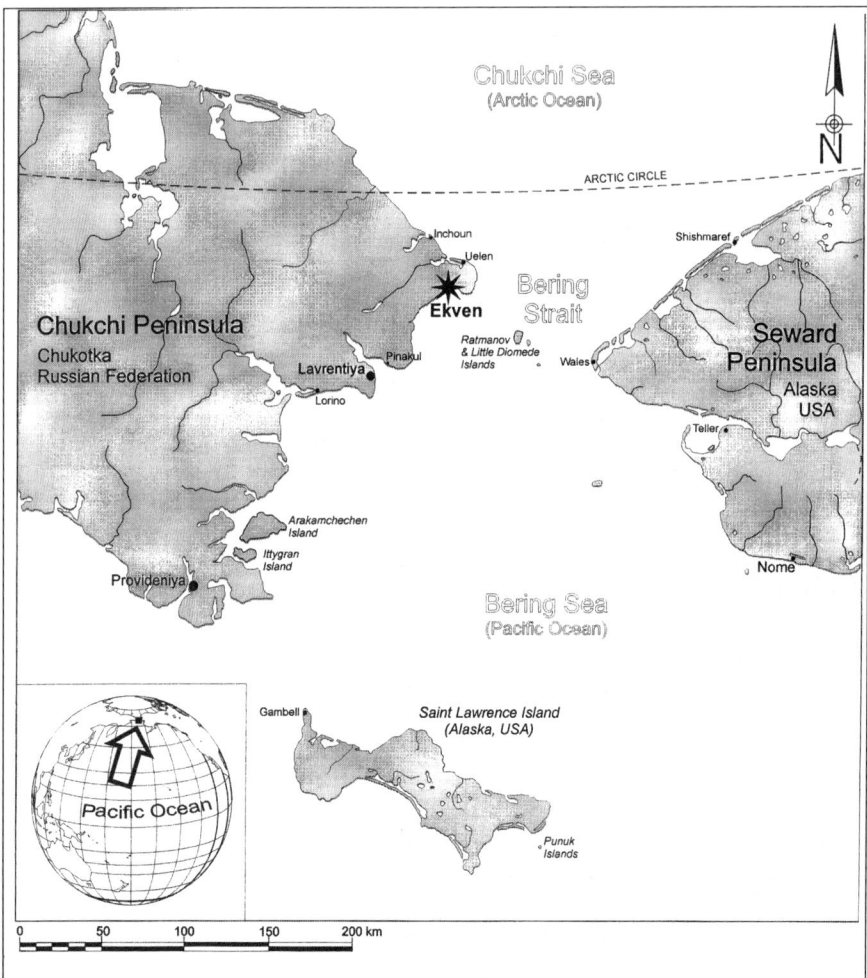

Figure 1. *Regional map showing the location of Ekven.*

Ekven has not been touched by uncontrolled excavations, in contrast to neighboring sites in Dezhnevo and Uelen, and further in Wales, Alaska. However, the seaward face of the settlement is rapidly being eroded by storm-driven waves. It is not known how much of the village has already disappeared. Taking advantage of the opportunity offered by the neat natural stratigraphic exposures, one of the Ekven subprojects, which will be reported upon here, is the study of the erosion face that transects the site and the abundant artifacts collected there (see als Csonka *et al.* 1999; Moulin and Csonka 2002).

PREVIOUS RESEARCH IN EKVEN AND IN THE REST OF THE ASIAN ESKIMO ZONE

The early manifestations of Neoeskimo cultures were first identified and studied on islands belonging to the United States but lying close to the Asian continent: Little Diomede where Jenness (1928) first identified the Old Bering Sea culture, but especially St. Lawrence Island and the neighboring Punuk Islets (Collins 1937; Geist and Rainey 1936; Rainey 1941). On the Siberian mainland, research started in 1945 with Rudenko's surveys of the eastern and southern coasts of the Chukchi Peninsula, between Uelen and Enmelen (Rudenko 1961). On the northern coast, settlements attributed to Birnirk and Thule were excavated in 1946 at Cape Baranov and on the Bear Islands, near the mouth of the Kolyma and more than 1000 km from Bering Strait (Okladnikov and Beregovaya 1971; Raushenbakh 1969).

The prehistoric cemetery at Uelen was excavated between 1957 and 1960 by Levin, Sergeev, and Arutiunov (Arutiunov and Sergeev 1969). Dikov also did some research work there between 1956 and 1963 (Dikov 1967). In 1961, Arutiunov and Sergeev moved to the newly discovered cemetery at Ekven, about 25 km southwest of Uelen, and excavated its 'eastern hill' until 1974 (Arutiunov and Sergeev 1975, 1983, 1990). In the meantime, Alekseev studied the physical anthropology of ancient and modern inhabitants of the Chukchi Peninsula (e.g., Alekseev 1972). Not far to the south of Ekven, other excavations took place at the cemetery of Chini in 1963-1965 (Dikov 1974). Dikov later surveyed many sites in the interior of the Chukchi Peninsula (e.g., Dikov 1997). The islands Ittygran and Arakamchechen, including the so-called Whale Alley, were surveyed in 1976-1979 (Arutiunov *et al.* 1982). The zooarchaeological studies that were conducted on the Chukchi Peninsula coasts since the 1970s are mentioned by Dinesman and Savinetsky (this volume).

Excavations at the Ekven cemetery ('western hill') were resumed in 1987 by a team from the State Museum of Oriental Arts in Moscow led by Mikhail Bronshtein and Kirill Dneprovskyi. The material collected through 1991 has been partially published in an exhibition catalogue (Leskov and Müller-Beck 1993). This project was halted at the end of the 1995 season when the local Native population finally managed to turn its discontent into a ban on the disturbance of graves (Plumet 1993; Bronshtein and Plumet 1995:43-44; Csonka 1998:68-69). By that time, several hundred graves had been excavated, but it was estimated that at least as many remained untouched (Bronshtein 1993:73). In recent years, a number of sites in the vicinity of Ekven were surveyed by a team from the Russian

Research Institute for Cultural and National Heritage, but detailed results from these surveys have neither been published nor been made available to us (there are some allusions to it in Gusev *et al.* 1999). During the course of our own work in Ekven, we were able to visit many of these sites.

STRENGTHS AND WEAKNESSES OF THE CURRENT STATE OF KNOWLEDGE

The excavations of cemeteries yielded large quantities of well-preserved artifacts of high aesthetic value, without the difficulty of having to dig through permafrost. They also provided material for physical anthropological studies. Arutiunov and Bronshtein (1985) refined but also modified (Dumond 1998:98-103) the stylistic distinctions in the ornamentation of Okvik and the different aspects of Old Bering Sea Culture (OBS), which had originally been proposed by Collins (1937) and Rainey (1941) on the basis of St. Lawrence Island materials. In the cemeteries of Uelen and Ekven near East Cape, the Russians identified patterns in the combinations of artifacts of different styles that were recovered in graves (Bronshtein and Plumet 1995:33-38). In both places, the chronocultural affiliation of artifacts ranged from Okvik and the three subdivisions of OBS, to Birnirk and Punuk. The main higher-level conclusions of these publications consisted in establishing that these cultures partly overlapped one another chronologically, and that their distinct traditions of art and decoration corresponded to the need for cultural differentiation among ethnic groups occupying the same territory simultaneously. It was also hypothesized that social differentiation was present from the beginning, ranging from slaves to shamans and secular leaders (for further elaborations see Mason 1998:255-259).

Twenty-eight radiocarbon dates were obtained on human bones from the Ekven cemetery; they range from 2200 to 600 years B.P. (Dinesman *et al.* 1999, Fig. 4.6.A: 81, Table 4.15 p. 114, and Appendix 2).[2] Their distribution suggests that the cemetery may have been used without interruption. It has generally been assumed that the corpses were brought from the settlement situated a few hundred meters distant. Since the first excavations of the Ekven cemetery,

---

[2] These dates were left uncalibrated by the authors; they give their reasons for this in their introduction (Dinesman *et al.* 1999:2-3). Mason proposed conservative calibrations for these dates with a reservoir effect of 400 years (1998:225-226, 263), with 56% (n=15) of the graves belonging to the interval of A.D. 800-1400. Dumond (1998:118, 134) suggests other coefficients for the correction of the reservoir effect, which would make these dates about a century younger; see below. The determinations from Dinesman *et al.* are not reported according to current conventions, as they are not adjusted for any known or assumed d13C.

archaeologists have collected artifacts in the rubble at the foot of the eroding bank of the settlement. Arutiunov and Sergeev (1975:6-7) affirm that the same cultures are represented in the village and in the cemetery, but only two or three of the harpoon heads from the settlement, which they use to illustrate this point, are of the OBS or Okvik types; most are Birnirk–looking, a few more Punuk-like, and some Western Thule or late prehistoric in appearance.

In another vein, Krupnik and his colleagues had already established that there was an area on the northeastern coast of the Chukchi Peninsula where gray whale bones predominated in archaeological contexts, in contrast with the southern zone centered around Provideniya Bay, where remains of bowheads were most numerous (Bogoslovskaya *et al.* 1982; Krupnik *et al.* 1983). From Dinesman (1999; this volume) and others, it was also known that the majority of the whale bones found at the Ekven settlement and cemetery belonged to (juvenile) gray whales. In order to better integrate and synthesize these zooarchaeological studies and the sociocultural interpretations derived from cemetery excavations, a research program focused on settlements is clearly a priority.

Despite the large number of graves excavated, and the surveys and test excavations conducted, much still remains to be learned about the prehistory of the Chukchi Peninsula. Survey data from the northern coast, as well as from the southern coast west and south of Nunligran, are scanty, and those from the interior are even more limited. Traces of the passage of early humans to America are so far almost nonexistent, and the links between inland cultures and coastal ones cannot yet be firmly established. The direct ancestors to the first fully developed Neoeskimo cultures—Okvik and Old Bering Sea—have not been found. Even the better-known aspect of the region's prehistory, that of the Neoeskimo cultures up to the historic Yupiget, is based on surveys and limited test excavations, and on data from a few cemeteries. The fact that the second Ekven expedition was conducted out of an art museum prolonged the 'long-standing Russian bias towards cemetery remains and the more elaborate Old Bering Sea styles' (Mason 1998:286), which had already been pointed out by Larsen (1968:88). The late prehistoric period remains poorly known (Schweitzer and Golovko 1995:140). A few recent studies have begun to shed light on the history of the region (Krupnik and Chlenov, n.d.; Schweitzer 1990; Schweitzer and Golovko 1995).

Foremost among the lacunae in our current knowledge are detailed data from settlement archaeology. At a time when no scientific excavations are being conducted on St. Lawrence Island, and while the prehistoric heritage is rapidly being destroyed and scattered, increasing our knowledge about the nearby Asian mainland becomes even more significant. Beside the need for new data, much could also be gained, at the level of theory and interpretation, from a patient reconciliation of the Russian and the American taxonomies for the region's prehistory, which diverged during the Cold War (see Bronshtein and Plumet 1995:18-20).

THE EKVEN SETTLEMENT INTERNATIONAL PROJECT, 1995-1998

Figure 2. *Coastal plain looking north from the summit of Mount Tunitlen. The Ekven erosion bank and cemetery hill are the areas in the shade, in the center; to the far right is Sphinx Point (Verbliuzhi or Camel Point); in the foreground is the Bering Sea; in the background to the left is the Arctic Ocean, and to the right, the Dezhnevo Mountain massif.*

As the foregoing summary indicates, Ekven was already, prior to our project, one of the best-known sites on the Chukchi Peninsula–at least its cemetery was. Extrapolations based on cemetery data suggest that Ekven must have been a large settlement, occupied perhaps continuously during a great part of the Neoeskimo period. The conclusions of Arutiunov and Bronshtein also attest to a considerable blending of different cultural traditions. Residents must have been attracted to the settlement and kept there by the abundant marine resources. The location of the settlement, in sight of Big Diomede Island (Ratmanova in Russian) and the Alaskan mainland, probably contributed to its cosmopolitanism and also to its prosperity; we may presume that intercontinental contact and trade always existed (Fig. 2). It will be one of our goals during the analyses of the new data to try to identify traded materials. The portion of the settlement remaining today is extremely well preserved in permafrost, but erosion is destroying it at a rapid pace. All these conditions encouraged the Swiss team to begin excavations there, along with the Russian team directed by Kirill Dneprovskyi and Mikhail Bronshtein, who had previously worked in the nearby cemetery, and a German crew from the University of Tübingen under the direction of Hansjürgen Müller-Beck.

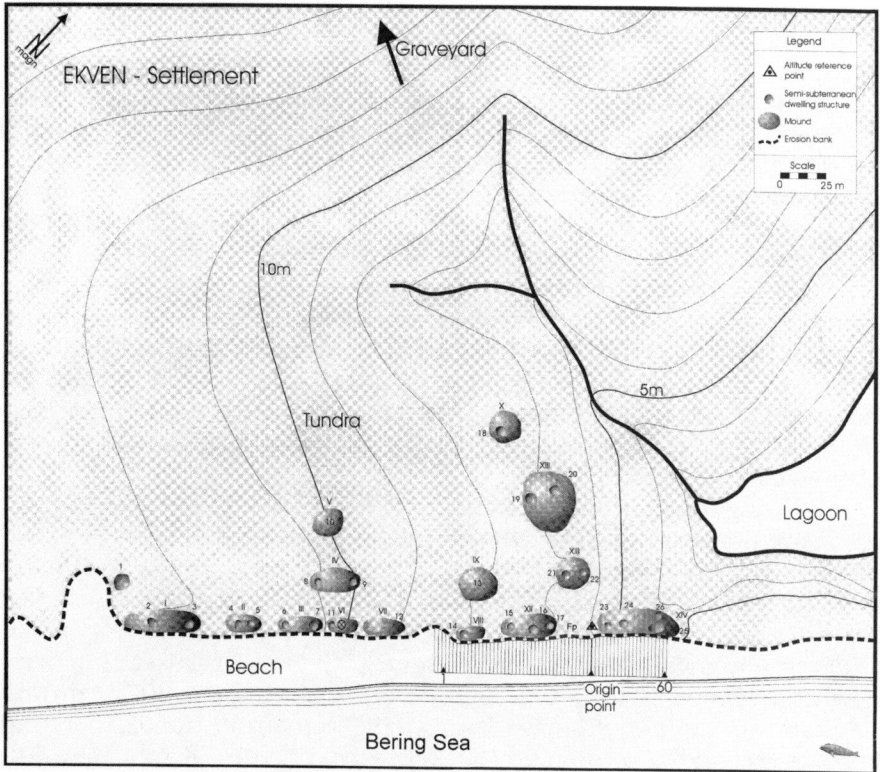

Figure 3. *Sketch map of the Ekven settlement. House depressions are marked by arabic numbers, mounds by roman numbers; the grid used for the collection of artifacts in the erosion bank is illustrated on the beach.*

The excavations at Ekven have been subdivided into several subprojects. One of them, initiated in 1995 by a team composed of Danes, Germans, Russians, and Swiss, and taken over by German and Russian teams in 1997, consists of the detailed excavation of a single house mound, designated EH-18 (Blumer 1996, 1997; Dneprovsky 2002). In 1997 and 1998, the Swiss team extensively studied different areas of the settlement through test excavations, but especially through the documentation of the erosion face (Blumer and Csonka 1998; Csonka *et al.* 1999). The data presented below stem mostly from this subproject.

LOCATION OF THE SITE, HISTORICAL TRADITION OF WHALING, AND AVAILABILITY OF WHALES

Ekven lies approximately 20 km west of Cape Peek, which marks the southeastern corner of the Dezhnevo Mountain massif. It is situated in the middle of a rolling plain that extends along approximately 12 km of coast from the foothills of that massif southwest to Mount Tunitlen. Sand and gravel bars demarcate several lagoons. The Ekven settlement is located near the shore of one of the lagoons. The presence of ruins is indicated by a series of mounds (see Fig. 3), most of them aligned parallel to the beach, along a stretch of approximately 250 m. Twenty-six depressions in the mounds were interpreted as collapsed houses.[3] Over the course of time, more dwellings probably existed, but their remains are now buried in the large mounds where the thickness of anthropic layers reaches up to 3 m. EH-27 (description below) is one such case, and it could not be identified on the basis of surface topography. Since the seaward-facing row of mounds has been eroded away by the sea, we may suppose that the village once extended farther in that direction. We have no way of knowing how much larger the settlement could have been, and we cannot determine how many of the dwellings were occupied simultaneously.

During our summer stays at Ekven, we noted the almost constant presence of gray whales leisurely moving along the coast, often only a few meters from the beach. These were primarily females with calves or, judging by their small size, other young animals.[4] In early Soviet times, Native hunters in the vicinity of Bering Strait followed an older tradition and caught gray and, in lesser numbers, bowhead whales (Krupnik *et al.* 1983:560). During the 20th century, Naukan was the only Eskimo (Yupiget) settlement in the district, and it was surrounded by coastal Chukchi villages, until it was closed in the late 1950s. However, according to Russian ethnographers, it is probable that many of the much more numerous coastal Chukchis were 'chukchified' Yupiget. In any case, their way of life and subsistence activities were very similar to those of the Yupiget

---

[3] The identification, numbering, and sketch mapping of mounds and depressions were performed in 1995 by Hans-Christian Gulløv and Hans-Christian Kapel, with the aid of a GPS.

[4] We noticed year-to-year variability in sightings of gray whales and were informed about it by local hunters. Current worldwide populations of gray whales are considered equivalent to what they were in precontact times (Yarborough 1995:65).

(Chichlo 1981; Krupnik and Chlenov n.d., Chapter 2).[5] We cannot date the beginning of this process of Chukchi expansion and assimilation, but we agree with Krupnik and Chlenov that it probably intensified in the 18th century. At any rate, Ekven and the surrounding coastal plain now belong to the *sovkhoze* (State farm) of Uelen, and are considered Chukchi land. The abandoned Ekven site was surrounded by two Chukchi settlements, Kaniskun (Dezhnevo), a few kilometers to the northeast, and Eleikei, the same distance to the southwest, which were closed in the 1950s.

During the period 1959–1993, aboriginal whaling was replaced by a whaling vessel operating every summer out of Vladivostok. Gray whale carcasses were landed in the villages from this vessel. Gray whale hunting by locals resumed immediately after the whaling vessel stopped operating. Nowadays, residents from Uelen come to Ekven only in the fall to hunt bearded seals and walrus. Due to the difficulty of towing them, gray whales are caught closer to Uelen. As in former times, hunters today target smaller, subadult gray whales. Their equipment does not allow them to pursue larger, more dangerous animals. The village of Lorino, whose large fox farm consuming sea mammal products has operated longer than in neighboring villages, takes the most whales. But whaling also occurs in Lavrentiya, Pinakul, Uelen, and a few other settlements. On the basis of current and historical data, one may assume that gray whales were abundant at different spots near the Asian coast of Bering Strait and that hunting them was a traditional activity in that area. Bowhead whaling was also practiced, to a lesser extent.

Of course, such extrapolation, in the vein of Steward's (1942) 'direct historical approach' can only be done with the support of evidence. Recent studies have documented climatic and ecological change, and associated consequences in the area (e.g., Mason and Gerlach 1995). Krupnik (1988, 1993a, b) has shown that these consequences, coupled with human influence on the animal populations they hunted, led to considerable fluctuations in the availability of their prey species. In any case, the seas close to the Asian shore of Bering Strait are particularly rich in nutrients and are therefore attractive to sea mammals feeding on them, due to relatively favorable ice conditions and the perennial formation of some polynyas (Mason 1998:245-249; Mason and Gerlach 1995; Stringer and Groves 1991).

That changes in the coastline took place at Ekven from the time it was first settled is evident from the fact that interiors of houses have been exposed by wave action. Dinesman *et al.* (1999, Chapter 2)

---

[5] Linguistic data suggest that the Naukan people may have been relatively recent Yupik immigrants from Alaska (Schweitzer and Golovko 1995:50-3). And indeed, their cranial characteristics set them apart from the people interred in the Ekven and Uelen cemeteries (Utermohle 1988:40-43).

indicate that the underwater coastal slope and the shoreline must have changed considerably over the last few millennia. According to them, the lagoon itself may have been a shallow bay until the bar was formed some two millennia ago. Until a few centuries ago, its connection with the sea remained more widely open. One may imagine that it was possible to land *kayaks* and *umiaks* in the lagoon near the houses in the northern corner of the settlement. It may also have been possible to drive small whales into the lagoon or simply to prevent them from retreating with the lowering of the tide, which would have made the hunt much easier than in open water (see Dumond 1995).

These considerations about site location, temporal variations in coastline, and ecological fluctuations remain partly conjectural. Nevertheless, when coupled with those about recent availability and hunting of gray whales, they reinforce the case that Ekven was indeed situated in a favorable spot for whaling during the times it was inhabited.

FAUNAL REMAINS

The diversity of species identified from excavated animal remains found in different features of the Ekven settlement makes it clear that the diet must have been varied during all of the occupation.[6] Among mammal bones, sea mammals clearly predominate (depending on the feature of the site, between 78% and 94% of the NISP). Ringed seal is by far the most abundant marine mammal species, followed by bearded seal and walrus. In all features, dog bones dominate the assemblage of land mammal bones; arctic fox make up one-fifth to one-third of that subcategory, while other land mammals are very rare. The avifauna contributes between 5 and 40% of the total number of identified bones, with up to 30 species represented. Molluscs and fish remains were found. Fishing is well indicated by ad hoc gear such as net fragments, net weights, fishing lures, line sinkers and hooks, and fishing line of baleen. The mixture of species in the diet probably developed over time, as Dinesman *et al.* (1999) show for the nearby Dezhnevo settlement.

---

[6] During the excavation of EH-18 as well as three test pits in other parts of the settlement, bones and shells were collected by layer and .25 m²; bones were also collected while cleaning the profile of the erosion face; they have been identified by Arkady Savinetsky. Whale bones were not counted since they were clearly elements of the architecture and would not necessarily reflect subsistence during the period when the dwellings were inhabited; they are, however, documented on the drawings by layers. The percentages in this paragraph refer to the different features, all layers confounded.

We have good indications about whale bone frequencies and species distribution through the work of Dinesman *et al.* (1999). Because of their bulkiness, we presume that a number of whale bones were not carried to the camp but were abandoned at the place of flensing.[7] But whale bones and baleen are abundant, which convinces us that whale products constituted a significant part of the diet. The presence of bones in a house, however, does not necessarily imply that its inhabitants consumed the meat that once covered them, but may have been stored manufacturing material. It is important to point out that bones were probably not used until they were naturally cleaned and bleached, which can take several decades, and that clean bones could have been reused more than once. Whatever the case, the faunal remains recovered indicate that besides an unspecified amount of whale meat, blubber, and skin, the diet was composed of many other animal species. However, a study of isotopic fractionation in human bones from the Ekven cemetery (Chu 1998) indicates that the inhabitants must have been almost totally dependent on marine protein, and that species of higher trophic level, especially the ringed seal, were eaten in much higher quantities than others such as walrus and baleen whales.

ARCHITECTURE

Most whale bones in Ekven were found in architectural contexts. In the cemetery, they were used to line graves and according to Krupnik (1983:559), the majority were from gray whales. The practice of lining graves with whale bones was also common on St. Lawrence Island (Bandi *et al.* 1984). We observed a number of meat caches containing whale and other bones in the profile of the erosion face. Some of them may have been part of the cache structure, while others may have been stored with the meat.

---

[7] Ethnographically, one observes that the larger the sea mammal, the greater the proportion of its useless parts—especially bones—that will be left on the beach and eventually disappear in the sea. When examining differential proportions of remains of species in archaeological contexts, one should remember possible distortions in the record induced by this effect. Ringed seals can easily be dragged into a house and flensed there, but this is less the case with bearded seals, and walrus are definitely butchered on the shore. The abundance of walrus ivory as raw material belies the small percentage of walrus bones found in archaeological contexts. It may well be that bowhead bones are underrepresented compared to gray whale, for a similar reason.

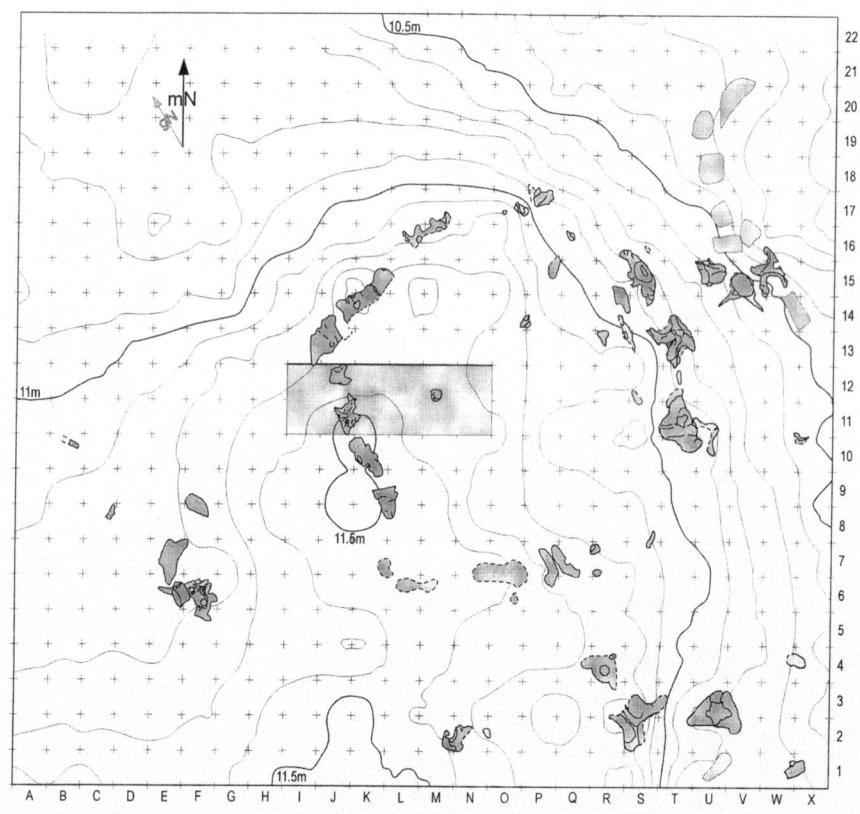

Figure 4. *Surface map of whale skull ring, showing location of test pit. Grid unit: 0.5 m.*

There are two rings of whale skulls on the surface of the Ekven settlement, near the edge of the erosion face. One of them is made of small gray whale skulls (Fig. 4; the location of the ring is marked with an X on mound VI in Fig. 3). A test excavation within the ring did not reveal any significant remains on the level associated with the base of the skulls (Blumer and Csonka 1998:102-107). The second ring is situated at the southwestern edge of the settlement, and a third ring can be seen 2 km to the northeast, at Sphinx Point. Bones from these features were dated by Dinesman *et al.* (1999), but we do not think the whales were necessarily freshly killed when their skulls were emplaced. In any case, whale skull rings are a common occurrence in the area. Arutiunov *et al.* (1982:64) describe and illustrate such rings from the Seniavin Strait Islands, and Krupnik (1993a:4) from the Mechigmen Bay area. They interpret them as ritual structures dating from late Punuk times. Indeed, whatever the measured age of the bones, the fact that they are found above ground, on top of the earlier houses, indicates that the rings cannot be very old.

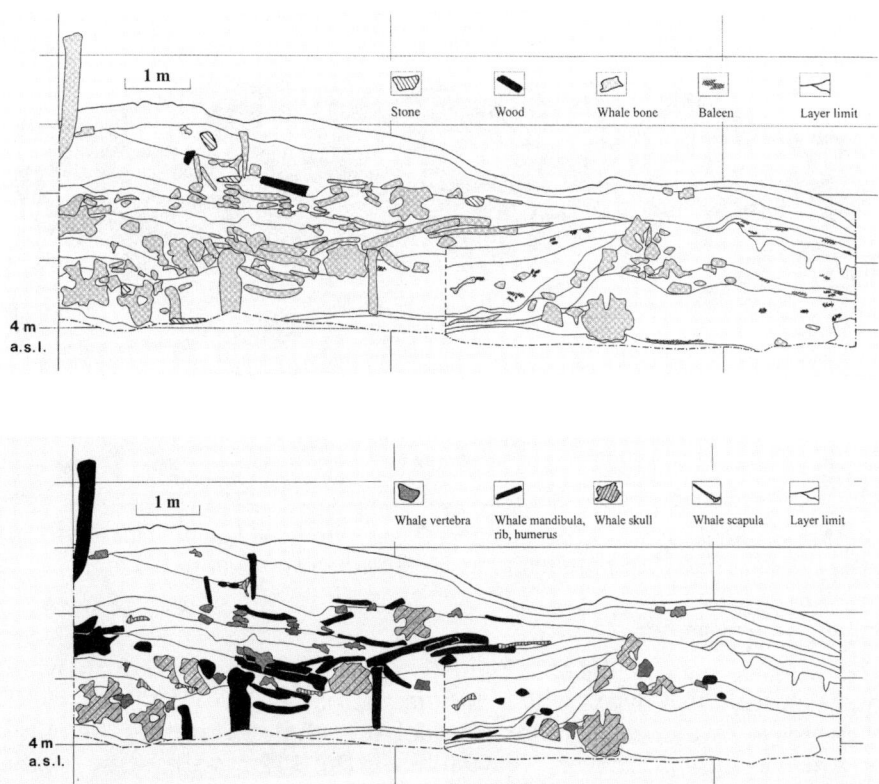

Figure 5. *Section of the erosion bank showing structural elements associated with the profile of EH-27 in Mound XII. Horizontal meters along the front are counted from the origin point (see Fig. 3).*

In the houses, whale bones overwhelmingly predominate as structural elements. This is apparent from the map of EH-18 (Dneprovsky 2002: Fig. 2), from our test pits, and especially in the seafront. Figures 5 to 8 illustrate the profiles of two dwellings as they appeared in the erosion face during the summer of 1998. The shapes of the mounds suggest that the erosion has cut approximately to the middle of the houses (see Csonka *et al.* 1999:Fig. 8). The upper part of the illustrations shows the relationships of whale bones to other structural elements of wood and stone. It also indicates the concentrations of baleen, usually consisting of unworked strips. The lower part of the illustrations differentiates between types of whale bones: skulls, vertebrae, bones of elongated shape (ribs, mandibulae, humeri), and flat bones (scapulae).

Figure 6. *Profile of Mound XII (see Fig. 3) seen from the beach, with EH-27 in the center (compare with Fig. 5). Length of surveyor rod: 3 m.*

Dwelling EH-27

The interior part of EH-27 (Figs. 5, 6) visible today is 7-8 m wide, between the two gray whale skulls visible at the left and right. Mandibles were used as pillars supporting whale rib rafters. Remains of a thick sod wall cover are visible on the right, with some of the sods resting on a bed of baleen strips that separated them from the sterile ground below. After its abandonment, several smaller structures seem to have been erected on top of the dwelling, without penetrating its roof. Part of the interior of EH-27 is filled with ice that resisted compression. Two radiocarbon dates were obtained from twigs. The first one, from an occupation layer, is calibrated to A.D. 963 (1022) 1191 (Ua-14903). The second one, from just above the roof, A.D. 1151 (1217) 1281 (Ua-14902).[8] These two assays bracket the period of occupation. Not enough diagnostic artifacts were recovered in the occupation layers to allow a chronotypological assessment of the occupation. Artifacts found in the rubble downslope from the dwelling in displaced deposits are fairly numerous, however, and most probably fell from the lower occupation floors. Nine harpoon heads were collected there: four of antler and five of ivory. The six less fragmented specimens are tentatively ascribed to OBS III (1 specimen, reworked), Late Birnirk (2), Punuk? (1), and Early Western Thule (2).

---

[8] Our calibration is with 2 sigma intervals, here as elsewhere in this paper, using OxCal v.2.15 r:4 sd:123 prob (chron.), (Stuiver *et al.* 1998).

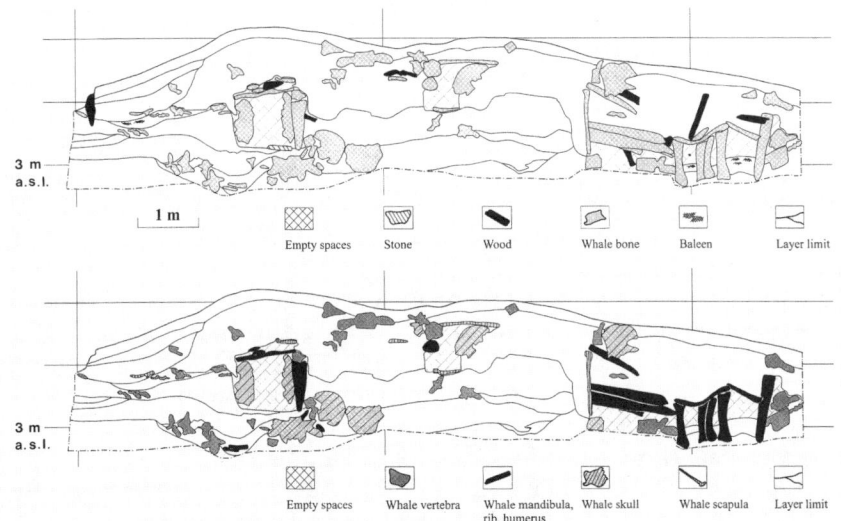

Figure 7. *Section of the erosion bank showing structural elements associated with the profile of EH-25 in Mound XIV. Horizontal meters along the front are counted from the origin point (see Fig. 3).*

Figure 8. *Profile of part of Mound XIV (see Fig. 3) showing house EH-25. Photo taken after a storm had removed some of the elements seen on Fig. 7.*

Dwelling EH-25

House EH-25 (Figs. 7, 8) was located on the northeastern edge of the settlement, near the western end of the sand and pebble bar separating Ekven lagoon from the sea. Of all the Ekven ruins, it lies lowest above sea level. Over the past decade, we have observed its rapid destruction. The bowhead whale skulls surrounding a sleeping niche (Dneprovsky 2002:177-178) were removed by the sea during a storm in 1995, and since then, waves have exposed the inner part of the dwelling. This mound is simpler than that in which EH-27 is situated, and the inner empty spaces have been preserved. The whale skulls situated at the bottom left of the illustration belong to an earlier phase of occupation. Several samples of local wood, taken in the occupation layers at the bottom of the empty spaces on the left and on the right, have been radiometrically assayed. The five assays range from A.D. 1300–1523 (Ua-14905) to 1441–1669 (Ua-14901) calibrated. Chronotypologically, the artifacts recovered agree with these dates and can be attributed to the late Punuk–Western Thule. Of all the areas of the settlement that were investigated, this is the most recent, and it corresponds roughly with the abandonment of the cemetery as dated by Dinesman and Savinetsky (this volume). Dinesman *et al.* (1999:22, 46) had initially attributed a date of 1272 ± 127 B.P. to this house, based on a sample from a whale skull. A comparison with our results suggests the following:

(a) part of the difference may be due to the reservoir effect affecting the whale bone measurement; Dumond (1998:116-17) currently estimates that a correction factor of between 510 ± 57 and 720 ± 53 years must be added to measurements on material of marine origin in order to make them comparable with those on material of terrestrial origin; this would bring their date closer to ours;

(b) dates obtained from whale bones used as architectural elements should not be considered as good indicators of the date of construction nor of the period of occupation of a dwelling.[9]

Dwelling EH-18

EH-18 was a large dwelling with a whale bone superstructure. Most of its interior and an outdoor activity area (73 m²) were excavated during the first four field seasons (Blumer 1996, 1997).[10] Massive mandibles stuck in the sterile substrate were used as main pillars. Other pillars were made of piles of vertebrae. Large schistose stone slabs lined the floor in the central part of the interior, and there was also a paving of small whale scapulae outside the house. Remains of five humans were found on the sleeping platform. Two radiocarbon assays of human hair taken from the same individual gave results of 1655±50 B.P. and 1645±45 B.P., calibrated to A.D. 530–680 and A.D.

---

[9] EH-25, the youngest house in the settlement, being situated closest to sea level, confirms Dinesman *et al.*'s (1999) contention that until three centuries ago, the Ekven beach must have been protected from storms by an offshore reef or sandbar.

[10] After this paper was completed, the Russian team excavated EH-18 during two more seasons, in 2000 and 2001 (Kirill Dneprovsky, pers. comm. 2002).

670–890 (AAR-2776-7, Gulløv and Kapel, pers. comm.; see Blumer 1997:63-4). Averaging the two measurements and recalibrating them using Dumond's (1998) parameters yields a two-sigma range of A.D. 808-983 (alternative interpretation see Dneprovsky 2002:171,179). Most of the diagnostic artifacts can be attributed to Birnirk, but a bundle of seven typical Punuk whaling harpoon heads, some of them unfinished, were found cached under one of the stone slabs. We thus have in EH-18 another case of association of Birnirk and Punuk artifacts, such as those found on St. Lawrence Island, across Bering Strait at Wales, and along the northwest coast of Alaska near Barrow (e.g., Ackerman 1961; Ford 1959; Stanford 1976; Yama'ura 1984).

Figure 9. *Abundance of whale bone in the Ekven erosion front.*

The above presentation makes it clear that whale bone was the predominant element of architectural structures during the settlement period we have documented, beginning about A.D. 600 cal. and ending no later than about A.D. 1600 cal (see also Moulin and Csonka 2002). Driftwood is also used in the constructions, but in limited amounts. Nowadays, driftwood is found in relative abundance on the beaches near Ekven. There are also large stone slabs in all the houses, but in small numbers. Such slabs can only be obtained quite far from the settlement, and they were no doubt difficult to transport. Whale scapulae were used as pavement in some areas perhaps for lack of stone. Our colleague geoarchaeologist,

Bernard Moulin, counted 280 whale bones in the stratigraphy of the erosion exposure that he documented: this represents 88% of the large elements used in construction, the rest being driftwood (8%) and stone (4%). One-third of all the whale bones are vertebrae—not necessarily structural elements, but many of the 26 whale skulls visible in the profiles seemed to be part of the architecture.

We are not certain that the dependence on whale bone for construction was due to a lack of other suitable materials. Vertical piles of whale vertebrae, functioning as supporting members in EH-18, could indicate that no better material, especially no better type of whale bone was handy. For construction purposes, large bones of bowheads or adult gray whales (mandibles, ribs) were needed. However, as the data from Dinesman *et al.* (1999; this volume) show, bones of smaller individuals predominate. We may hypothesize that the reliance on whale bones for structures was a reflection of a compelling cultural choice. This interpretation can be applied to the presence of whale bones in human graves, and we are inclined to extend it to the dwellings of the settlement as well.

ARTIFACTS FROM THE EROSION FACE

During two seasons, we collected over 2000 artifacts from the erosion face of the settlement–in addition to about 750 pottery sherds. Most of the artifacts were found in crumbled sediments at the foot of the bluff and could only be assigned a two-dimensional provenience, but close to 300 were collected in situ with documented stratigraphic context. We gathered material in the northeastern half of the erosion face, along a 120 m stretch (see Fig. 3). The two dwellings described above, EH-25 and EH-27, are situated along this stretch, the second one at its northeastern edge. Chronotypologically, most of the diagnostic objects can be assigned to Birnirk, Punuk, and Western Thule, in roughly comparable proportions. Only a handful of objects have Old Bering Sea II or III shape and decoration and, when found *in situ*, they were side by side with Birnirk and Punuk material. Despite the sporadic presence of late OBS material and a date of 390 B.C.–A.D. 10 cal.[11] in one of the lowermost anthropic layers, we think that most archaeological remains found in this part of the erosion face, as well as other dated features of the settlement (EH-18 and test pits) fall in the interval A.D. 600–1600 cal. Circumstantial evidence also comes from the scarcity of chipped stone artifacts. In 1998, for instance, we found only one chipped stone artifact in contrast to 70 items of polished stone. The proportion of chipped stone is much higher in the graves containing OBS material in the nearby cemetery. It may be that later inhabitants all but obliterated remains from their

---

[11] Arkady Savinetsky, written comm., 1999: 2158 ± 85 B.P. (Iemae-1233).

ancestors when constructing their own shelters directly inside earlier dwellings. Just as likely is the possibility that older houses were situated closer to the shore, in the part of the settlement that has already been engulfed by the sea. We may also wonder if the early graves at the Ekven cemetery were actually those of people from the nearby Sphinx Point settlement, where OBS I and Okvik material has been found.

Baleen and Whale Bone

Baleen is abundant in all occupation layers. We left behind a few large unworked sections (some with incisions for lashing may have been parts of toboggans), and numerous small strips and bundles of thread embedded in organic remains in the occupation layers. Of the artifacts we collected in the erosion front, 139 have baleen as their main or only component, or close to 7% of the total. The great majority of baleen artifacts are small sections, strips, coils, and bundles of thread, many with knots. Three of the more complex assemblages could have been snares. There are three fragments of nets, and two nets on loops (diameters 10 and 20 cm), interpreted as ice scoops. We also found three miniature bows, one pair of snow goggles, and one bullroarer, all made of baleen. Among the larger artifacts were 18 baleen composite vessels' walls, some of them with their oval wooden bottoms attached. Since we found only one small such vessel with a wooden rim, it is clear that the baleen wall form was more popular. In addition, 37 artifacts are composites with baleen as the second material, usually for lashings.

Out of 174 bone artifacts, only ten were of whale bone: three mattock blades, one (undecorated) pottery paddle, four polished sections of unknown function, an unidentified artifact on a rib, and a fishing net weight. To that may be added a number of vertebrae whose epiphyses had been trimmed. We know of the use of such elements as counterweights for the skin cover of tent-like structures in more recent contexts, but we cannot ascertain their function in the semisubterranean houses covered with sod.

Because baleen does not remain preserved on beaches as long as whale bones, its presence in archaeological contexts is a more reliable indicator of whaling. Thus, the abundance of baleen at Ekven, often left unworked, confirms the wide availability of whale products in the village.

Artifacts Related to Whaling

Artifactual evidence of whale hunting and processing is present but sparse in the settlement.[12] Whaling harpoon heads are less frequent

---

[12] Sheehan (1990:278) and Dekin (1990:545) both remark that in Utqiaġvik, artifactual evidence of whaling was less pervasive than one might

than smaller ones. Of 40 socketed harpoon heads found in the erosion front, only two were of sufficient size to have been used in whaling. One of them has Punuk decoration, the other was probably deposited there after the abandonment of the site (Blumer and Csonka 1998:98-99). An additional whaling harpoon head was found in an occupation layer in the test excavation of EH-21. And as mentioned above, a bundle of seven Punuk whaling harpoons, 21 to 23 cm long, was discovered in EH-18 (see illustration in Dneprovsky 2002:Fig.3, p. 194)). Other evidence of ice and open-water sea mammal hunting is present: smaller harpoon heads, foreshafts, socket pieces and finger rests, proximal ice picks, throwing boards, float handles and float plugs, wound plugs, parts and models of kayaks and umiaks, and so on. Some of the larger fragmentary polished slate blades could have been used as lance heads or flensing knives in whaling. There are more complete examples from the cemetery (Leskov and Müller-Beck 1993:134). Judging from their shape, some of the larger ivory 'ice picks' could have been used as lances. It is conceivable that if gray whales were somehow trapped in shallow waters, they could be dispatched with lances, harpoon heads thus being unnecessary. The few specimens found may have been used in the less frequently practiced bowhead hunt. Finally, none of the zoomorphic sculptures and iconographic decorations found so far represent anything connected with whales.

DISCUSSION AND CONCLUSION

With so much circumstantial evidence of whaling, it surely was practiced to some degree. The association of the Punuk culture, traits of which are definitely present in Ekven, and an increase in whaling is well established by previous publications (e.g., Ackerman 1984:112) and it is not our intention here to refine this discussion. Our data do not add anything new to the question of whaling during Okvik and OBS times. They do provide evidence that since about A.D. 600, the earliest times we could document in the settlement so far, quantities of whale bone were used as architectural elements in large dwellings despite the probable availability of alternate materials. Baleen must also have been abundant since an unworked surplus of it was present despite its popularity and usefulness as a raw material. During the perhaps one millennium of Birnirk, Punuk, and Western Thule occupation, we see in the archaeological record a continuous use of whale products. The older dwellings are as replete with whale remains as the more recent ones.

Whales could not provide for all human needs; diets were supplemented by many different species. Foremost among them were

---

expect. To account for it, they provide ethnographic examples of ritual disposal of implements and reshaping and reuse of others.

ringed seals. These pinnipeds must have also furnished skins for clothing and perhaps bedding, especially in the quasi-absence of caribou. In recent times, meat and the skins of these terrestrial mammals were obtained only through trade from inland people. The rather high percentage of dog bones in dwellings may reflect a taste for their meat; by ethnographic analogy they may also indicate periods of hunger when other food was not available (see Mason 1998:293). In any case, subsistence was almost exclusively provided by the sea. As to the question of the proportions of whale species hunted and evidence of whale size selection, Dinesman and Savinetsky's paper in this volume furnishes adequate answers. Scavenging on stranded carcasses does not seem to have been a major source of whale products.[13]

Only additional data and analyses will allow us to reconstruct the culture history of Ekven and its surroundings with detail and certainty. But we may now say with confidence that the picture offered by the settlement differs considerably from that derived from the cemetery. Ekven was probably, at some time, an important link in the middle of a chain of settlements with predominant Birnirk traits extending east to west minimally between Point Barrow in Alaska and the mouth of the Kolyma in Siberia. The bearers of the Birnirk culture have been portrayed by Mason (1998:281-288, 311) as a marginal group threatened by the Punuk people, but at the same time Mason postulates that Uelen and Ekven were centers of political power in the region. The once strong Birnirk presence at Ekven, established in spacious houses replete with whale bones and baleen artifacts, contradicts this rather destitute image of their culture. Instead, it suggests that this site may have been an important center of Birnirk culture. And since most agree that Birnirk is essentially an outgrowth of the Okvik-Old Bering Sea cultures found only in Asia (e.g., Collins 1973:24n. 30), we will someday have to address the question of where this development took place. The Ekven region, with its predominantly OBS II and III cemeteries contemporary with a then predominantly Birnirk settlement, may eventually illuminate this transition.[14]

---

[13] The other main predator of gray whales in Bering Strait is the killer whale, which often only partly consumes its prey. Nowadays, gray whales freshly killed by orcas are retrieved and consumed by Natives. Such scavenging may also influence the mortality curves based on the size of the bones, since killer whales also do select smaller individuals.

[14] Interestingly enough, Utermohle (1988:40-43) found that the characteristics of crania at Uelen and Ekven were very close to those of Birnirk people and to Inupiaq-speaking people generally, as opposed to other Eskaleuts.

Punuk traits are present in Ekven from early Birnirk times, and they persist alongside Western Thule artifacts until the abandonment of the settlement in the late prehistoric period. From its geographical position, one may thus postulate that Ekven was also a bridgehead for the diffusion of Punuk traits and/or people to the New World. A cursory comparison with the Wales sites just across Bering Strait makes apparent how similar—yet with nuances—the Ekven settlement assemblages are to the mixed 'Kurigitavik culture' (Yama'ura 1984, following Collins 1940; Jenness 1928; Morrison 1991). Obviously, the clarification of the nature and extent of the interactions and influences between Punuk and Birnirk, and the rise of (Western) Thule, must be based on detailed comparative data from both sides of the strait and the intervening islands (Csonka 2000). In Ekven, we are faced with a more intricate situation than the 'trait unit intrusion' (Ackerman 1962:33) of one culture in a site of another.

That hostilities were part of the equation can be deduced from artifactual evidence. In Ekven, elements of archery are abundant. Several dozen Punuk 'war' type ivory arrowheads with sharp triangular sections (Bandi 1995:170-171) were found at the erosion front. Many of them were distal fragments typically broken along the deep incisions underlining the only barb, as if they had been designed to detach on impact and remain in the wound.[15] EH-18 has also yielded antler armor elements (Dneprovsky 2002:Fig 12, p. 203)).

After at least a millennium of existence, the village of Ekven was abandoned a few centuries ago. Did whaling fail because of the onset of the Little Ice Age? There may be other answers. The more recent houses may have been periodically inundated during storms, following the disappearance of a hypothetical barrier bar offshore, forcing their inhabitants to seek drier ground. The late prehistoric period also saw the territorial pressure of the Chukchi in all areas of what is now Chukotka, including the coast, and their gradual assimilation or replacement of Eskimos. Did the last inhabitants abandon Ekven, situated at the foot of a low hill and difficult to defend, to move to protected locations in the Dezhnev Mountain massif, at a time when it may already have been colonized by immigrated Alaskan Yupik? Or were the survivors simply assimilated into the neighboring populations?

Acknowledgments. I am pleased to acknowledge the special contributions of Reto Blumer (SLFA, Zurich) and Bernard Moulin (Lozeron, France) to the collection of data in the field as well as to some elaborations presented in this paper. They are the authors of all the figures except the photographs. Bones were identified by Arkady

---

[15] Some from Wales are illustrated in Morrison (1991:Pl. 5 j, k). In 1999, while participating in the excavations at Wales at the invitation of Roger Harritt, I saw similar arrowheads among the excavated artifacts.

Savinetsky. Financial support for this project was provided by two concurrent subsidies from the Swiss National Science Foundation (SNSF 1997–2000) and by the Swiss-Liechtenstein Foundation for Archaeological Research Abroad (1992–2001). Our work was conducted under a license from the Russian Academy of Sciences attributed to Kirill Dneprovskyi from the Museum of Oriental Arts in Moscow. We also held authorizations from the Department of Culture of the Autonomous Region of Chukotka, from the Chukotka Regional Beringia Park, from the Council of Elders of the Chukotskyi District, and had approval from the Council of Elders and from the State farm of Uelen. We are particularly grateful to the Native persons and institutions in Chukotka who allowed us to work in their land and who offered us invaluable help and support. This text has greatly benefited from comments by Roger Harritt, Allen McCartney ; and an anonymous reviewer. The responsibility for opinions, errors, and omissions remains mine.

REFERENCES

Ackerman, Robert E.
  1961   *Archaeological Investigations into the Prehistory of St. Lawrence Island, Alaska.* Philadelphia: University of Pennsylvania, Ph.D. dissertation.
  1962   'Culture Contact in the Bering Sea: Birnirk–Punuk Period,' pp. 27-34 in J.M. Campbell, ed., *Prehistoric Cultural Relations Between the Arctic and Temperate Zones of North America.* Montreal: Arctic Institute of North America, Technical Paper 11.
  1984   'Prehistory of the Asian Eskimo Zone,' pp. 106-118 in D. Damas, ed., *Handbook of North American Indians 5: Arctic.* Washington, D.C.: Smithsonian Institution.
Alekseev, Valeri P.
  1972   Results of Historico-Ethnological and Anthropological Studies in the Eastern Chukchi Area. *Inter-Nord* 12: 234-244.
Arutiunov, Sergei A. and Mikhail M. Bronshtein
  1985   'The Problem of Distinguishing between Old Bering Sea and Okvik Ornamental Styles,' pp. 17-21 in R. Fellmann *et al.*, eds., *Jagen und Sammeln* (Festschrift for H.-G. Bandi). Berne: Stämpfli and Co.
Arutiunov, Sergei A., Igor I. Krupnik, and Mikhail A. Chlenov
  1982   *Kitovaya Alleya: Drevnosti Ostrovov Proliva Seniavina* (Whale Alley: Antiquities of the Seniavin Strait Islands). Moscow: Nauka.
Arutiunov, Sergei A. and Dorian A. Sergeev
  1969   *Drevnie Kultury Aziatskikh Eskimosov: Uelenskii Mogilnik* (Ancient Cultures of the Asian Eskimos: the Uelen Cemetery). Moscow: Nauka.

1975   *Problemy Etnicheskoi Istorii Beringomoria (Ekvenskii Mogilnik).* (Problems of the Ethnic History of the Bering Sea: the Ekven Cemetery). Moscow: Nauka.

1983   'Nauchnie rezultaty rabot na Ekvenskom drevneesskimoskom mogilinike (1970-1974 gg.) (Scientific Results of the Work in the Old Eskimo Cemetery at Ekven 1970-1974),' pp. 200-229 in V. Alekseev ed., *Na styke Chukotki i Aliaski.* Moscow: Nauka.

1990   Issues in the Ethnic History of the Bering Sea [Translation of Chapters 1 & 8 of Arutiunov and Sergeev 1975]. *Soviet Anthropology & Archeology* 28(4): 50-77.

Bandi, Hans-Georg

1995   'Siberian Eskimos as Whalers and Warriors,' pp. 165-183 in A.P. McCartney, ed., H*unting the Largest Animals: Native Whaling in the Western Arctic and Subarctic:* Edmonton: Canadian Circumpolar Institute (CCI) Press, University of Alberta, Occasional Publication No 36, Studies in Whaling No. 3.

Bandi, Hans-Georg *et al.* (in collaboration with E. Anliker-Bosshard and A.B. Hofmann-Wyss)

1984   *Lorenz Insel-Studien: Berner Beiträge zur archäologischen und ethnologischen Erforschung des Beringstrassengebietes* 1: *Allgemeine Einführung und Gräberfunde bei Gambell am Nordwestkap der St. Lorenz Insel, Alaska.* Berne: Stuttgart, Haupt (Academica Helvetica 5/1).

Blumer, Reto

1996   'Première expédition archéologique internationale en Tchoukotka, Sibérie nord-orientale: Rapport de la contribution suisse aux travaux de l'été 1995,' pp. 110-150 in *Annual Report, 1995.* Vaduz and Bern: Swiss-Liechtenstein Foundation for Archaeological Research Abroad.

1997   'Seconde expédition archéologique internationale en Tchoukotka, Sibérie nord-orientale: Rapport de la contribution Suisse à la campagne de 1996,' pp. 57-78 in *Annual Report, 1996.* Vaduz and Bern: Swiss-Liechtenstein Foundation for Archaeological Research Abroad.

Blumer, Reto and Yvon Csonka

1998   'Archaeology of the Asian Shore of Bering Strait: Swiss Contribution to the Third International Expedition,' pp. 83-130 in *Annual Report, 1997.* Vaduz and Bern: Swiss-Liechtenstein Foundation for Archaeological Research Abroad.

Bogoslovskaya, L.S., Igor I. Krupnik, and L.M. Votrogov

1982   The Bowhead Whale Off Chukotka: Migrations and Aboriginal Whaling. *Report of the International Whaling Commission* 32: 391-399.

Bronshtein, Mikhail M.

1993   'Ekven: Einzigartige archäologische Fundstätte in Nordostasien,' pp. 73-83 in A. Leskov and H. Müller-Beck, eds., *Arktische Waljäger vor 3000 Jahren: Unbekannte sibirische Kunst.* Mainz-Munich: Hase & Koehler Verlag.

Bronshtein, Mikhail and Patrick Plumet
  1995   Ekven: l'art préhistorique béringien et l'approche russe de l'origine de la tradition culturelle esquimaude. *Etudes/Inuit/Studies* 19(2): 5-59.

Chichlo, Boris
  1981   Les Nevuqaghmiit ou la fin d'une ethnie. *Etudes/Inuit/Studies* 5(2): 29-47.

Chu, Pei Pei
  1998   *Dietary Variation Among the Prehistoric Asiatic Eskimo.* Burnaby, B.C.: Simon Fraser University, Unpublished M.A. thesis, Department of Archaeology.

Collins, Henry B.
  1937   *Archaeology of St. Lawrence Island, Alaska.* Washington, D.C.: Smithsonian Institution, Miscellaneous Collections 96 (1).
  1940   *Outline of Eskimo Prehistory.* Washington, D.C.: Smithsonian Institution, Miscellaneous Collections 100: 533-592.
  1964   'The Arctic and Subarctic,' pp. 85-114 in J.D. Jennings and E. Norbeck, eds., *Prehistoric Man in the New World.* Chicago: University of Chicago Press.
  1973   'Eskimo Art,' pp. 1-131 in H.B. Collins, F. De Laguna, E. Carpenter and P. Stone, eds., *The Far North: 2000 Years of American Eskimo and Indian Art.* Washington, D.C.: National Gallery of Art.

Csonka, Yvon
  1998   'La nécessité d'une approche nuancée,' in: Debate: Should Ethnographic Collections be Returned? *Tsantsa* 3: 67-70.
  2000   'Archaeology of Bering Strait: Short Report on a Contribution to the Excavations in Wales, Alaska, in the Summer of 1999,' pp. 59-66 in *Annual report 1999.* Vaduz and Zurich: Swiss-Liechtenstein Foundation for Archaeological Research Abroad.

Csonka, Yvon, Reto Blumer and Bernard Moulin
  1999   'Archaeology of the Asian Side of Bering Strait: Swiss Contribution to the Fourth International Fieldseason,' pp. 83-130 in *Annual Report 1998.* Vaduz and Zurich: Swiss-Liechtenstein Foundation for Archaeological Research Abroad.

Dekin, Albert A., Jr.
  1990   'Summary and Conclusions,' pp. 534-545 in A.A. Dekin Jr. *et al.*, eds., *The 1981 Excavations at the Utqiaġvik Archaeological Site, Barrow, Alaska*, Vol. 1. Barrow: The North Slope Borough Commission on Iñupiat History, Language, and Culture.

Dikov, Nikolai N.
  1967   'Uelenskii mogilnik po dannym raskopok b 1956, 1958 i 1963 godakh (The Uelen Cemetery According to Data from the Excavations in 1956, 58 and 63),' pp. 45-79 in *Istoria i Kultura Narodov Severa Dalnego Vostoka.* Moscow: Nauka.
  1974   *Chininski Mogilnik* (The Chini Cemetery). Nauka, Novosibirsk.
  1997   *Asia at the Juncture with America in Antiquity: The Stone Age of the Chukchi Peninsula.* Translation by Richard Bland. National Park Service, Anchorage [1st edition in Russian 1993, St. Petersburg: Nauka].

Dinesman, L. G., N. Kiseleva, A. Savinetsky, and B. Khassanov
    1999   *Secular Dynamics of the Coastal Zone Ecosystems of the Northeastern Chukchi Peninsula.* Tübingen: Mo Vince Verlag [1st edition in Russian 1996. Moscow: Argus.]

Dneprovsky, Kirill A.
    2002   'Ekven House H-18: A Birnirk– and Early Punuk– Period Site in Chukotka, pp. 167-206 in D.E. Dumond and R.L. Bland, eds. *Arcaheology on the Bering Strait Region: Research on Two Continents.* University of Oregon Anthropological Papers, No. 59.

Dumond, Don E.
    1987   *The Eskimos and Aleuts.* Revised edition. London: Thames and Hudson.
    1995   'Whale Traps on the North Pacific?,' pp. 61-61 in A.P. McCartney, ed. H*unting the Largest Animals: Native Whaling in the Western Arctic and Subarctic:* Edmonton: Canadian Circumpolar Institute (CCI) Press, University of Alberta, Occasional Publication No 36, Studies in Whaling No. 3.
    1998   The Hillside Site, St. Lawrence Island, Alaska. *University of Oregon Anthropological Papers* No. 55.

Ford, James A.
    1959   Eskimo Prehistory in the Vicinity of Point Barrow, Alaska. *Anthropological Papers of the Museum of Natural History* 47(1).

Geist, Otto and Froelich Rainey
    1936   *Archaeological Excavations at Kukulik, St. Lawrence Island, Alaska: Preliminary Report.* University of Alaska Miscellaneous Publications 2. Washington, D.C.: U.S. Government Printing Office.

Gusev, Sergey, Andrei Zagorulko and Aleksey Porotov
    1999   Sea Mammal Hunters of Chukotka, Bering Strait: Recent Archaeological Results and Problems. *World Archaeology* 30(3): 354-369.

Jenness, Diamond
    1928   Archaeological Investigations in Bering Strait, 1926, pp. 71-80 in *Canada Department of Mines Bulletin 50, Annual Report for 1926.* Ottawa: National Museum of Canada.

Krupnik, Igor I.
    1983   'Gray Whales and the Aborigines of the Pacific Northwest: The History of Aboriginal Whaling,' pp. 103-120 in M.L. Jones, S.L. Swartz and S. Leatherwood, eds., *The Gray Whale: Eschrichtius robustus.* Orlando: Academic Press.
    1988   Asiatic Eskimos and Marine Resources: A Case of Ecological Pulsations or Equilibrium? *Arctic Anthropology* 25(1): 94-106.
    1993a   Prehistoric Eskimo Whaling in the Arctic: Slaughter of Calves or Fortuitous Ecology? *Arctic Anthropology* 30(1): 1-12.
    1993b   *Arctic Adaptations: Native Whalers and Reindeer Herders of Northern Eurasia.* Hanover and London: University Press of New England.

Krupnik, Igor I., L.S. Bogoslovskaya, and L.M. Votrogov
    1983   Gray Whaling off the Chukotka Peninsula: Past and Present
            Status. *Report of the International Whaling Commission* 33:
            557-562.
Krupnik, Igor and Mikhail Chlenov
    n.d   *Survival in Contact: Yupik (Asiatic Eskimo) Transitions, 1900-
            1990.* unpublished.
Larsen, Helge
    1968   Near Ipiutak and Uwelen-Okvik. *Folk* 10: 81-90.
Leskov, Alexander M. and Hansjürgen Müller-Beck, eds.
    1993   *Arktische Waljäger vor 3000 Jahren: Unbekannte sibirische
            Kunst.* Mainz-Munich: Exhibition catalogue. v. Hase & Koehler
            Verlag.
Mason, Owen K.
    1998   The Contest Between the Ipiutak, Old Bering Sea, and Birnirk
            Polities and the Origin of Whaling During the First Millennium
            A.D. along Bering Strait. *Journal of Anthropological
            Archaeology* 17: 240-325.
Mason, Owen and C. Gerlach
    1995   Chukchi Hot Spots, Paleo-Polynias, and Caribou Crashes:
            Climatic and Ecological Dimensions of North Alaska
            Prehistory. *Arctic Anthropology* 32(1): 101-130.
Morrison, David
    1991   *The Diamond Jenness Collections from Bering Strait.* Ottawa:
            Archaelogical Survey of Canada, Mercury Series Paper 144.
Moulin, Bernard and Yvon Csonka
    2002   'The Erosion Front at Ekven: A Stratigraphic and
            Geoarchaeological Approach,' pp. 227-259 in D.E. Dumond
            and R.L. Bland, eds., *Archaeology in the Bering Strait Region:
            Research on Two Continents.* University of Oregon
            Anthropological Papers No. 59.

Okladnikov, Aleksei P. and Nina Beregovaya
    1971   *Drevnie poselenia Baranova mysa* (Ancient Settlements of Cape
            Baranov). Novosibirsk: Nauka.
Plumet, Patrick
    1993   'Le patrimoine archéologique au Kamchatka et en Tchoukotka,'
            pp. 299-309 in Chichlo, B. (ed), *Sibérie 3: Questions
            sibériennes: les peuples du Kamchatka et de Tchoukotka.* Paris:
            Institut d'études slaves.
Rainey, Froelich
    1941   Eskimo Prehistory: The Okvik Site on Punuk Islands.
            *Anthropological Papers of The American Museum of Natural
            History* 37(4).
Raushenbakh, V.M.
    1969   *Novye nakhodki na Chetyrekhstolbovom ostrove* (New Finds on
            the Four-Columns Island). Moscow: Sovietskaïa Rossia.
Rudenko, S. I.
    1961   *The Ancient Culture of the Bering Sea and the Eskimo Problem.*
            Arctic Institute of North America, Translations from Russian
            sources 1. Toronto: University of Toronto Press [Original in
            Russian 1947].

135

Schweitzer, Peter
    1990   *Kreuzungspunkt am Rande der Welt: Kontaktgeschichte und soziale Verhältnisse der sibirischen Eskimo zwischen 1650 und 1920.* Vienna, Austria: University of Vienna: Ph.D. dissertation.

Schweitzer, Peter and Evgeniy Golovko
    1995   *Contacts Across Bering Strait, 1898-1948.* Report to the U.S. National Park Service. Alaska: US National Parks Service.

Sheehan, Glenn W.
    1990   'Excavations at Mound 34,' pp. 181-325 in A.A. Dekin Jr. *et al.*, eds., *The 1981 Excavations at the Utqiaġvik Archaeological Site, Barrow, Alaska*, Vol. 2. Barrow: The North Slope Borough Commission on Iñupiat History, Language, and Culture.

Stanford, Dennis J.
    1976   *The Walapka Site, Alaska: Its Place in the Birnirk and Thule Cultures.* Washington, D.C.: Smithsonian Institution, Contributions to Anthropology 20.

Steward, Julian
    1942   The Direct Historical Approach to Archaeology. *American Antiquity* 7(4): 337-343.

Stringer, W.J. and J.E. Groves
    1991   Local and Areal Extent of Polynyas in the Bering and Chukchi Seas. *Arctic* 44 (suppl. 1): 164-171.

Stuiver, M., P.J. Reimer, E. Bard, J.W. Beck, G. S. Burr, K.A. Hughen, B. Kromer, F.G. McCormac, J. v.d. Plicht, and M. Spurk Stuiver
    1998   INTCAL98 Radiocarbon Age Calibration, 24000-0 cal B.P. *Radiocarbon* 40: 1041-1083.

Utermohle, Charles J.
    1988   'The Origin of the Inupiat: The Position of the Birnirk Culture in Eskimo Prehistory,' pp. 37-46 in R.D. Shaw, R.K. Harritt and D.E. Dumond, eds., *The Late Prehistoric Development of Alaska's Native People.* Anchorage: Alaska Anthropological Association, Aurora Monograph Series 4.

Yama'ura, Kiyoshi
    1984   Toggle Harpoon Heads from Kurigitavik, Alaska. *Bulletin of the Department of Anthropology Tokyo* 3: 213-262.

Yarborough, Linda F.
    1995   'Prehistoric Use of Cetacean Species in the Northern Gulf of Alaska,' pp. 63-81 in A.P. McCartney, ed., *Hunting the Largest Animals: Native Whaling in the Western Arctic and Subarctic.* Edmonton: Canadian Circumpolar Institute (CCI) Press, University of Alberta, Occasional Publication Series No. 36, Studies in Whaling No. 3.

Secular Dynamics of the Prehistoric Catch and Population Size of Baleen Whales off the Chukchi Peninsula, Siberia

As Based upon the Study of Historical Whale Bone from Ancient Coastal Sites

Lev G. Dinesman and Arkady B. Savinetsky
Translated by Alexandr N. Kuznetsov
Edited by Igor I. Krupnik

Abstract. *The paper discusses the results of recent surveys of ancient coastal sites along the northeastern shores of the Chukchi Peninsula, with large concentrations of prehistoric whale bones. The majority of some 1200 identified whale skulls and scapulae belonged to gray and bowhead whales. Of those, more than 100 bones were sampled for radiocarbon dating. It is argued that active whaling had started along the eastern (Bering Strait) shore of the Chukchi Peninsula around 2300–2100 BP and along the arctic coast, around 1800–1700 BP. Yearlings of both gray and bowhead whales were primarily targeted between 2300 and 1300 BP; later hunters started to harvest bigger and older whales. Two periods of high abundance of gray whales in ancient site deposits occurred around 2100–1900 and 700–500 BP: The reasons for such secular trends remain unclear. The share of bowhead whales was relatively low between 2300 and 1300 BP.; however, the role of bowhead whaling increased dramatically after 1300 BP, most probably because of the heavy summer ice conditions in the Bering Strait.*

INTRODUCTION

The utilization of whale bone for dwelling construction by the Natives of the Chukchi Peninsula, Siberia (commonly known as Chukotka), has been documented since the mid-1600s, after Semen Dezhnev's voyage of discovery in 1648 along the shores of northeastern Asia (Berg 1946:28, 32). However, the first special surveys of historical whale bone in the ruins of ancient sites in Chukotka have only been conducted fairly recently (see Arutyunov *et al.* 1982; Bogoslovskaya *et al.* 1981, 1982; Krupnik *et al.* 1983; Chlenov and Krupnik 1984; Krupnik 1982, 1984, 1987, 1993b; Dinesman *et al.* 1996). Judging from the size of whale bone found as well as from (some historical evidence) the ancient residents of Chukotka—from the Gulf of Anadyr to Cape Dezhnev ('East Cape')—primarily hunted the gray whale (*Eschrichtius gibbosus* (Erxleben, 1777)) and the Greenland whale (*Balaena mysticetus* (Linnaeus, 1758)), mainly their immature individuals. Occasionally, they hunted the Humpbacked (*Megaptera novaeangliae* (Borowski 1781)) and the White Whale (*Delphinapterus leucas* (Pallas, 1776)).

In 1989-1992, a team of the Severtsov Institute of Ecology and Evolution of the Russian Academy of Sciences studied several ancient coastal sites on the Chukchi Peninsula, from the mouth of the Chegitun River on the arctic shore to Mechigmen Bay, south of Bering Strait. The full outcome of our study of prehistoric ecosystem dynamics in Chukotka were published elsewhere (Dinesman *et al.* 1996; 1999). This paper presents an extended, updated summary of the related survey of the historical whale bone deposited in the ruins of ancient dwellings and in abrasion slopes of the coastal sites. The aim of our study was to identify the biological species hunted by ancient whalers of Chukotka and to reveal whether long-term 'secular' dynamics, if any, could be established in the Native whale catch and in the size of whale populations targeted by prehistoric hunters.

All radiocarbon dating cited in this paper was processed by the Group of Historical Ecology of the Severtsov Institute of Ecology and Evolution in Moscow, using two scintillation spectrometers. Age was registered from 1950. The benzene standard of the Geological Institute of the Russian Academy of Sciences was used, which is equivalent to the International Standard (Arslanov *et al.* 1987). The radiocarbon dates obtained are not expressed in terms of calendar time because of the absence of correlation curve for sea samples for this region.

WHALE BONES FROM THE NORTHERN AND EASTERN SHORE OF THE CHUKCHI PENINSULA

During three field surveys in 1989, 1990, and 1992, we studied ancient whale bone located in the ruins of dwellings and on erosion taluses of collapsing settlements along the northern (Arctic) and eastern (Bering Strait) shores of the Chukchi Peninsula. Several abandoned Native villages and ancient sites were surveyed, including those near the mouth of the Chegitun River, at Cape Dezhnev, at the prehistoric Ekven cemetery and settlement, at Eleikei Creek, Tunytlen Mountain, Yandogai, and at Masik in the Mechigmen Bay (Fig. 1).

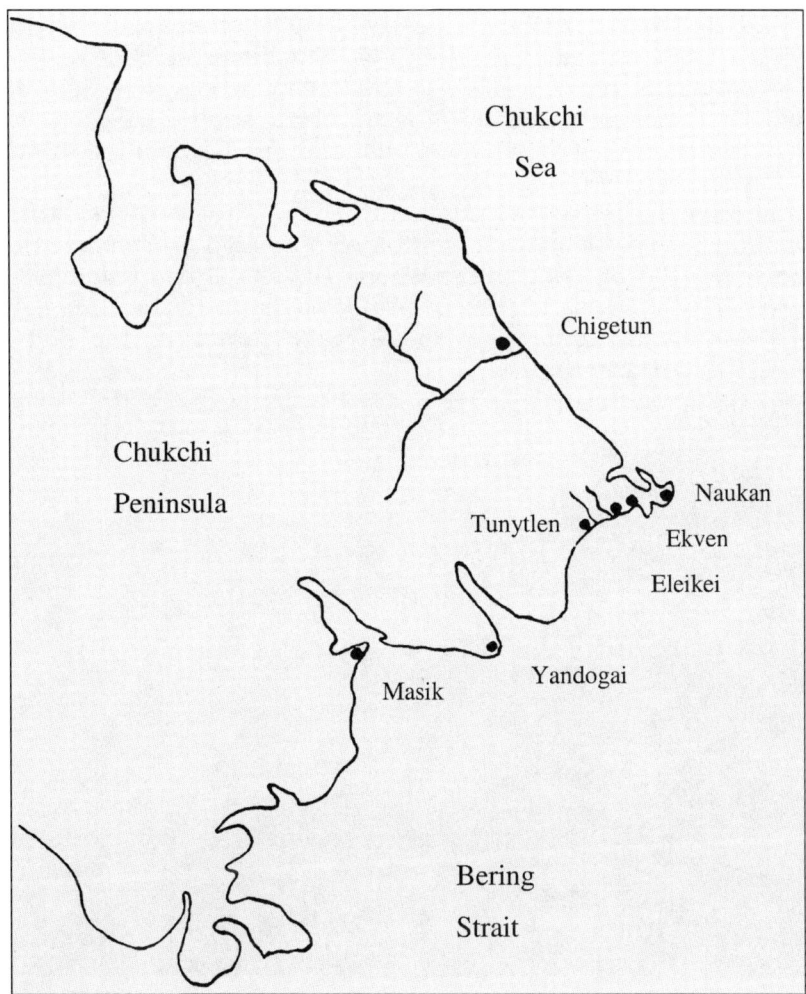

Figure 1. *Area of survey for historical whale bone in Chukotka, 1989-1992.*

Sites C: near the mouth of the Chegitun River. Five ancient sites were studied.The first of the Chegitun settlements is situated on a hill on the western coast of Chegitun Lagoon. Ruins of six dwellings and several meat pits were found here. Altogether, 109 skulls and scapulae of whales were located in these ruins.

The second Chegitun site is situated approximately 1.5 km to the west, on a scarp of a very high rocky shore. It is heavily damaged by erosion, with only the ruins of two or three dwellings remaining. Here, the fragments of three gray whale skulls and nine scapulae were found.

The ruins of the third Chegitun site are located at the eastern shore of Chegitun Lagoon. Several whale skulls in very bad condition were identified.

The fourth of the Chegitun settlements is situated to the east of Chegitun Lagoon, near the mouth of Ekichuvervehem Creek. The site's position is on a seaside ledge of 4–5 m elevation above the beach. It is very damaged by erosion. Here, several ribs and lower jaws of whales, two gray whale skulls, and one bowhead skull whale were found.

Farther to the east, at the mouth of the Chutpen River, there is a fifth Chegitun settlement. It is situated on a high scarp and extends for some 80–100 m along the seashore; the site is badly damaged by erosion. The thickness of the cultural layer is up to 3 m. Altogether, 100 whale skulls and 56 scapulae were found in the ruins of dwellings and meat pits.

Figure 2. *'Whale alley' built from mandibles of mature bowhead whales in the Naukan settlement.*

Naukan and *Nunak:* Two abandoned sites were surveyed in the Cape Dezhnev area. Both are located on high cliffs of the Dezhnev headland. Eskimo (Siberian Yupik) hunters lived in Naukan until 1958 and at Nunak until the late 1800s (Leont'ev and Novikova 1989). Four whale skulls and 21 scapulae were recovered in the ruins of both settlements.

Ekven: The ancient Ekven settlement is situated on sand hills to the west of Cape Verblyuzhii, between the lagoon and the mouth of the Kytylynvehem stream. Its seaside part was destroyed by erosion. In the ruins and talus of this settlement, 102 gray and bowhead whale skulls and scapulae were found.

Eleikei: An ancient settlement located on a hill above Eleikei Lagoon, preserved in very good condition. Ruins of its dwellings are covered by sod. Only some whale ribs emerge above the ground. Near the settlement, small, probably ritual assemblages of large stones and 68 gray and bowhead whale skulls are located.

Tunytlen: A settlement on a slope of Tunytlen Mountain; is heavily damaged by erosion and is half-buried in a talus of large-sized debris. A total of 37 gray and bowhead whale skulls and 67 scapulae were recorded here.

Yandogai: The ruins of the abandoned Yandogai settlement are situated on the south side of St. Lawrence Bay. A total of 51 skulls and scapulae were found in the ruins.

Masik: The ruins of the Masik site are located at the southern portion of the long spit that separates Mechigmen Inlet from Mechigmen Bay. The spit is formed by subparallel, interconnected bars of sand and pebbles, terminating at a group of sand hills, about 10 m high. The house ruins are in two rows, found on parallel bars and on hills and occupy 1.5–2 km of the beach. The site was first surveyed by Bogoslovskaya, Chlenov, and Krupnik in 1981 (see photos in Chlenov and Krupnik 1984—*ed.*). Over 550 gray and bowhead whale skulls were recorded there.

Table 1. Bones of the Gray and the Greenland Whales from ruins of ancient settlements.

Settlements	Number of bones		Percentage of Gray Whale
	Bowhead Whale	**Gray Whale**	
Chegitun	35	245	87
Naukan	16	9	36
Ekven	31	71	69
Eleikei	6	62	91
Tunytlen	11	93	90
Yandogai	27	24	47
Masik	17	544	97

Altogether, 1196 identifiable fragments of whale skeletons were reported at the above-mentioned sites. Of these, only five sets of bone remains (less than 0.5%) belong to humpback whale and *belukha* (white whale). They were found at the Tunytlen Mountain, at Eleikei Creek, and at the Ekven sites. All other bone remains belong to gray and bowhead whales. Considering the ratio of bones of these two species, ancient hunters almost everywhere preferred to harvest gray whales (Table 1). Only at Yandogai and at two Cape Dezhnev sites, particularly in Naukan, did the species ratio shift in favor of the bowhead whale. Here, the share of gray whale bones decreased from 69–97% to 36–47%.

Figure 3. *The skull of a bowhead whale on the beach near the Dezhnevo site.*

Size and Status of the Whale Bone Remains

While collecting and measuring prehistoric whale bones in the ruins of ancient settlements, special attention was paid to the crania (skulls) and scapulae of whales. In scapulae, the length and width of blades were measured, following procedures first introduced in McCartney 1976 (see also Gerlach *et al.* 1993). In crania, the length and width of *os supraoccipitale*, the zygomatic, and the condylar widths were measured—which, again, is a common practice in whale bone research (see Tomilin 1957; McCartney 1976; Savelle and McCartney 1994; Arutyunov *et al.* 1982) (Fig. 3).

Os supraoccipitale and the condylar width were selected for measuring due to the common position and the status of preservation of whale skulls. In most cases, the exact measurement of the condylobasal length and the zygomatic width was either impossible or required extensive digging. The rostral parts were most often truncated, destroyed, or were positioned deeply into the ground.

According to our measurements of the gray whale skulls *in situ* with the best status of preservation (Knyazev and Savinetsky 1995), the usual ratio of zygomatic to supraoccipitale width is 1.30–1.38. Several earlier studies by Tomilin (1957), Zimushko (1970), Yablokov and Bogoslovskaya (1984), and Krupnik (1984, 1987, 1993b) have demonstrated that scapulae of up to 70 cm wide and of skulls with *os supraoccipitale* of up to 65 cm wide belong to juvenile gray whales (that is, to calves and yearlings—*ed.*). Their body length was estimated from 4.3 to 8.1 m, which corresponds to animals one year or under one year of age. Larger skulls and scapulae belong to whales from 8.4 to 12.1 m in length. This range comprises both immature (subadult) gray whales that are older than one year and mature (adult) whales (above 3–5 years—*ed.*).

Based on our measurements of *in situ* bowhead whale crania and the earlier measurements of some 50 skulls by Krupnik in 1977 (Krupnik 1982:165-166), we estimated that the zygomatic and condylar width are connected by the equation: ZW = 6.94 x CW – 42 (where ZW is the zygomatic width and CW is the condylar width; see Knyazev and Savinetsky 1995) (Figs. 4 and 5).

Figure 4. *Scapula of gray whales at the Tunytlen site.*

Using the data on bowhead whale size published by Tomilin (1957), we assumed that in calves' crania, the zygomatic width does not exceed 120 cm and that among immature whales of older age it varies from 120 to 230 cm. For mature or adult bowhead whales, the zygomatic crania width could be even larger, as shown by several adult crania measured by Krupnik at the 'Whale Alley' site (Arutyunov *et al.* 1982:165-166; Knyazev and Savinetsky 1995).[1]

Figure 5. *Scapula of a bowhead whale at the Tunytlen site*

The relationship of the body length (BL) and the length of the scapulae (SL) in the bowhead whale is described by the equation (Gerlach *et al.* 1993): BL = 11.943 x SL + 212.586

Considering that bowhead whales reach maturity at the body length 12–14 m (Savelle and McCartney 1994:290), it is possible to calculate that scapulae longer than 83 cm belong to mature animals. Gray whale crania and scapulae were measured at the Chegitun, Eleikei, and Tunytlen prehistoric sites (Knyazev and Savinetsky 1995) and also at the Ekven (Dinesman *et al.* 1996) and Masik sites. Everywhere, the size distribution pattern was very similar. At Chegitun, Ekven, and Tunytlen sites, the width of gray whale scapulae varied from 40 to 95–103 cm. However, smaller scapulae with a width of less than 70 cm predominated, representing 74–95% of the overall sample. Among the gray whale skulls at the Chegitun sites, the overall width of *os supraoccipitale* ranged from 50 to 95 cm; but here again, 80% of the measured crania widths fell within a fairly narrow range of 50–65 cm. Of 26 gray whale crania measured at the Eleikei site, 25 (96%) had an *os supraoccipitale* width between 50 to 68 cm (Fig. 6). Finally, over 90% of gray whale crania at the Masik site belonged to calves and juveniles, which is very close to the

---

[1] A far more detailed analysis of the statistical relation(s) between bowhead whale size, age, and the basic crania measurements has been presented by Savelle and McCartney (1994)—*ed.*

estimate by Bogoslovskaya and Krupnik after their initial survey at Masik in 1981 (Krupnik *et al.* 1983; Chlenov and Krupnik 1984). These measurements clearly prove that at all prehistoric sites surveyed in Chukotka, juvenile gray whales were selectively targeted and killed.

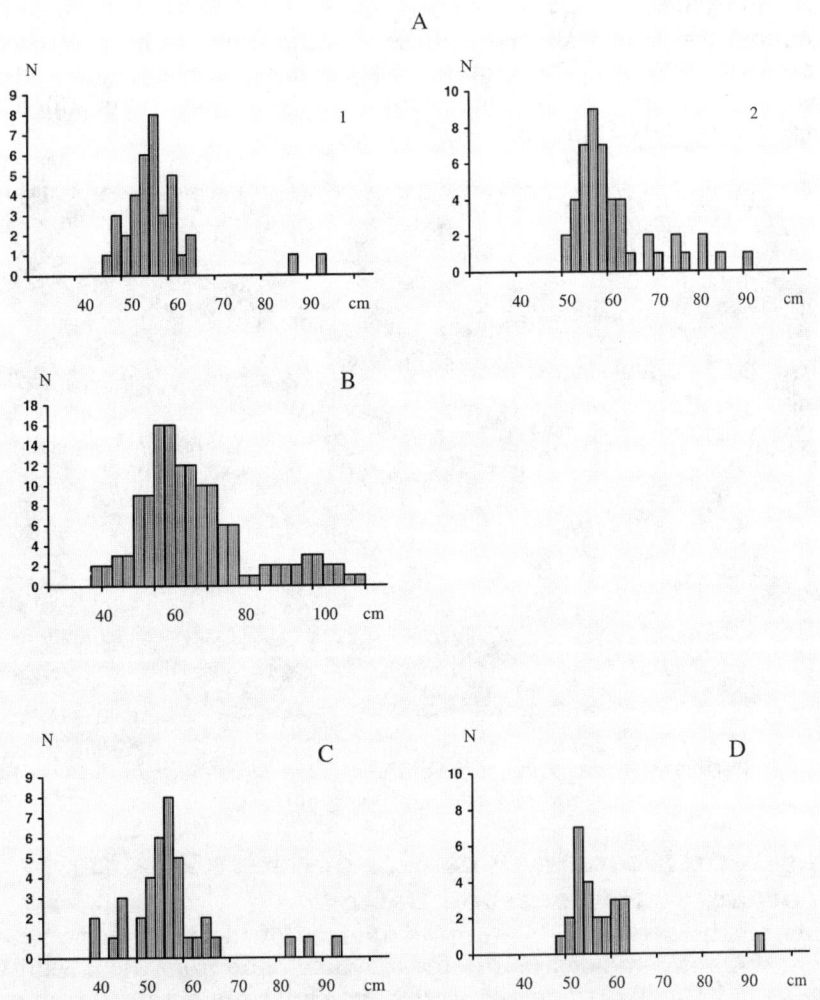

Figure 6. *The width of scapulae (I) and of supraoccipital bones (II) of gray whales from Ancient sites: a—Chegitun, b—Ekven, c—Tunytlen, d—Eleikei.*

The distribution of bowhead whale bones at ancient sites followed a different pattern. Of 19 crania measured at the Chegitun sites, 12 skulls represented immature whales (older than one year) and seven skulls belonged to adults (Knyazev and Savinetsky 1995). Based upon the scapulae size, four scapulae belonged to adult

145

whales and one to an immature whale of over one year. Thus, at the Chegitun sites, both adult and subadult bowhead whales were killed.[2] Judging from the whale bone surveyed and measured at Chegitun, the remains of neither calves nor yearlings are represented in our sample.

The situation was quite different at the Ekven site. Of 17 bowhead whale crania measured at Ekven, eight (47%) had a zygomatic width of less than 100 cm, while nine skulls (53%) were larger than 120 cm. Thus at the Ekven site, both bowhead whale calves and older whales (subadults and adults) were killed in approximately equal proportion (Fig. 7).

Figure 7. *The erosion slope of the ancient Ekven site. In the foreground are skulls of young gray whales.*

Long-Term (Secular) Dynamics of Native Whaling According to Radiocarbon Dating

To reveal the secular (long-term) dynamics of the prehistoric whale catch, the serial radiocarbon dates of whale bone from the Chegitun, Ekven, and Masik settlements were run. The dates obtained from the prehistoric whale bone were plotted against the time scale subdivided into subsequent intervals of 200 years (Fig. 8a,b.c). Counting the 200-year intervals began by shifting toward the middle of the 19th century, to counter potential anthropogenic change in isotopic comp-

---

[2] The authors do not address the issue of possible use of the bones of stranded whales nor that of the selective collection and/or recycling of larger bones that belonged to mature whales for house construction, human burials, and/or other needs—*ed.*

position of natural carbon. This excludes a possibility of inaccurate dating due to the radioactivity of organisms that died in the second half of the 19th century and during the 20th century.

The change in the number of radiocarbon dates within one interval would reflect the secular dynamics of the number of whales (*Nbs*) landed by hunters of a given settlement. In fact, the number of killed animals is more accurately measured by the parameter *Nbs/t*, i.e., by the relative accumulation rate of whale bone at the site area.

A total of 27 gray and bowhead whale bones were dated at the five Chegitun sites on the arctic coast of Chukotka (Knyazev and Savinetsky 1995). As the date distribution illustrates, the earliest bones of subadult bowhead whales (i.e., those older than one year) were deposited at 1900–1700 B.P. (Fig. 8a). During the following period, the number of whale bones at the settlement gradually increased, reaching its maximum at about 900–700 B.P. Then accumulation of bones of subadult whales dropped dramatically, and it ceased completely about 500 B.P. The latest date available from the Chegitun series is 465±89 B.P. (IEMAE-868). The only recent date of a large bowhead skull of an adult whale (<73 B.P.–IEMAE-871) clearly belongs to the recent period and is not connected to ancient whaling. Except for this late date, all other dates obtained from the bones of adult bowhead whales in Chegitun fall within the range of 1430±83 (IEMAE-877) to 332±82 B.P. (IEMAE-880). Their maximum number is within the 200-year period between 900 and 700 B.P. (Fig. 8a).

The earliest date for the gray whale juvenile bones at the Chegitun sites is 1809±137 B.P. (IEMAE-910); for the bones of older subadult whales, 1420±81 B.P. (IEMAE-936). The latest date on ancient gray whale bones available at the Chegitun sites is 626±71 B.P. (IEMAE-939). Here, native hunting for gray whales increased significantly between 1500 and 1300 B.P. and remained at a stable high level. The bulk of gray whale bones was accumulated within the interval of 1500–700 B.P., with the maximum recorded around 1100–900 B.P.

From the Ekven site in the Bering Strait area, a total of 17 bowhead and 32 gray whales were dated. The oldest date obtained on the bowhead whale bone is 2166±68 (IEMAE-831), the latest date is 450±83 B.P. (IEMAE-807). Prior to 1300 B.P., the bones of juvenile whales predominated, though the overall number of bones was relatively low (Fig. 8b). During the period of 1300–500 B.P., the share of adult whales increased, with the peak of accumulation about 1100–900 B.P.

The earliest date obtained on gray whale bone at the Ekven site is 2505±129 B.P. (IEMAE-845). The number of accumulated bones of this species increased steadily, reaching its maximum at about 1900–1700 B.P. (Fig. 8b). It ebbed significantly during the next 200–year interval of 1700–1500 B.P. It then increased again at 1500–1300

B.P., and underwent a gradual decline after that. The youngest date obtained on gray whale bone from the Ekven site is 575±139 B.P. (IEMAE-843). The bones of juvenile gray whales largely predominated at the Ekven settlement.

The earliest date obtained on gray whale bone at the Ekven site is 2505±129 B.P. (IEMAE-845). The number of accumulated bones of this species increased steadily, reaching its maximum at about 1900–1700 B.P. (Fig. 8b). It ebbed significantly during the next 200–year interval of 1700–1500 B.P. It then increased again at 1500–1300 B.P., and underwent a gradual decline after that. The youngest date obtained on gray whale bone from the Ekven site is 575±139 B.P. (IEMAE-843). The bones of juvenile gray whales largely predominated at the Ekven settlement.

From the Masik site, located about 100 km southeast of the Ekven settlement, bones of 17 bowhead whales and 25 gray whales were dated. All dates on bowhead whale bones fall within the 1000 year range from 1628±93 (IEMAE–1098) to 617±81 B.P. (IEMAE–1102). However, most of the dates fall within the period 1300–500 B.P. (Fig. 8c). About half of all bones accumulated during this period belonged to subadult whales older than one year.

The earliest date on gray whale bones obtained at the Masik site is 1251±100 B.P. (IEMAE–1125). During the period of 1100–500 B.P., the number of accumulated bones gradually increased (Fig. 8c) and then dropped abruptly. For the later period, only two dates are available: 479±75 (IEMAE–1123) and 227±67 years B.P. (IEMAE–1134). Except for a few rare cases, all bones of gray whales at the Masik site belonged to calves and yearlings.

The pattern of distribution of radiocarbon dates on prehistoric whale bones described above leads to the following scenario. At the Chegitun settlements, whaling for grays and bowheads started at about 1900–1700 years ago, but it was of minor intensity during its first 400 years. It increased significantly between 1500 and 1300 B.P., although the peak in gray whaling was not reached until 1100–900 B.P. and in bowhead whaling, around 900–700 B.P. Local gray whale hunting at Chegitun ceased about 500 years ago, but hunting for bowheads continued for 200 more years, until roughly 300 B.P.

At the Ekven site, the difference in the earliest available dates on gray whale *versus* bowhead whale bones is nearly 500 years. However, it would be more reasonable to assume that whaling for both species started here almost simultaneously, in the middle of the first millennium BC. This is strongly supported by the additional date of 2683±149 (IEMAE-829) obtained on a juvenile bowhead whale bone at the nearby site at Cape Verblyuzhii. In any case, sustainable hunting for gray and bowhead whales at Ekven did not start until 2300–2100 B.P. (Fig. 8b) and it ceased completely about AD 1500. During this period of 1600–1800 years, local hunters at Ekven harvested primarily juvenile gray whales, mostly calves and yearlings.

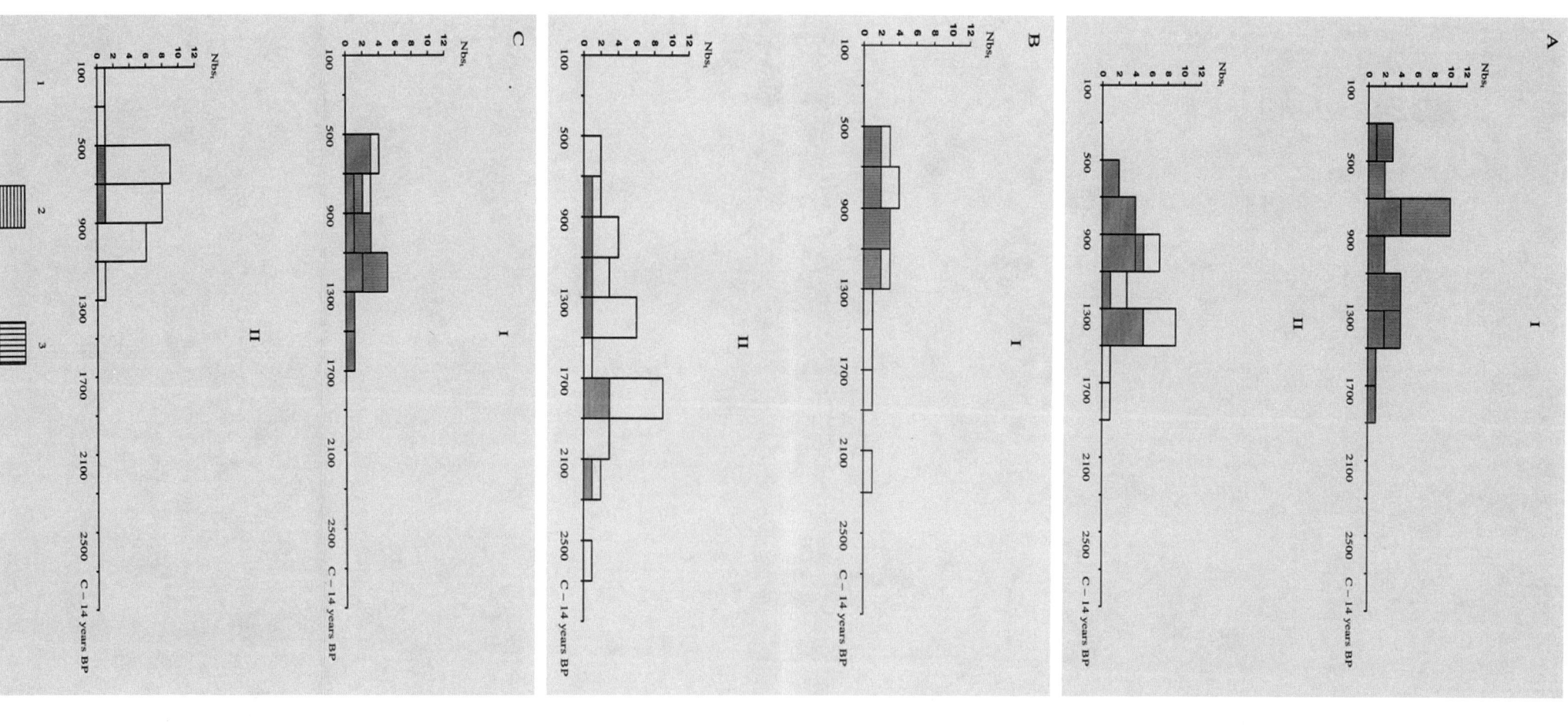

Figure 8. *Distribution of the radiocarbon dates of whale bone from ancient sites:*
a—Chegitun, b—Ekven, c—Masik. I—bowhead whale; II—gray whale;
1—yearlings and calves; 2—juveniles (older than one year); 3—adult animals.

Sustainable hunting for gray whales at Ekven developed rapidly and reached its maximum at about 1900–1700 B.P. It ebbed abruptly at about 1700-1500 B.P., restored again at 1500–1300 B.P., and then decreased gradually, until its termination at A.D. 1500.

Between 2300 and 1300 B.P., hunters at Ekven targeted, or at least managed to land, bowhead juveniles and calves only. Later, subadult and even adult whales were hunted regularly, with a slight peak in the adult whale harvest at about 1100–900 B.P. Hunting for bowhead whales at Ekven ceased at about 500 B.P., that is, simultaneously with the end of gray whaling.

At the Masik site, whaling for bowhead whales started almost 1000 years later, at about 1700-1500 B.P. This may be related to a comparatively late formation of the settlement established on a young stretch of accumulative shore (pebble spit—*ed.*). The ruins of older settlements in the Masik area and along the shores of Mechigmen Bay, in general, might be located on higher cliff outcrops and earlier, loose coastal deposits. During its first 400 years, hunting for bowhead whales at Masik remained at a very low level. It increased substantially between 1300 and 1100 B.P., and proceeded without noticeable change during several subsequent centuries. The catch consisted mainly of subadult (immature) whales that were older than one year. About 500 B.P. hunting for bowhead whales at Masik ceased completely.

Based upon the age of surveyed whale bones, the first gray whales were killed at Masik about 1300–1100 B.P. The catch increased rapidly, reaching its maximum at about 700–500 B.P., then it dropped sharply and continued at a very low level until recently (the small native whaling village at Masik was abandoned in 1952; see Krupnik 1984). During this time, calves and yearlings of gray whales were selectively targeted and killed.

Normally, the chronology of ancient whaling at each individual site depends primarily upon the time of its establishment and its duration. Thus, it is not surprising that it varies significantly in different areas. Also, the ratios of bowhead *versus* gray whales harvested and the share(s) of the predominant age/size group(s) in the Native catch differ significantly at the various sites. The latter may be related to the geographical and local factors in whale migration routes, to different subsistence specialization of individual coastal communities, and, last but not least, to the varying human needs for whale products (generally, sea mammals) through time. Still, all three coastal areas surveyed in 1989–1992 demonstrated a significant and almost simultaneous increase in local bowhead whale hunting that took place beginning at about 1500–1300 B.P.

Long-term (Secular) Dynamics of Whale Populations near the Ekven Site

It should be noted that the data cited above illustrate the long-term dynamics of the Native whale catch,[3] but not of the whale population under human exploitation. In order to determine the trends in the whale population(s), some additional methods are used. Various modifications of the method first introduced by DeLury (1947) are commonly applied to estimate the size of the exploited populations of vertebrates. They are widely used in analyzing the status of modern cetacean and pinniped stocks and for the establishment of catch quotas (Allen 1966; Bockstoce and Botkin 1983; Allen and Kirkwood 1988). According to the modification of DeLury's method suggested by Chapman (1974), the average number of animals (N) during a certain period (t), the value of their commercial catch (C_t), and the hunting effort (f) employed in securing this catch are related by the following equation:

$N_t = C_t/qf_t$, where q is the catch efficiency whose value is determined by the nature of the hunting methods used.[4]

The ratio C_t/f_t in this equation estimates the amount of catch per unit hunting effort or, in other words, the efficiency of the harvest technology. In modern ecological literature, it is commonly designated by the abbreviation *CPUE* (Catch per Unit Effort). Under a stable level of catch efficiency q, the *CPUE* value changes proportionally to the abundance of the exploited population, irrespective of its recruitment, its natural mortality and its losses through hunting (Ricker 1975). Thus, it may well serve as a relative indicator of the abundance of the exploited population.

Obviously, the reconstruction of long-term changes in the relative abundance of hunted animals by use of osteological data collected at ancient sites turns into a (cumulative) estimate of the amount of actual catch, of hunting efforts employed, and of catch efficiency of prehistoric hunting teams. A credible way to approach this complicated task was considered elsewhere (see Dinesman 1996). In the case of prehistoric whaling, the amount of bones of animals (*Nbs*) accumulated during the period t at a certain site may serve as a relative indicator of the number of animals killed by ancient whalers during this period. This number may be calculated

---

[3] The data presented illustrate the trends in whale bone site accumulation; it may reflect both the active hunting for whales and the collection (or recycling) of beached whale bone—*ed.*

[4] The 'catch efficiency' of a certain hunting method is the ratio of the number of animals procured under this method versus all expenditure (efforts) in attempts at detecting and harvesting these animals off the exploited population.

fairly easily whenever a sufficiently representative sample of whale bone with adequate radiocarbon dating is available.

The 'hunting effort' is commonly understood as the combination of a certain number of hunting gear and of the duration of its use during the hunting season. For example, in modern modeling of catch quotas for marine mammal stocks, 'hunting efforts' used in obtaining the catch are usually estimated by the numbers, the size (displacement), and the overall term of activities of the hunting vessels, or by the time used in search for the animals. In modern fisheries and and/or commercial hunting, the hunting effort is commonly measured through the number of fishermen engaged in fishing, number of hunting licenses sold, number of animals killed per one rifle, or by the overall number of 'rifle–days' during a given hunting season (Ricker 1975; Caughley 1977). It is easy to note that these and similar methods take into account both the level of human hunting activity (i.e., number of people exploiting a certain game stock within a given territory; the number of fishing vessels, etc.), and the technology used in this activity (size of vessels, type of equipment, etc.). It would be senseless to compare the catch of two teams, one armed with bows and arrows, the other with rifles with optical sights, or, likewise, of two hunting vessels of different displacement and different equipment. Therefore, in studies of modern game populations all hunting efforts are carefully standardized according to their technological or 'catch' efficiency.

It is hardly possible to use similar methods in estimating 'hunting efforts' of the prehistoric coastal communities, since most of the required data (parameters) are not available. Thus, the only reasonable strategy is to simplify the task by focusing on secular changes in relative size of the exploited game populations.

According to modern reconstruction, ancient sea mammal hunters of Chukotka live in small permanent villages (communities). These were relatively autonomous economic, social, and reproductive units, which exploited certain hunting areas, with well-known and defined boundaries (Krupnik 1993a:29-31). All adult males and teens capable of working were hunters. Thus, the level of the 'hunting activity' of each coastal community largely depended upon the total number of active hunters, i.e., males roughly between ages 15 and 60.

Judging from the annual subsistence cycle of the historical Yupik (Asiatic Eskimo) and Chukchi communities in Chukotka, their 'hunting activity' was highly seasonal in its focus and in the overall 'hunting effort' alike. Based upon Krupnik's description of the Siberian Yupik subsistence cycle in the early 1900s (Krupnik 1993a:73), winter (from December to March–April) was the main period for individual hunting of seals at breathing holes and along ice-cracks and leads. In spring (April–June), men were engaged primarily in collective hunting for bearded seals (April–May), walruses, and bowhead whales (April–June) from large skin boats. The short summer season (June–August) was dominated by more individualized fishing, bird hunting, and catching of small seals by netting, and shooting at the shore resting sites. In the Mechigmen Bay area, summer was also the main season for catching small gray whales feeding close to the shoreline. In the fall (September–November), hunters once again turned to collective pursuit of bowhead whales and walruses in large skin boats. Those biggest game animals were normally inaccessible to individual hunters and even to solitary hunting teams. Thus, whaling and walrus hunting, and the following processing of catch always required cooperation and simultaneous participation of all able-bodied residents of the settlement (Krupnik 1993a:31, 45).

Due to such clearly expressed seasonality in hunting activity, all or most of the males in the community were simultaneously engaged in hunting for one or other (often for several) game species, using highly seasonal hunting techniques specially devised for certain animals and/or environmental conditions. Obviously, in any given time (t), the number of able-bodied males may be used as a reliable though indirect indicator of both the level of hunting activity of the community and of the fluctuating level of 'hunting efforts' used by its members in harvesting each of the key game species hunted. Of course, this will be a fair generalization of the different hunting efforts and methods used, but their long-term dynamics should be manifested quite clearly.

The secular changes in the number and share of able-bodied males in the ancient community of Ekven have not been addressed in any previous study of the Ekven settlement. However, it is clear that they had to be related (correlated) to the changes in the overall number(s) of residents at the settlement. Therefore, any indicator of the changing overall population size of the Ekven prehistoric village would be suitable for the general estimate of its hunting activity in relative units.

Several methods have been suggested in modeling the structure and absolute numbers of ancient human populations based upon the study of the burial samples at the prehistoric cemeteries (Acsadi and Nemeskeri 1970; Ubelaker 1975). However, these methods normally require complete excavation of all graves at the cemetery site and a thorough identification of age and sex of the skeletons recovered. At the same time, the changing hunting activity of small ancient whaling settlements, like the Ekven site in Chukotka, may be estimated in relative units, without the complete investigation of nearby burial grounds.

Judging from the basic postulates of paledemography, the average size of the prehistoric human population (P_t) during the period t is directly proportional to the number of burials at the cemetery site (*Ngr$_t$*) and inversely proportional to the level of human mortality (*M$_t$*).

According to the model introduced by Krupnik (1993a:219-230), ancient indigenous communities in the High Arctic shared a special type of demographic structure and population dynamics adapted to the extreme conditions of their environment. It was characterized by high birthrates and overall capacity for rapid population growth (in the time of adequate food supply only—*ed.*). The population increase, however, was limited by high mortality of both children and adults, and by a relatively short life span of able-bodied adults and elders. Even in the early 1900s, death rates in the Siberian Yupik communities in Chukotka, particularly for adults, could surge dramatically during the years of famines and epidemics. Of course, in the long run, such extreme and short-term aberrations in the community age-sex structure would be leveled off. Thus, if the duration of the period *t* selected for our estimate of the community hunting activity surpassed by several times the average life span of one reproductive generation (i.e., about 20–25 years), such short-term fluctuations in adult mortality could be all but overlooked. As a result, the number of burials accumulated at the village burial ground during certain time intervals may be used as a rough measure of the potential hunting activity of the community. This parameter can be best measured in person-year(s). It reflects the total pressure of the human community on the exploited game stocks expressed in the 'hunting efforts' of male hunters. Thus, one could simply resort to the analysis of temporal (long-term) distribution of a sufficiently representative random sample of burials at the cemetery site as a useful and practical approach, if backed by adequate radiocarbon dating of a relatively large number of burials.

By using the same time intervals in our assessment of catch unit(s) of the exploited species (animals/years) and of the human hunting activity invested to secure the game (persons/years), it is possible to estimate the amount (volume) of catch per unit effort (*CPUE*) from the historical skeletal remains as:

$CPUEbs_t = Nbs_t/Ngr_t$., where the index is expressed in relative units (animals/persons).

The dynamics of prehistoric catches of both bowhead and gray whales by the ancient hunters at the Ekven site for the period 2500–500 B.P. (measured by 200–year intervals) was considered above (Fig. 8). Accordingly, one may restore the changes in the relative numbers of residents of the Ekven prehistoric village based on radiocarbon dating of burials at the Ekven cemetery. The cemetery is located some 800 m further inland from the Ekven site, slightly above it on the coastal slope. Archaeologists argue that both sites were established simultaneously and were built and used by the same community (Arutyunov and Sergeev 1975). The Ekven burial ground is split into two major sections, with the eastern and western hills separated by a narrow depression. Excavation at the eastern hill was started in 1961–1965 by a team headed by Dorian Sergeev and Sergei Arutyunov (Arutyunov and Sergeev 1975). Since 1987, the burials on the western hill have been excavated by a team led by Michael Bronshtein (1991) and Cyrill Dneprovskii. Altogether, it was possible to provide radiocarbon dates for 27 excavated burials on the western hill-based bone collagen. Previously, two burials were dated by radiocarbon at the Smithsonian Institution in Washington, D.C.: one from the eastern and one from the western hill (Bronshtein 1986). To compare the dating, we checked the same grave on the western hill and received almost the same date as the one obtained at the Smithsonian.

In addition, we tested historical whale bone excavated from graves on the eastern hill and left *in situ* by Arutyunov and Sergeev in the 1960s and 1970s. Twelve radiocarbon samples were obtained from these historical bones in 1989 (Dinesman 1996). The dates shed light on the time of burials at this section of the Ekven prehistoric cemetery. A potential recycling of whale bones in graves from the abandoned ancient dwellings cannot be excluded (and it was surely practiced—*ed.*); but it may hardly distort the overall results. Altogether, we presently have a sample of 39 tested burials at the Ekven prehistoric cemetery dated either by human bones or by historical whale bone; this is, however, a small percentage (less than 10%—*ed.*) of the total number of graves excavated at the cemetery.

The soil cover on the burial hills is highly diverse, due to the presence of well-defined small patches of sand and rock debris. Ancient residents clearly preferred to use sandy plots for burial, as digging the graves there was much easier (M. Bronshtein, pers. comm.). Sites with heavier ground and/or rock debris were normally avoided or used quite rarely. This irregular combination of plots that were more or less suitable for burial produced highly peculiar and geometrically random use of the overall cemetery area. However, the archaeologists (Sergeev and Arutyunov in the 1960s and the 1970s, and Bronshtein's team after 1987) excavated the cemetery field systematically, along sequential squares. This pattern made preferential selection of the excavated graves by archaeologists according to the time of burial logistically and technically difficult. Our statistical evaluation of the radiocarbon dates obtained from the burials (see Dinesman 1996) confirmed that the grave sample was indeed obtained by the archaeologists by a more or less random procedure.

The earliest burials at the Ekven prehistoric cemetery were made at 2300–2100 B.P., as seen from the distribution of the available radiocarbon dates (Fig. 9). This dating is very close to that obtained from the oldest bone remains of the bowhead whales and walruses at the nearby Ekven settlement. However, the oldest evidences of the successful gray whale hunting at Ekven appeared somewhat earlier (Fig. 8). Thus we may argue that the first inhabitants of the Ekven settlement either did not bury the dead on the nearby hills (later to become the long-term village grave site) or that the earliest graves have not yet been discovered. The dates for the latest (most recent) remains of whales and walruses at the Ekven settlement and of the latest burials at the Ekven burial ground almost coincide (Table 2). Obviously, the use of both the village and the graveyard by the prehistoric residents ceased almost simultaneously. Therefore, we may rely on the skeletal remains from the Ekven burial ground and on the long-term dynamics of the number of burials to estimate the change(s) in hunting activity of the residents of Ekven during almost the entire time of the site's existence.

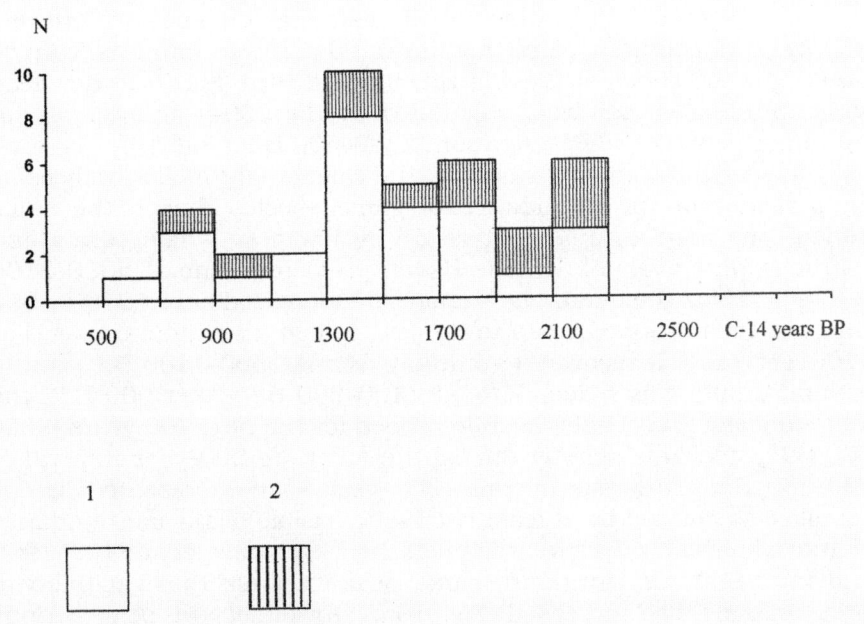

Figure 9. *Distribution of the radiocarbon dates of burials at the Western (1) and Eastern (2) Burial hills at the Ekven Burial site.*

If we are to measure the level of potential hunting activity by the number of burials by 200-year intervals (Fig. 9), the fluctuations were quite remarkable during some 1800 years of the site occupation (2300–500 B.P.). During the first 800 years (from the end of the first millennium BC until roughly AD 500), the level of hunting activity was relatively stable and rather low. It surged about 1500–1300 B.P., and then sharply decreased about 1300–1100 B.P., to remain again quite low and fairly stable until about 500 B.P. (A.D. 1400). Then, the village evidently was abandoned by its residents and no more burials were made at the cemetery hills.

Table 2. Margin radiocarbon dates of burials from the Ekven burial ground and whale bones from the cultural layer and ruins of the Ekven settlement.

Object of dating	Number of dates	Radiocarbon age (years BP)	
		Maximum	Minimum
Bones from the Ekven settlement:			
Gray Whale	32	2505+/-129	575+/-139
Bowhead Whale	17	2166+/-68	450+/-83
Burials from the Ekven burial ground	39	2220+/-65	612+/-88

This scenario of the overall historical changes in hunting activity and potential catch volume of the Ekven hunters can be backed by the series of the *CPUEbs* index for the gray and bowhead whale hunting. Those were calculated by the 200–year intervals for the entire period of site occupation (2300–500 B.P.; Fig. 10).

Taking the changing values of the calculated *CPUEbs$_t$* indices as an indicator of the abundance of game species during the same period, one may assume that relatively few gray whale calves and yearlings were available off the Ekven prehistoric site at 2300–2100 B.P. (Fig. 10a). Their numbers reportedly increased about 2100–1900 B.P. and then dropped again to the low, in fact, the minimum level at 1700–1500 B.P. It recovered gradually during 1500–1100 B.P., and a new maximum was reached about 1100–900 B.P. After 900 B.P. the abundance of gray whale juveniles ebbed for the next 200 years (900–700 B.P.), to surge slightly during the following 200–year interval of 700–500 B.P. Thus the overall 1800–year span represented in our sample was marked by at least two highly visible peaks in abundance of gray whale calves and yearlings off the Ekven village: at 2100–1900 and 1100–900 B.P. During the same 'peak intervals' the availability of older animals also increased (Fig. 10a). Obviously, the peak periods listed above were favorable for all age groups of the gray whales.

The long-term changes in the abundance of bowhead whale off the Ekven prehistoric village site can also be tested by the availability of young (juvenile) animals. Local hunters obviously were able to harvest small bowhead whales as early as the middle of the first millennium B.C. (Fig. 10b). Hunting for larger whales, however, did not start until 1300–1100 B.P., and any possible change(s) in their abundance can be evaluated after this time only. Bowhead whales were (are) available to local hunters on the Chukchi Peninsula during seasonal, spring and fall migrations *via* Bering Strait. Traditionally, most of the whales were killed in the fall when they approached the coast at close distance. Bowhead whale calves are born in February–May, and the lactation period extends for about a year (Tomilin 1957;

Sokolov and Arsen'ev 1994). Thus, the newborn calves and yearlings move northward in spring *via* Bering Strait accompanied by their mothers, since they are unable to survive independently. Here, the availability of whale sucklings and yearlings may be a good indicator to the presence of adult female bowhead whales participating in reproduction. Based on the *CPUEbs$_t$* values calculated for the 200–year intervals (Fig. 5b), the number of the bowhead whale juveniles and calves off the Ekven settlement was at the low level during 2300–1100 B.P. It surged and ebbed during 1100–500 years B.P. when the abundance of larger whales was also quite unstable (Fig. 10b).

Taking the changing values of the calculated *CPUEbs$_t$* indices as an indicator of the abundance of game species during the same period, one may assume that relatively few gray whale calves and yearlings were available off the Ekven prehistoric site at 2300–2100 B.P. (Fig. 10a). Their numbers reportedly increased about 2100–1900 B.P. and then dropped again to the low, in fact, the minimum level at 1700–1500 B.P. It recovered gradually during 1500–1100 B.P., and a new maximum was reached about 1100–900 B.P. After 900 B.P. the abundance of gray whale juveniles ebbed for the next 200 years (900–700 B.P.), to surge slightly during the following 200–year interval of 700–500 B.P. Thus, the overall 1800–year span represented in our sample was marked by at least two highly visible peaks in abundance of gray whale calves and yearlings off the Ekven village: at 2100–1900 and 1100–900 B.P. During the same 'peak intervals' the availability of older animals also increased (Fig. 10a). Obviously, the peak periods listed above were favorable for all age groups of the gray whales.

The long-term changes in the abundance of bowhead whale off the Ekven prehistoric village site can also be tested by the availability of young (juvenile) animals. Local hunters obviously were able to harvest small bowhead whales as early as the middle of the first millennium B.C. (Fig. 10b). Hunting for larger whales, however, did not start until 1300–1100 B.P., and any possible change(s) in their abundance can be evaluated after this time only. Bowhead whales were (are) available to local hunters on the Chukchi Peninsula during seasonal, spring and fall migrations *via* Bering Strait. Traditionally, most of the whales were killed in the fall when they approached the coast at close distance. Bowhead whale calves are born in February–May, and the lactation period extends for about a year (Tomilin 1957; Sokolov and Arsen'ev 1994). Thus, the newborn calves and yearlings move northward in spring *via* Bering Strait accompanied by their mothers, since they are unable to survive independently. Here, the availability of whale sucklings and yearlings may be a good indicator to the presence of adult female bowhead whales participating in reproduction. Based on the *CPUEbs$_t$* values calculated for the 200–year intervals (Fig. 5b), the number of the bowhead whale juveniles and calves off the Ekven settlement was at the low level during 2300–

1100 B.P. It surged and ebbed during 1100–500 years B.P. when the abundance of larger whales was also quite unstable (Fig. 10b).

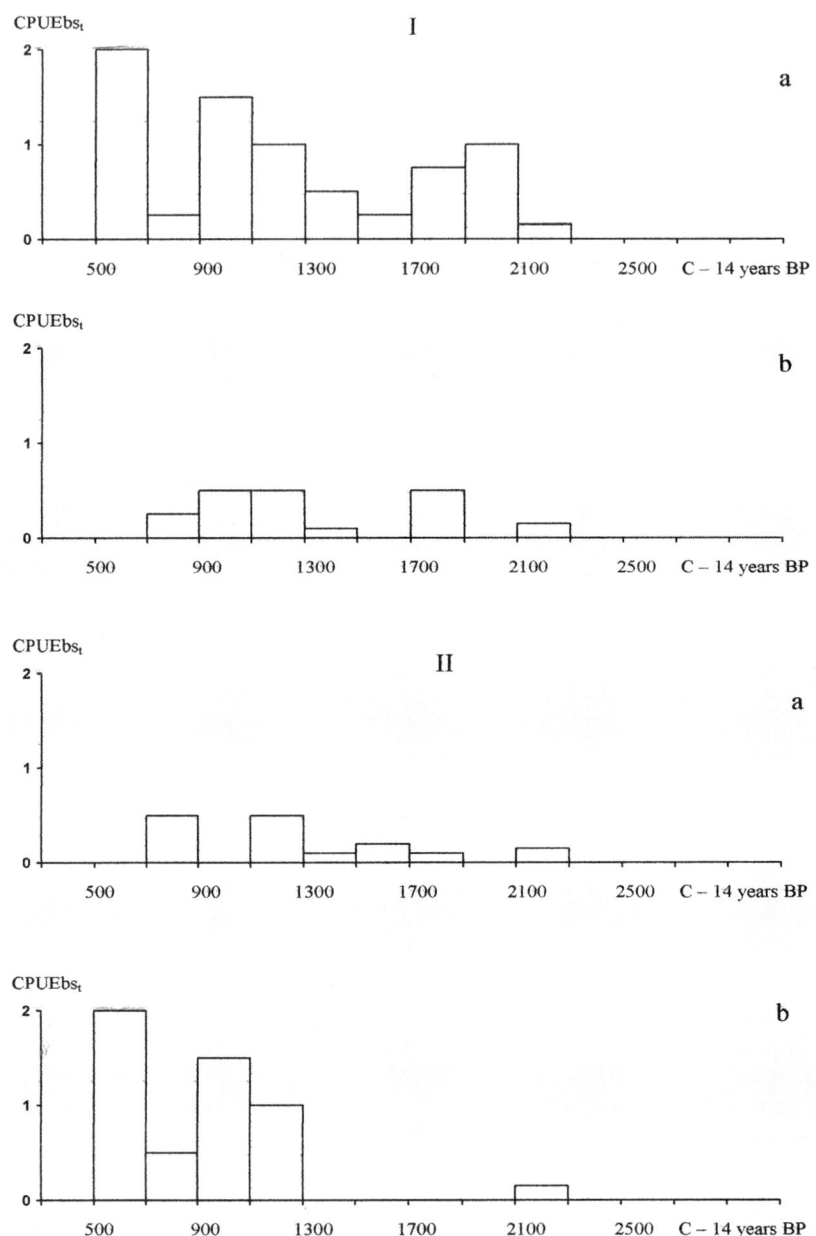

Figure 10. *Secular dynamics of the relative abundance of bowhead and gray whales off the Ekven ancient whaling site: I—gray whale, II—bowhead whale; (a) —yearlings and calves; (b) —animals older than one year.*

We may look for the explanation of such long-term dynamics in the available palaeoenvironmental record. About 1200 B.P., the summer ice cover of the surf and coastal zone in the Bering Strait area increased significantly, as judged by humus accumulation in beach deposits in the study area (Dinesman *et al.* 1996). Obviously, during this period the ice conditions in the Bering Strait region were quite favorable to the bowhead whale as well as to the walrus population. This pattern was sustained for the next several centuries, and it clearly coincided with the increase of bowhead whales off the Ekven hunting settlement (as seen from the bone accumulation at the Ekven site—*ed.*). Most probably, the increase in bowhead whale abundance was caused by the redistribution of animals across their usual range due to the change in ice conditions. However, the possibility of the overall 'population peak' (maximum) of the bowhead whale stock in the Chukchi and Bering Sea cannot be excluded.

Gray whales which spend summer months in the coastal waters off the Chukchi Peninsula normally prefer the areas where less than one-tenth of total water surface is covered with ice (Moore *et al.* 1986). However, they can visit areas with drifting ice and stay there in lanes and small patches of open water (Sleptsov 1955). Off the ancient Ekven whaling village, one secular peak in gray whale numbers (abundance) took place during the period of decreased ice coverage; the second peak, however, coincided with the time of increased summer ice about 1200 B.P. Thus, no clear relationship can be established so far between the summer abundance of gray whale and the sea–ice situation.

During the summer, the gray whales normally move to the coastal waters off the Chukchi Peninsula; their winter areas for feeding and reproduction are located far down south, in the coastal lagoons of southern California (Tomilin 1957). Without reliable information on the environmental history in this latter area, it is hardly possible to discuss the factors of secular dynamics in population number and whale distribution in the Bering Strait region.

CONCLUSIONS

Our data show that active coastal whaling by the prehistoric residents of Chukotka originated more than 2000 years ago. It was started along the Bering Strait shore of the Chukchi Peninsula about 2300–2100 B.P.; along the arctic (Chukchi Sea) shore it began about 1800–1700 B.P. Our analysis of historical whale bone tested at ancient coastal sites confirm the scenario that was presented earlier by Krupnik, Bogoslovskaya, and coauthors (Krupnik 1984; 1987; 1993b; Krupnik *et al.* 1983; Bogoslovskaya *et al.* 1982; Arutyunov *et al.* 1982). Ancient Chukotka whalers, indeed, hunted for both gray and bowhead whales; but the role as well as the abundance of each whale species, were subjected to remarkable secular changes. During 2300–1300 B.P. hunters targeted primarily (if not exclusively) calves and yearlings of both whale species. Later, they were able to kill larger (older) animals. Two historical peak periods of high abundance of gray whales can be established: at about 2300–1700 B.P., with a maximum at 2100–1900 B.P., and 1500–500 B.P., with a maximum at 700–500 B.P. The cause of such long-term fluctuations is not clear. During the whole period of 2300–1300 B.P., the local population (or availability?) of the bowhead whale was at a low level. The abundance of bowhead whales off Chukotka increased dramatically at about 1300 B.P., presumably, due to the heavier summer ice conditions in the Bering Strait. So far, we can trace this period of high abundance of bowhead whales until 500 years B.P. only, because of the subsequent abandonment of the Ekven whaling settlement.

The data obtained on the historical whale bone at the ancient Chukotka whaling sites illustrate that the long-range fluctuations in local whale stock size and abundance was a common phenomenon, at least during the last 2500 years. However, it is hard to determine whether such changes were local trends or whether they affected the overall whale population(s) in the Bering Strait region, until more data from other prehistoric sites and other coastal areas become available.

Acknowledgments. The study of the long-term ecological changes on the Chukchi Peninsula by environmental surveys at prehistoric whaling sites was supported by several agencies. Major grants were awarded under the research programs, 'Problems of the World Ocean, of the Arctic and Antarctic,' and 'Biodiversity' and by the Russian State Committee on Scientific and Technological Progress and by the Russian Foundation for Basic Research (Grants NN 97–04–48350, 00-04-48150, and 03-04-49323). During our field studies and the following discussions, we received continuous advice, criticism, and cooperation from several people. Contributions by Michael M. Bronshtein, Lyudmila S. Bogoslovskaya, and Igor Krupnik should be particularly acknowledged. The authors are grateful to Igor Krupnik and D. West for checking and editing the translated draft of the original Russian version. Special thanks go to Allen P. McCartney for his encouragement of our work on this summary paper, which made our study available to a much broader professional audience, and to two anonymous reviewers for their critical comments on the initial draft.

REFERENCES

Acsadi, G. and J. Nemeskeri
  1970   *History of Human Life Span and Mortality* Budapest: Academical Kiedo.
Allen, K.R.
  1966   Some Methods for Estimating Exploited Populations. *Journal of Fisheries Research Board of Canada* 23(10): 1553-1574.
Allen, K.R. and G.P. Kirkwood
  1988   'Man Impacts on Marine Mammals,' pp. 151-269 in *Fish Population Dynamics*. Chichester: John Wiley & Sons.
Arslanov, Kh.A., S.B. Chernov and T.V. Tertychnaya
  1987   'Kalibrovka vsesoyuznogo radiouglerodnogo standarta po standartu NBS i kol'tsam sosny 1846-1855 gg. (Calibration of the All-Union Radiocarbon Standard by the NBS Standard and Pine Annual Rings, 1846-1855),' pp. 254-255 in *Metody radioizotopnoi geologii (Methods in Radioisotope Geology): Tesisy dokladov Vesoyuznoi shkoly-seminara*, Part 2. Moscow: Nauka.
Arutyunov, Sergei A., Igor I. Krupnik, and Mikhail A. Chlenov
  1982   *Kitovaya alleya. Drevnosti ostrovov proliva Senyavina* (The Whale Alley. The Antiquities of the Senyavin Strait Islands). Moscow: Nauka
Arutyunov, Sergei A., and Dorian A. Sergeev
  1975   *Problemy etnicheskoi istorii Beringomorya (Ekvenskii mogilnik)* [Issues in Ethnic History of the Bering Sea Region (The Ekven Burial Site)]. Moscow: Nauka
Berg, L.S.
  1946   *Otkrytie Kamchatki i Ekspeditsii Beringa* (Discovery of Kamchatka and Bering's expeditions), 3rd Ed. Moscow-Leningrad (in Russian)
Bockstoce, John R. and Daniel B. Botkin
  1983   The Historical Status and Reduction of the Western Arctic Bowhead Whale Population by the Pelagic Whaling Industry, 1848-1914. *Report of the International Whaling Commission*, Special Issue 5: 107-146.
Bogoslovskaya, L.S., L.M. Votgorov, and I.I. Krupnik
  1982   The Bowhead Whale of Chukotka Migrations and Aboriginal Whaling. *Report of the International Whaling Commission* 32: 391-399.
Bogoslovskaya, Lyudmila S., Leonard M. Votrogov, and Tatyana N. Semenova
  1981   Distribution and Feeding of Gray Whales in Chukotka in the Summer and Autumn of 1980. *Report of the International Whaling Commission* 31: 507-510.
Bronshtein, Mikhail M.
  1986   Typologicheskie varianty drevneeskimosskogo graficheskogo ornamenta (Typological Variations of the Ancient Eskimo Graphic Ornamentation). *Sovetskaya etnografiia* 6: 46-58.
Bronshtein, M.M.
  1991   *Drevneeskimosskoe iskusstvo kak istoriko-etnograficheskii istochnik* (Ancient Eskimo art as an historical-ethnographical

source), Avtoreferat dissertatsii, cand. Hist. Sci. Moscow. (in Russian).

Caughley, G.
1977   *Analysis of Vertebrate Populations.* London: Wiley (Russian translation [1979] under the title: Analiz populyatsii pozvonochnykh). Moscow: Mir.

Chapman, D.G.
1974   Estimation of Population Size and Sustainable Yield of Sei Whales in the Antarctic. *Report of the International Whaling Commission* 24: 82-90.

Chlenov, Michael and Igor Krupnik
1984   Whale Alley. A Site on the Chukchi Peninsula, Siberia. *Expedition* 26(2): 6-15.

DeLury, D.B.
1947   On the Estimation of Biological Populations. *Biometrics* (5).

Dinesman, Lev G.
1996   Vekovaya dinamika chislennosti morzha, serogo kita i grenlandskogo kita na chukotskom poberezh'e Beringova proliva (Secular Dynamics of the Abundance of the Walrus, the Gray Whale, and the Bowhead Whale along the Chukotkan Coast of the Bering Strait) *Uspekhi sovremennoi biologii* 116 (2): 255-271.

Dinesman, L.G., N.K. Kiseleva, A.B. Savinetsky, and B.F. Khassanov
1996   *Vekovaya dinamika pribrezhnykh ekosistem severo-vostoka Chukotki* (Secular Dynamics of the Coastal Ecosystems of North-Eastern Chukotka). Moscow: Argus.

Dinesman, L.G., N.K. Kiseleva, A.B. Savinetsky, and B.F. Khassanov
1999   Secular Dynamics of Coastal Zone Ecosystems of the Northeastern Chukchi Peninsula. Chukotka: Cultural Layers and Natural Deposits of the Last Millennia. Tübingen: Mo Vince Verlag.

Gerlach C., J.C. George, and R. Suydam
1993   Bowhead Whale (*Balaena mysticetus*) Length Estimations Based on Scapula Measurements. *Arctic* 46(1): 55-59.

Knyazev, A.V. and A.B. Savinetsky
1995   Drevneeskimosskii promysel kitov na poberezh'e Chukotskogo morya i Beringova proliva (Ancient Eskimo Whaling along the Coast of the Chukchi Sea and the Bering Strait). *Bulleten' Moskovskogo obshchetva ispytatelei prirody* 100(3): 22-33.

Krupnik, Igor I.
1982   'Izmereniya cherepov grenlandskikh kitov s 'Kitovoi allei' (Measurements of the Bowhead Whale Crania from the 'Whale Allee),' pp. 165-167 in Arutyunov, S.A., Krupnik I.I., and M.A. Chlenov. *Kitovaya alleya. Drevnosti ostrovov proliva Seniavina.* Moscow: Nauka.

1984   'Gray Whales and the Aborigines of the Pacific Northwest: The History of Aboriginal Whaling,' pp. 103-120 in M.L. Jones, S.L. Swartz, and S. Leatherwood (eds.), *The Gray Whale, Eschichtius robustus.* Orlando: Academic Press

1987   Bowhead *vs.* the Gray Whale in Chukotkan Aboriginal Whaling. *Arctic* 40(1): 16-32.

1993a   *Arctic Adaptations. Native Whalers and Reindeer Herders of Northern Eurasia.* Hanover and London: University Press of New England.

1993b   Prehistoric Eskimo Whaling in the Arctic: Slaughter of Calves or Fortuitous Ecology. *Arctic Anthropology* 30(1): 1-12.

Krupnik, Igor I., Lyudmila S. Bogoslovskaya, and Leonard M. Votrogov

     1983   Gray Whaling off Chukotka Peninsula: Past and Present Status. *Report of the International Whaling Commission* 33: 557-562.

Leont'ev, V.V., and K.A. Novikova

     1989   *Toponimicheskii slovar' severo-vostoka SSSR* (Toponymic dictionary of the north-east USSR). Magadan. (in Russian)

McCartney, Allen P.

     1976   *Study of Archeological Whale Bones for Reconstructions of Canadian Arctic Bowhead Whale Stocks and Use by Prehistoric Inuit.* Preliminary Report, Northern Environmental Protection. Ottawa: Department of Indian and Northern Affairs.

Moore, S.E. and T.J. Clarke

     1983   A Comparison of Gray Whale and Bowhead Whale Distribution, Abundance, Habitat Preference and Behavior in the Northeastern Chukchi Sea, 1982-1984 *Report of the International Whaling Commission* 36: 273-279.

Moore, S.E., T.J. Clarke and D.K. Ljungblad

     1986   A Comparison of Gray Whale and Bowhead Whale Distribution, Abundance, Habitat Preference, and Behavior in the Northeastern Chukchi Sea, 1982-1984. *Report of the International Whaling Commission* 36: 273-279.

Ricker, W.E.

     1975   *Computation and Interpretation of Biological Statistics of Fish Populations.* Ottawa: Department of the Environment, Fisheries, and Marine Service.

Savelle, James M. and Allen P. McCartney

     1994   'Thule Inuit Bowhead Whaling: A Biometrical Analysis,' pp. 281-310 in D. Morrison and J-L. Pilon, eds., *Threads of Arctic Prehistory: Papers in Honour of William E. Taylor Jr.* Ottawa: Canadian Museum of Civilization, Mercury Series. Archaeological Survey of Canada Paper No. 149.

Sleptsov, Mikhail M.

     1955   *Biologiya i promysel kitov Dalnevostochnykh morei* (Biology and Catches of Whales in the Far Eastern seas). Moscow: Pishchepromizdat. (in Russian)

Sokolov, V.E. and V.A. Arsen'ev

     1993   *Usatye kity* (Whalebone whales). Moscow: Nauka. (in Russian)

Tomilin, Avenir G.

     1957   *Kitoobraznye (Cetacea).* Moscow: Akademiia Nauk USSR.

Ubelaker, Douglas H.

     1975   *Human Skeletal Remains.* Chicago: Aldine.

Yablokov, Alexey V. and Lyudmila S. Bogoslovskaya

     1984   'A Review of Russian Research on the Biology and Commercial Whaling of the Gray Whale,' pp. 465-486 in M.L. Jones, S.L Swartz and S. Leatherwood, eds., *The Gray Whale, Eschichtius robustus.* Orlando: Academic Press.

Zimushko, V.V.
  1970   K voprosu opredeleniya vozrasta serogo kita (To the Determination of Age of the Gray Whale). *Izvestiya Tikhookeaskogo instituta rybnogo khozyaistva i okeanografii* 71.

Prehistoric Bowhead Whaling in the Bering Strait and Chukchi Sea Regions of Alaska:
A Zooarchaeological Assessment

James M. Savelle and Allen P. McCartney

Abstract: *Zooarchaeological data on bowhead whales* (Balaena mysticetus) *are presented for 17 archaeological sites of different ages from northwestern Alaska. Using a measurement schedule for bowhead bones, measurements were taken on 1681 bones from a total bone assemblage representing at least 1294 individual animals. Bowheads dominate cetacean species found at these archaeological sites, but gray whales are represented in significant numbers at St. Lawrence island sites. The majority of these bones derived from juvenile (yearling) bowheads, which, based on regression analyses, measured 7–9 m in length when alive. This pattern of small whale selection conforms to previously discovered patterns of small bowhead hunting by ancient Eskimos. A few bowhead mandibles are from large subadults and adults, and probably derive from stranded animals. Further, we find insufficient evidence to support the conclusion that Old Whaling culture was based on hunting of baleen whales.*

INTRODUCTION

There is a considerable literature dealing with prehistoric and historic Inuit whaling in the Bering Strait region (e.g., Collins 1937; Larsen and Rainey 1948; Ford 1959; Stanford 1976; Bogoslovskaya *et al.* 1982; Giddings and Anderson 1986; Sheehan 1985, 1997; various papers in McCartney 1995a, and various papers in this volume). Typically, these studies concentrated primarily upon identification of whaling technology and the development of local and regional culture histories, and, more recently, on various social and economic contexts within which whaling was embedded. However, attempts to investigate the zooarchaeology of prehistoric and historic Eskimo whaling in a systematic fashion are few (e.g., Gerlach *et al.* 1993; Krupnik 1993; Yarborough 1995).

In this paper, we contribute to the zooarchaeology of Alaskan whaling societies by summarizing multiple site data relating to bowhead whale harvesting and whale bone use by prehistoric and historic Eskimo whalers at various localities in the Bering Strait and Chukchi Sea regions, and provide preliminary interpretations. The present study follows on from similar research conducted by us in the Canadian Arctic on prehistoric Thule Eskimo bowhead whaling societies (see e.g., McCartney 1980; McCartney and Savelle 1993; Savelle and McCartney

1994, 1999), preliminary bowhead size selection assessments among prehistoric and historic Alaskan Eskimo societies (McCartney 1995b), and allied zooarchaeological assessments of gray whale selection and use (Savelle and McCartney 2002a).

STUDY AREA AND FIELD DATA

A total of 17 coastal sites or site clusters were visited in 1996 and 1998 (Fig. 1). The type and amount of data collected at each site varied, depending upon logistical considerations. At Nuwuk, Birnirk, Utqiagvik, Walakpa, Nunagiak, Cape Krusenstern, and Point Hope, all surface whale bone was identified and counted, and measurements taken on crania, mandibles, scapulae, and cervical vertebrae, using a standard medical anthropometer. At the Point Franklin and Wales sites, data collection was restricted to measurable crania, mandibles, scapulae, and cervical vertebrae only. At the Gambell sites of Old Gambell, Siglugaghyaget, Ayveghyaget, Mayughwaaq, and Muruktah, investigations were restricted to examining measurable crania and mandibles only. At most sites, bone elements were typically incorporated as structural materials within the features, although occasionally they occurred as isolated elements scattered around the sites. Note that we do not include in this study whale bones that, based on provenience and/or information provided by local residents or guides, could be attributed to recent whaling activities.

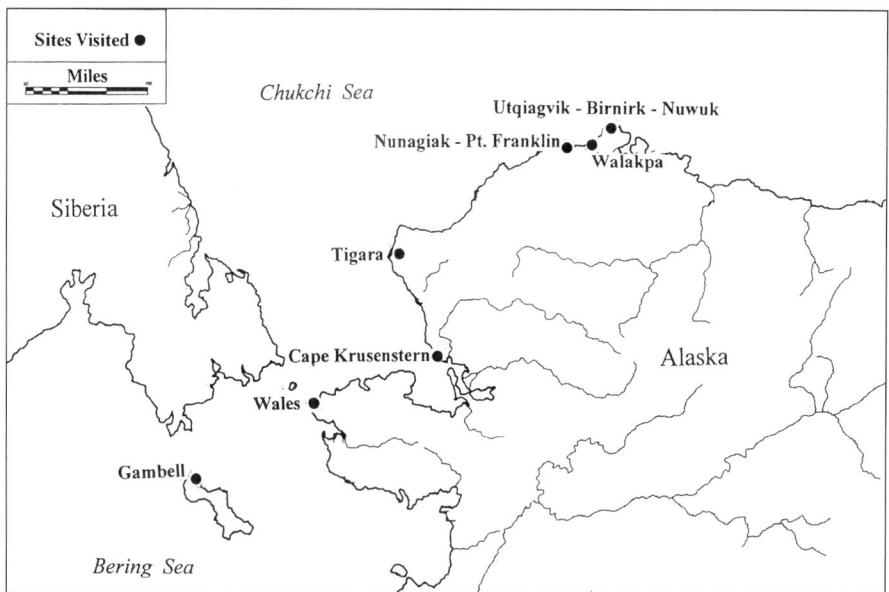

Figure 1. *Sites at which whale bones were recorded in 1996 and 1998. From Savelle and McCartney 2002a.*

The 17 sites cover a large temporal portion of the cultures that comprise the Northern Maritime tradition (or the Thule tradition) that begins at approximately A.D. 1 or contemporary cultures (we treat the old whaling culture separately below). They range in their whaling affiliation between strongly associated (such as the St. Lawrence Island sites and Tigara) to not associated (Ipiutak). These cultures include at various locations the Okvik/Old Bering Sea, Punuk, Birnirk, Western Thule, and Ipiutak cultures or combinations thereof. These range in age from two thousand years B.P. (such as Okvik/Old Bering Sea) to the 19th century (Tigara; for an overview of Western Arctic radiocarbon dates, see Gerlach and Mason 1992).

The measurements taken on the various bone types were based on a measurement schedule originally designed for bowhead whales by Dr. Edward Mitchell, then of the Department of Fisheries and Oceans, Canada. Osteometric regression models for bowheads using these measurement schedules were then developed by McCartney (1980) and subsequently modified by McCartney *et al.* (n.d.); we use the latter models in this paper.

RESULTS

Minimum Number of Bowheads Represented

Minimum number of individual animals (MNI) at each site are summarized in Table 1. MNIs are based on site-level bone totals to minimize the effects of aggregation (see e.g., Grayson 1984). Note that in several instances, mandibles were originally recorded as 'unidentified.' These were typically mandible mid-shaft fragments, but were identical in all respects to bowhead mandibles, and have thus been assigned to the bowhead category. The total MNI for all sites examined is 1294. It should be stressed that these are absolute minimums for the sites visited. As noted above, bone counts and measurements were restricted at many of the sites visited, and no attempt has been made to factor in the effects of the lack of visibility of subsurface bones, subsequent bone removal following site abandonment, and site erosion; an absolute MNI 10 times that recorded by us is certainly well within the possible range for original bowhead totals.

We also recorded bone elements for other species at the various sites. The total gray whale MNI for the regions examined is 88, with the majority of these recorded in the Gambell area of St. Lawrence Island (Table 1; see Savelle and McCartney 2002a). Otherwise, one minke whale mandible, and three right mandibles and a cranium from unidentified species were recorded in the Gambell area.

Composition of Bone Assemblage

A total of 2844 bowhead bones were recorded during the investigations at the various sites, and these are summarized by site in Table 1. Note that these totals do not include small fragments (primarily from maxillae and ribs). Savelle and McCartney (2002b) have recently investigated in detail these and Canadian Thule bowhead assemblages in the context of architectural utility, to which the reader is referred for a much more detailed discussion. Assemblages from sites or site areas for which bone counts were complete are combined and converted to minimal animal units (MAUs) in Table 2 (see Lyman 1994 for a discussion of the procedures and rationale for the use of MAUs in zooarchaeological analyses).

Table 2. Minimal animal units (MAUs) for all bowhead sites combined.

	Site	Crania	mand	maxillae	cerv	vert	ribs	scap
1	Nuwuk	78.00	24.00	3.00	.00	.00	3.00	.00
2	Birnirk	58.00	47.00	12.00	2.00	10.00	10.00	2.00
3	Utqiagvi	19.00	78.00	7.00	.00	4.00	22.00	2.00
4	Walakpa	4.00	1.00	.00	.00	.00	.00	.00
5	Nunagiak	22.00	96.00	13.00	2.00	8.00	1.00	3.00
6	Tigara	3.00	1769.00	6.00	.00	1.00	25.00	55.00
7	Ipiutak	.00	6.00	.00	.00	5.00	.00	.00
8	CK-PWC	1.00	.00	.00	.00	4.00	.00	.00
9	CK-Thule	.00	5.00	1.00	.00	11.00	3.00	.00
10	CK-Other	1.00	4.00	.00	.00	8.00	2.00	.00
11	TOTAL	115.00	2030.00	42.00	4.00	51.00	66.00	62.00
12	MAU	115.00	1015.00	10.50	4.00	1.06	2.54	31.00

For the present analysis, and following Savelle and McCartney (2002b), we assign all vertebrae to one category (with a final architectural utility rank of 5), and exclude flipper elements, sternum, and hyoid, since these are primarily food-related as opposed to architecturally-related when found in archaeological sites. Admittedly a 'coarse-grained' approach, the combined assemblage is nevertheless consistent with bone selection and use in a primarily architectural context. In general, the overall bone assemblage composition follows the expectations based on bone architectural utility (Savelle 1997). That is, those elements with the highest 'frame' utility (mandibles, maxillae, scapulae) and highest 'bulk' utility (crania, cervical vertebrae) predominate over other bone elements with lower utility values, and there is a strong positive correlation between bone element MAU and architectural utility (r_s = 0.764, P = 0.046). The one notable exception is the site of Tigara, at which crania are highly underrepresented. This is consistent with the well-documented symbolic return of whale skulls to the sea following butchering (see e.g., Larsen and Rainey 1948; VanStone 1962; Worl 1980; Lowenstein 1993).

Table 1. Bowhead whale bone elements recorded in 1996 and 1998, and derived MNIs. Numbers in brackets indicate the total of each bone type measured at each site.

	Site	Crania	Mandleft	Mandrigh	Mandns	Maxillae	Cerv	Vert	Ribs	Scapleft	Scaprigh	Scans	Humerus	Ulnarad	Other	MNI
1	Nuwuk	7(6)	9(6)	10(9)	5	3			3							12
2																
3	Birnirk	58(39)	15(14)	15(7)	17	12	2(1)	10	10	2(2)				2	1 chevro	58
4																
5	Utqiagvik	19(9)	24(14)	19(14)	35	7		4	22	1(1)	1(1)	1				39
6																
7	Walakpa	4(4)		1(1)												4
8																
9	Point Franklin	4(4)	2(2)	4(4)			1(1)			2(2)						4
10																
11	Nunagiak	22(19)	28(19)	23(19)	45(1)	13	2(2)	8	1	2(2)	1(1)	1		1		48
12																
13	Tigara	3(1)	713(513)	696(513)	360	6		1	25	14(10)	17(16)	24	2			885
14	Ipiutak/Nearlpiutak		2(2)	3(3)	1			5						5		3
15																
16	Wales															
17	Kurigitavik		3(3)	6(6)	1						4(4)					6
18	Hillside	1(1)	33(33)	24(24)	1(1)					12(12)	4(4)					33
19	Beach	1	22(22)	32(32)						3(3)	1(1)					32
20																
21	Cape Krusenstern															
22	OWC	1(1)						4								1
23	Western Thule		1	2(1)	2	1		11	3				1	1		3
24	Other	1(1)	2(1)	2(1)				8	2				1			2
25																
26	Gambell															
27	Old Gambell	52(52)	29(21)	21(20)	9											52
28	Siglugaghyaget	69(67)	20(20)	19(19)	1(1)											69
29	Ayveghyaget	20(20)	8(6)	2(1)	1											20
30	Mayughwaaq	16(16)	7(7)	7(7)	2											16
31	Muruktuh	2(2)	7(7)	5(3)	2											7
32																
33	TOTALS	280(242)	925(690)	891(683)	482(3)	42	5(4)	49	66	36(32)	28(27)	26	4	9	1	1294

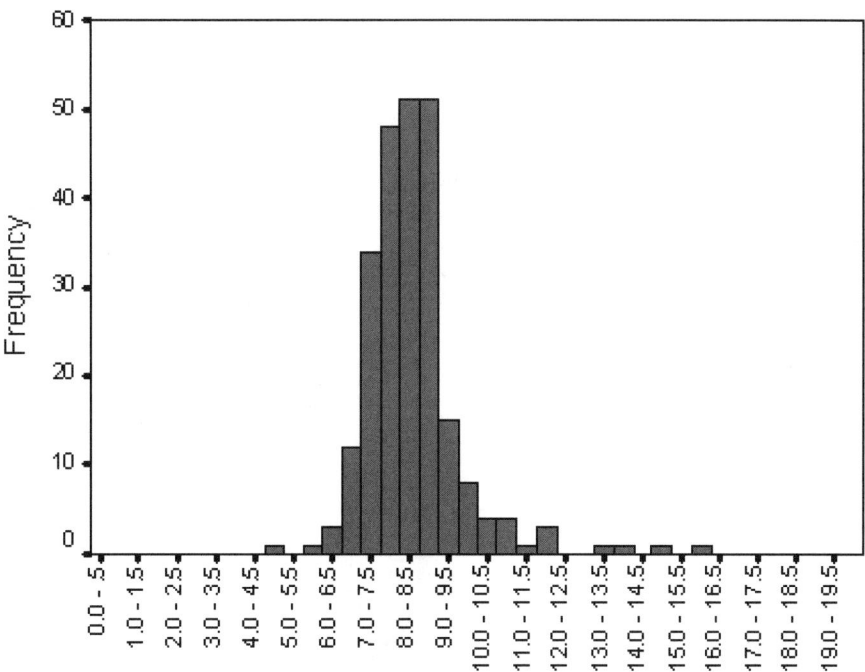

Figure 2. *Estimated bowhead whale lengths based on crania, all sites combined. Whale length (metres) N = 242.*

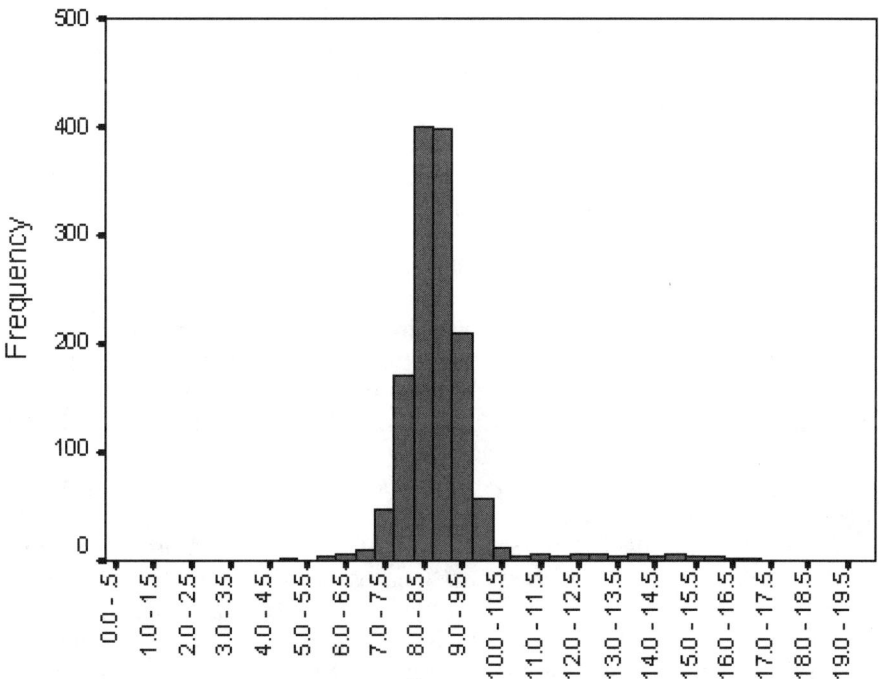

Figure 3. *Estimated bowhead whale lengths based on mandibles, all sites combined. Whale length (metres) N = 1376.*

Size Selection of Bowheads Represented

From the total assemblages, measurements appropriate for original live length determinations were obtained on 242 crania, 1376 mandibles, 59 scapulae, and four cervical vertebrae. Individual measured element numbers for each site are presented in Table 1. In this paper, we deal with mandibles and crania only in summarizing the biometrical data, since these elements far outnumber scapulae and cervical vertebrae at essentially every site (see Table 1). Length estimates based on the biometric analyses for crania and mandibles for all sites combined are summarized in Figures 2 and 3.

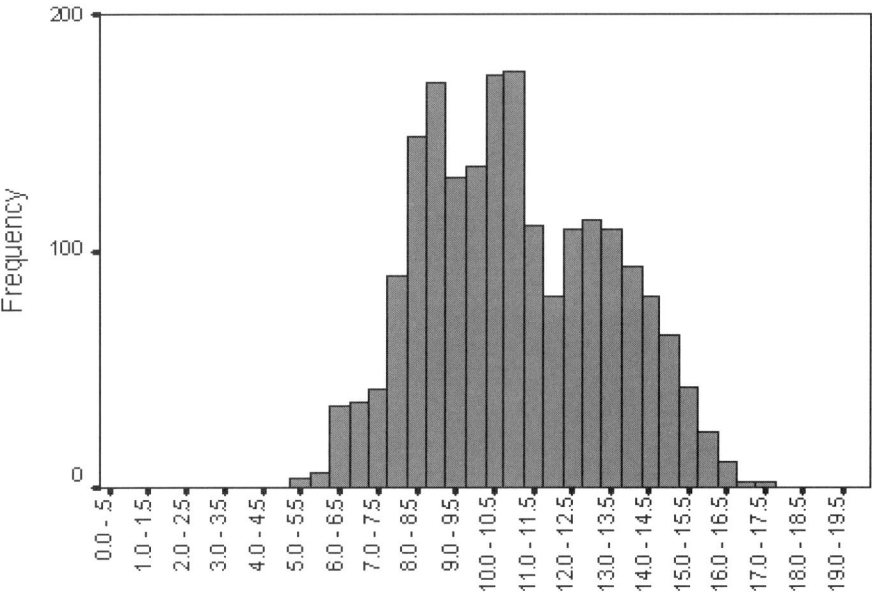

Figure 4. *Estimated whale lengths of live bowheads measured photogrammetrically in the Chukchi and Beaufort seas, 1981-1986 (estimated figures from Koski et al. 1988). Whale length (metres) N = 1988*

With very few exceptions (see below), the vast majority of bowhead whales represented at the various sites measured in the 7.0–9.0 m range, whether based on length estimates derived from mandibles or crania, and these size profiles differ markedly from a living population profile (Fig. 4). This is the general size range of yearling bowheads (see e.g., Nerini *et al.* 1984; Breiwick *et al.* 1984; Koski *et al.* 1993), although as demonstrated by Schell and Saupe (1993), the upper range may also include older juveniles. The Alaska size selection pattern is very similar to that of Thule Eskimo bowhead whales identified in the Canadian Arctic (see e.g., McCartney 1980; Savelle and McCartney 1994). Following our interpretation of the Canadian Thule size selection patterns, and in confirmation of McCartney's (1995b) initial compilation of archaeological and ethnographic data for Alaska, we suggest that these results indicate a definite selection for smaller whales, especially yearlings, in contrast to random hunting strategies, wherein animals would be harvested from a living population in direct proportion to their abundance. Further, we suggest that the pursuit of older (or at least larger) bowheads was rarely practiced because of the danger in hunting and/or the difficulty in processing these larger animals, while the pursuit of calves (4–6 m long animals) may have been especially dangerous given their close physical association with their mothers.

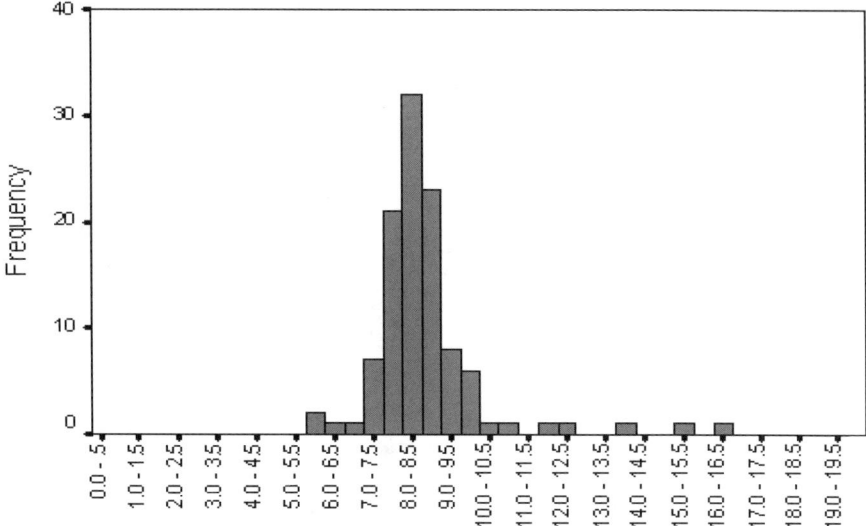

Figure 5. *Estimated bowhead whale lengths based on mandibles, Barrow area sites. Whale length (metres) N = 108.*

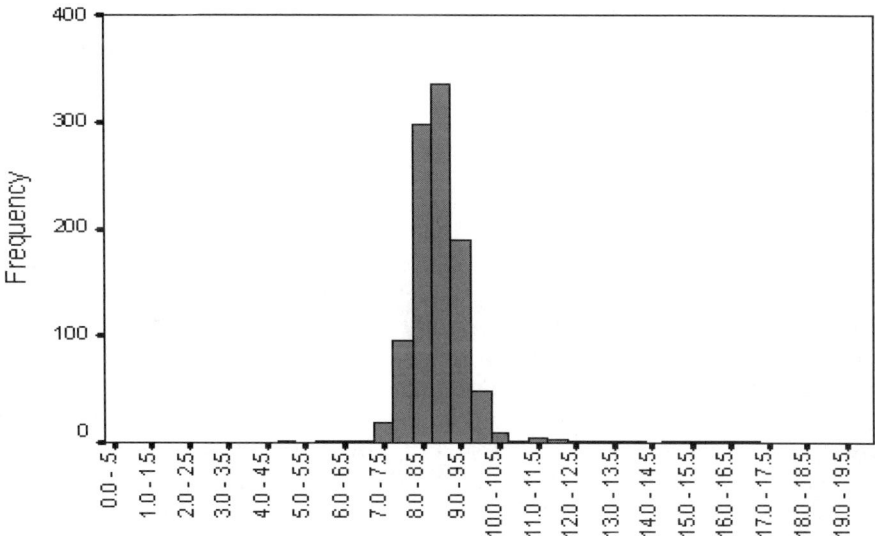

Figure 6. *Estimated bowhead whale lengths based on mandibles, Point Hope area sites. Whale length (metres) N = 1031*

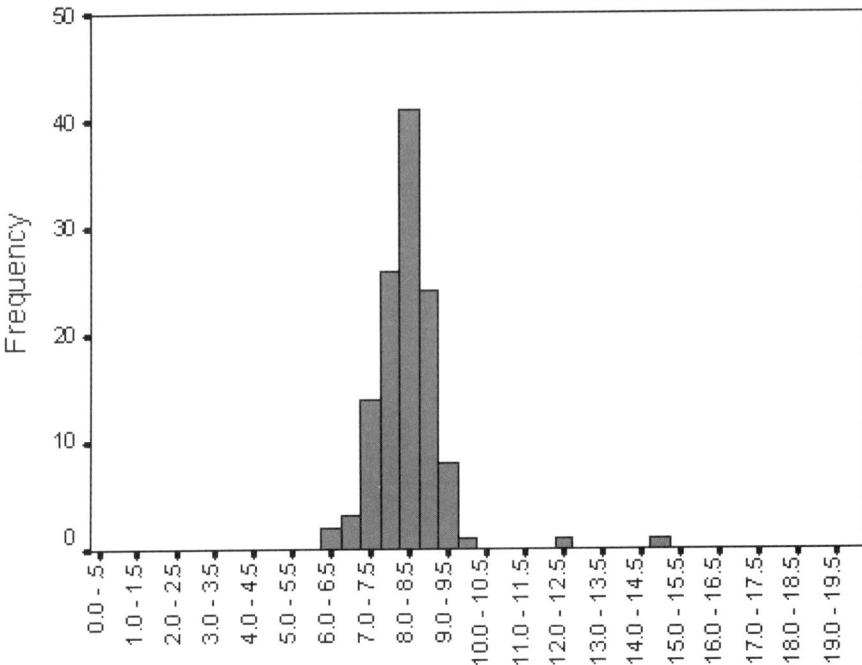

Figure 7. *Estimated bowhead whale lengths based on mandibles, Wales area sites. Whale length (metres) N = 121.*

When the length estimates are subdivided into different regions, however, local variations to the pattern of yearling selection become apparent. In the three major northern site groupings—Barrow area (Nuwuk, Birnirk, Utqiagvik, Walakpa, Point Franklin, Nunagiak), Point Hope area (Tigara, including surface material in the vicinity of the Ipiutak and Near Ipiutak sites), and Wales area (Kurigitavik, Hillside, Beach)—the yearling pattern generally holds for both mandibles (Figs. 5–7) and crania (Fig. 8), although the occasional larger subadult and adult whale is represented. Note that we do not present histograms for the Wales and Point Hope crania, since each of these measured samples consist of one element only; derived length for the Wales cranium is 6.68 m, and for the Point Hope cranium 7.45 m. In the Gambell area sites (Old Gambell, Siglugaghyaget, Ayveghyaget, Mayughwaaq, and Muruktah), on the other hand, while the crania follow the general yearling selection pattern (Fig. 9), there is a very definite secondary peak in the 12–15 m size range based on mandibles (Fig. 10). Note that we also recorded 26 'large' (probably in the 12–15 m original length range) bowhead mandibles for which 'minimum' value measurements only could be obtained at these sites, but which are not included in the histograms. Because there is not a gradual fall-off from the smaller size categories to the 12–15 m-long bowheads, we suggest that these larger mandibles were probably derived from naturally-stranded carcasses, with the mandibles only being transported back to the sites for construction purposes. The lack of an equivalent peak for crania would suggest that these elements from bowheads in the 12 m or larger range are probably too unmanageable for transportation and construction purposes (see e.g., Savelle 1997). This explanation can also be applied to the occasional large bowhead bones found at sites further north.

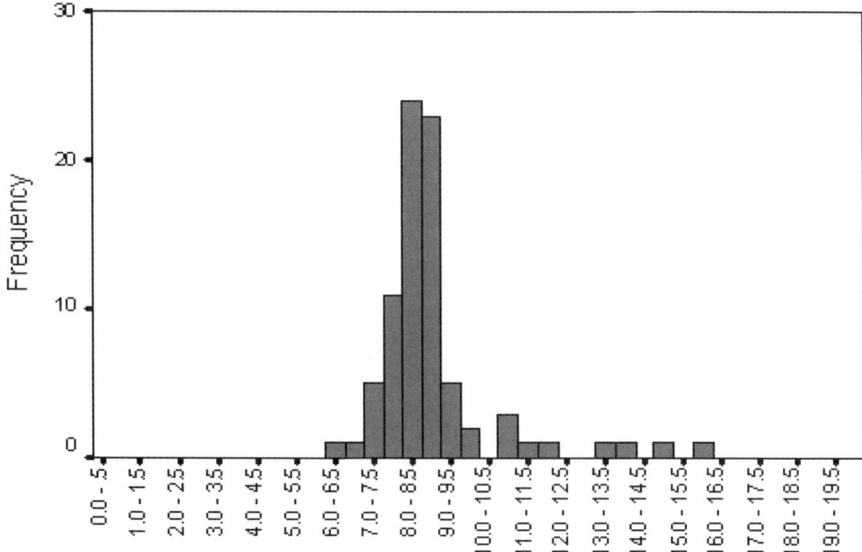

Figure 8. *Estimated bowhead whale lengths based on crania, Barrow area sites. Whale length (metres) N = 81*

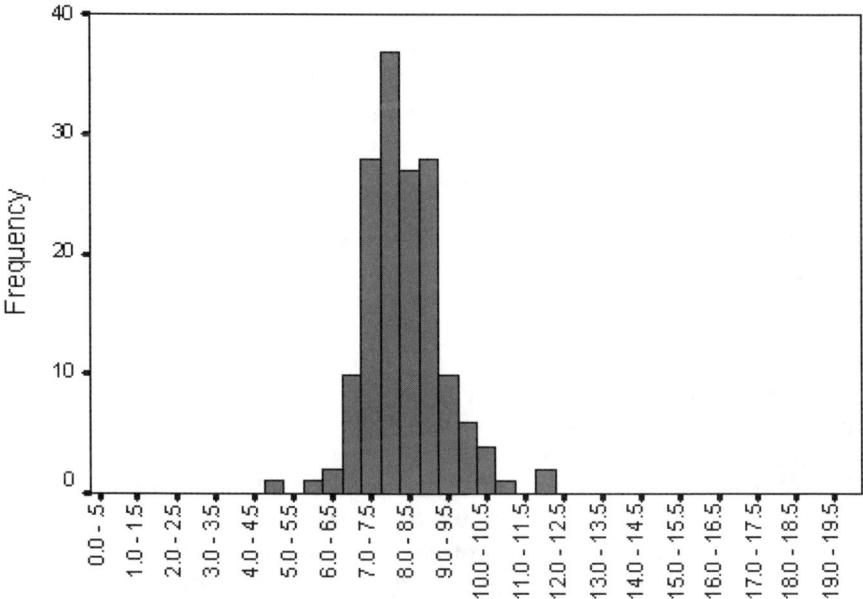

Figure 9. *Estimated bowhead whale lengths based on cramia, Gambell area sites. Whale length (metres) N = 157.*

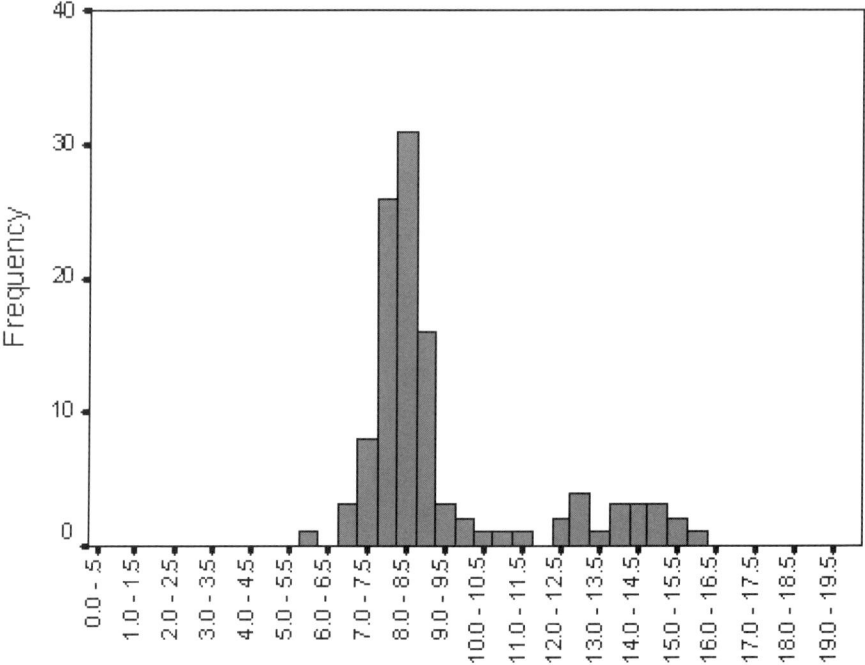

Figure 10. *Estimated bowhead whale lengths based on mandibles, Gambell area sites. Whale length (metres) N = 112.*

Bowhead Whaling at Cape Krusenstern

Given the importance of Cape Krusenstern in the history of study of the whaling origins in the Bering Sea/Chukchi Sea region, we deal with the data from sites in that area separately. While we had the opportunity to examine a number of sites on this series of beach ridges, we recognize that considerable disturbance, including bone removal, has taken place since the pioneering studies in this region by Giddings (1967) and Giddings and Anderson (1986). Furthermore, Mason and Gerlach (1995) and Ackerman (1998:252ff) have treated the issue in considerable detail, and we can do little more here than comment on whale bones recorded in the region and the presumed whaling pattern used there.

Whale bone elements were recorded at several features on the cape, and are listed in Table 1. Briefly, these include four worn vertebrae on the surface within several hundred meters on either side of the Old Whaling culture dwellings and one bowhead cranium adjacent to the Old Whaling culture site at beach 53 (the latter derived length is 7.95 m). No other whale bones were recorded anywhere near the site, and there was no evidence of storage pits. Given the very thin (2–3 cm) vegetation cover in the area, it is unlikely that either would not have been visible. Otherwise, one isolated cranium fragment (species unidentified) was noted about 1 km from the Old Whaling site, apparently in secondary context, and an isolated gray whale cranium (derived original length is 8.51 m) was recorded approximately 100 m from the modern beach.

Elsewhere at Cape Krusenstern, we recorded whale bones at several localities. Within a large pit dwelling (?) identified by a '1950' survey marker, two gray whale crania, one bowhead cranium, four bowhead mandibles (two right, two left), two ribs, five large (adult?), and three smaller vertebrae were recorded. Derived lengths for the bowhead crania is 9.63 m, for two measurable bowhead mandibles, 13.99 and 9.51 m long, and for the two gray whale crania, 8.65 and 9.03 m long.

At the excavated Western Thule dwellings H–25 and H–29 on beach ridge 10, five bowhead mandibles (three right, two left), one maxilla, one humerus, three ribs, eleven vertebrae, and five ulnae and radii were recorded. The one measurable mandible has a derived length of 12.60 m, and we noted that all other bone elements appear to have been derived from 'large' bowheads. Finally, at Western Thule dwellings 4, 5, 6 and 7 on beach ridges 10–11, one gray whale scapula, one rib fragment (species unidentified), and one bowhead humerus from a 'large' individual were recorded. Several shallow, probable cache depressions were noted in association with the dwellings.

The Old Whaling culture was defined by Giddings (1961, 1967; Giddings and Anderson 1986) as a result of his 1958-1959 inspection of winter houses on beach ridge 53. The 'uncommon number of whale bones there' (Giddings 1967:224) and his assumption that large chipped blades were whale harpoon end blades, led him to identify these houses as those of early whalers. Briefly, there are a number of reasons why they are not. Bowheads migrate northward across outer Kotzebue Sound during the spring migration (Braham *et al.* 1980) but do not occupy the inner sound near Cape Krusenstern due to ice conditions, shallow water, or lack of copepods on which they feed. Rather, belugas are common here. Compared to sites of intense whaling, such as Thule sites in the Canadian Arctic, few whale bones mark ridge 53 and the winter houses are constructed of wood rather than whale bones (Giddings and Anderson 1986:231 ff). Giddings describes whale bones partly buried in gravel near the houses, and they may have been naturally deposited as early Holocene-age carcasses during the formation of the beach ridge, similar to stranded carcasses found in the Canadian Arctic (Dyke and Morris 1990; Savelle and McCartney 1994). Some of the bones observed and illustrated at beach ridge 53 (Giddings 1967:Fig. 91) are clearly of adult whales, indicating stranded rather than hunted whales (McCartney 1995b). As pointed out above, there are no large cache pits associated with the supposed Old Whaling houses that might accommodate the large amount of whale meat and blubber produced by whaling. It is debatable whether the large stone blades found by Giddings were actually whaling harpoon end blades, especially since no bone or ivory harpoon heads were found at the site (Giddings and Anderson 1996:258-259). The fact that no baleen was found at the Old Whaling houses, that no other Old Whaling sites have been recognized since Giddings' identification of the beach ridge 53 site, and that the dates of Old Whaling precede the rest of the Alaskan whaling tradition (Northern Maritime or Thule tradition) by 1500 years, suggest that the Old Whalers were not engaged in whaling.

Even among later Western Thule societies in this region, the relatively small number of whale bones recorded suggest that whaling was not nearly as intensive as at sites at other locales dating to this period, and the relatively high proportion of bone elements derived from large bowheads would suggest a high incidence of bone scavenging.

CONCLUSIONS

1. Bowheads dominate cetacean assemblages throughout the regions examined, although gray whales were also hunted in significant numbers in the St. Lawrence Island region.

2. Bone assemblage composition at sites for which we have reasonably complete or representative bone counts demonstrate a definite selection for elements of high architectural utility over those with lower architectural utility.

3. The vast majority of bowheads represented at the various sites derive from animals in the 7.0-9.0 m size range—that is, primarily yearlings. This is consistent with McCartney's (1995b) earlier discussions of Alaska bowhead selection, and is similar to yearling-dominated bowhead size profiles throughout prehistoric Thule whaling sites in the Canadian Arctic. We interpret this pattern as one indicative of specific selection for smaller bowheads.

4. A number of bowhead bones, primarily mandibles, at several sites derive from larger subadults and adults in the 12.0–15.0 m range. As opposed to a gradual falloff with increasing size, these mandibles form distinct secondary peaks in the size profiles. We interpret this to be the result of scavenging of mandibles from larger, stranded bowhead carcasses.

5. Evidence for whaling at the Old Whaling culture site at Cape Krusenstern is minimal, and we agree with the assessment by Mason and Gerlach (1995) and Ackerman (1998) that Old Whaling culture was not a well-established whaling culture. Furthermore, evidence for whaling by later cultures at Cape Krusenstern is equivocal at best, and what limited size data we have from these later sites suggest at least some scavenging of larger, stranded bowheads, and perhaps gray whales.

Acknowledgments. We wish to recognize the National Science Foundation, Office of Polar Programs for grant OPP-9807051, which supported the research reported here. This study is part of a larger collaborative project entitled 'Whale Hunting Societies in the Western Arctic: A Regional Integration' which focused on Alaskan whaling communities between 1995-2001. We also wish to acknowledge the assistance offered by Amanda Crandall and Julia Clark and the Department of Wildlife Management, North Slope Borough, during the 1996 field season in North Alaska. We are especially indebted to Conrad Ozeeva, Betty Smith, Luke Koonook, Andrew Tooyak, Jr., Herbert Foster, Roger Harritt, Glenn Sheehan, Craig George, Douglas and Kathie Veltre, Bob Gal, Steve Klingler, Chris Young, and John and Chris Darwent for their field assistance.

REFERENCES

Ackerman, Robert E.
  1998   Early Maritime Traditions in the Bering, Chukchi, and East
          Siberian Seas. *Arctic Anthropology* 35(1): 247-262
Bogoslovskaya, Lyudmila, L.M. Votrogov, and Igor I. Krupnik
  1982   The Bowhead Whale Off Chukotka: Migrations and Aboriginal
          Whaling. *Report of the International Whaling Commission* 32:
          391-399.
Braham, Howard W., Mark H. Fraker, and Bruce D. Krogman
  1980   Spring Migration of the Western Arctic Population of Bowhead
          Whales. *Marine Fisheries Review* 42(9-10): 36-46
Breiwick, J.M., L. Eberhardt, and Howard Braham
  1984   Population Dynamics of Western Arctic Bowhead Whales
          (*Balaena mysticetus*). *Canadian Journal of Fisheries and
          Aquatic Sciences* 41: 484-496.
Collins, Henry B, Jr.
  1937   *Archaeology of St. Lawrence Island, Alaska.* Washington, D.C.:
          Smithsonian Institution, Miscellaneous Collections 96(1).
Dyke, Arthur S. and Thomas F. Morris
  1990   *Postglacial History of the Bowhead Whale and of Driftwood
          Penetration: Implications for Paleoclimate, Central Canadian
          Arctic.* Ottawa: Geological Survey of Canada Paper No. 89-24.
Ford, James A.
  1959   *Eskimo Prehistory in the Vicinity of Point Barrow, Alaska.*
          Washington, D.C.: Anthropological Papers of the American
          Museum of Natural History 47(1).
Gerlach, Craig, John C. George, and Robert Suydam
  1993   Bowhead Whale (*Balaena mysticetus*) Length Estimations
          Baised on Scapula Measurements. *Arctic* 46(1): 55-59.
Gerlach, Craig and Owen K. Mason
  1992   Calibrated Radiocarbon Dates and Cultural Interactions in the
          Western Arctic. *Arctic Anthropology* 29(1): 54-81

Giddings, J. Louis
    1961   Cultural Continuities of Eskimos. *American Antiquity* 27(2):
            155-173
    1967   *Ancient Men of the Arctic.* New York: Alfred A. Knopf.
Giddings, J. Louis and Douglas D. Anderson
    1986   *Beach Ridge Archaeology of Cape Kruenstern: Eskimo and Pre-*
            Eskimo Settlements Around Kotzebue Sound, Alaska. National
            Parks Service, Publications in Archaeology No. 20.
Grayson, Donald
    1984   *Quantitative Zooarchaeology.* New York: Academic Press.
Koski, W., G. Miller and R. Davis
    1988   The Potential Effects of Tanker Traffic on the Bowhead Whale
            in the Beaufort Sea. Ottawa: Indian and Northern Affairs
            Canada, Environmental Studies 58.
Koski, W.R., R.A. Davis, G.W. Miller, and D.E. Withrow
    1993   'Reproduction,' pp. 239-274 in J.J. Burns, J.J. Montague, and
            C.J. Cowles, eds., *The Bowhead Whale.* Lawrence, KS: Allen
            Press, Society for Marine Mammalogy, Special Publication No.
            2.
Krupnik, Igor I.
    1993   Prehistoric Eskimo Whaling in the Arctic: Slaughter of Calves
            or Fortuitous Ecology? *Arctic Anthropology* 30(1): 1-12.
Larsen, Helge and Froelich Rainey
    1948   *Ipiutak and the Arctic Whale Hunting Culture.* Washington,
            D.C.: Anthropological Papers of the American Museum of
            Natural History 42.
Lowenstein, Tom
    1993   *Ancient Land: Sacred Whale, The Inuit Hunt and its Rituals.*
            New York: Farrar, Straus and Giroux.
Lyman, Richard L.
    1994   *Vertebrate Taphonomy.* Cambridge: Cambridge University
            Press.
Mason, Owen K. and S. Craig Gerlach
    1995   'The Archaeological Imagination, Zooarchaeological Data, the
            Origins of Whaling in the Western Arctic, and 'Old Whaling'
            and Choris Cultures,' pp. 1-31 in A.P. McCartney, ed., *Hunting*
            the Largest Animals: Native Whaling in the Western Arctic and
            Subarctic. Edmonton: Canadian Circumpolar Institute (CCI)
            Press, University of Alberta, Occasional Publication No. 36,
            Studies in Whaling No.3.
McCartney, Allen P.
    1980   *Study of Archaeological Whale Bones for the Reconstruction of*
            Canadian Arctic Bowhead Whale Stocks and Whale Use by
            Prehistoric Inuit. Final report prepared for the Northern
            Environmental Protection Branch, Department of Indian and
            Northern Affairs, Ottawa.
    1995a ed., *Hunting the Largest Animals: Native Whaling in the Western*
            Arctic and Subarctic. Edmonton: Canadian Circumpolar
            Institute (CCI) Press, University of Alberta, Occasional
            Publication No. 36, Studies in Whaling No. 3.

1995b 'Whale Size Selection by Precontact Hunters of the North American Western Arctic and Subarctic,' pp. 83-108 in A.P. McCartney, ed., *Hunting the Largest Animals: Native Whaling in the Western Arctic and Subarctic.* Edmonton: Canadian Circumpolar Institute (CCI) Press, University of Alberta, Occasional Publication No. 36, Studies in Whaling No.3.

McCartney, Allen P. and James M. Savelle.
1993 Bowhead Whale Bones and Thule Eskimo Subsistence-Settlement Patterns in the Central Canadian Arctic. *Polar Record* 29(168): 1-12.

McCartney, Allen P., James M. Savelle, Edward M. Mitchell, Melanie De L'Etoile and John C. George.
n.d. *The Measurement of Bowhead Whale (Balaena mysticetus) Bones at Thule Eskimo Sites in the Central Canadian Arctic.* Unpublished manuscript, 37p.

Nerini, Mary K., Howard W. Braham, William M. Marquette, and David J. Rugh
1984 Life History of the Bowhead Whale, *Balaena mysticetus* (Mammalia: Cetacea). *Journal of Zoology* 204: 443-468.

Savelle, James M.
1997 The Role of Architectural Utility in the Formation of Zooarchaeological Whale Bone Assemblages. *Journal of Archaeological Science* 24: 869-885.

Savelle, James M. and Allen P. McCartney
1994 'Thule Inuit Bowhead Whaling: A Biometrical Analysis,' pp. 281-310 D. Morrison and J.L. Pilon, eds., *Threads of Arctic Prehistory: Papers in Honour of William E. Taylor Jr.* Ottawa: Archaeological Survey of Canada, Canadian Museum of Civilization, Mercury Series, Paper 149.

1999 Thule Eskimo Bowhead Whale Interception Strategies. *World Archaeology* 30(3): 437-451.

2002a Prehistoric Gray Whale Harvesting in the Bering Strait and Chukchi Sea Regions of Alaska: A Zooarchaeological Assessment. *University of Oregon Anthropological Papers* 59: 307-318.

2002b The Application of Bowhead Whale Bone Architectural Utility Indices to Prehistoric Whale Bone Dwelling Sites in Alaska and Arctic Canada. Osaka: *Bulletin of the National Museum of Ethnology* 27(2): 361-387.

Schell, Donald M. and Susan M. Saupe
1993 'Feeding and Growth as Indicated by Stable Isotopes,' pp. 491-509 in J.J. Burns, J.J. Montague, and C.J. Cowles, eds., *The Bowhead Whale.* Lawrence, KS: Allen Press, Society for Marine Mammalogy, Special Publication No. 2.

Sheehan, Glenn W.
1985 'Whaling as an Organizing Focus in Northwestern Alaskan Eskimo Society,' pp. 123-154 in T.D. Price and J.A. Brown, eds., *Prehistoric Hunters-Gatherers: The Emergence of Cultural Complexity.* New York: Academic Press.

Sheehan, Glenn W.
    1997   *In the Belly of the Whale: Trade and War in Eskimo Society.*
            Anchorage: Alaska Anthropological Association, Aurora
            Monograph Series VI.
Stanford, Dennis J.
    1976   *The Walakpa Site, Alaska: Its Place in the Birnirk and Thule
            Cultures.* Washington, D.C.: Smithsonian Institution,
            Contributions to Anthropology No. 20.
VanStone, James
    1962   *Point Hope: An Eskimo Village in Transition.* Seattle: University
            of Washington Press.
Worl, Rosita
    1980   'The North Slope Inupiat Whaling Complex,' pp. 305-320 in Y.
            Kotani and W.B. Workman, eds., *Alaska Native Culture and
            History.* Senri Ethnological Studies No.4.
Yarborough, Linda F.
    1995   'Prehistoric Use of Cetacean Species in the Northern Gulf of
            Alaska', pp. 63-81 in A.P. McCartney, ed., *Hunting the Largest
            Animals: Native Whaling in the Western Arctic and Subarctic.*
            Edmonton: Canadian Circumpolar Institute (CCI) Press,
            University of Alberta, Occasional Publication No. 36, Studies in
            Whaling No.3.

Ancient Whaling and the Biogeography of Bowhead and Gray Whales

Howard W. Braham

Abstract. B*owhead* (Balaena mysticetus) *and gray (*Eschrichtius robustus*) whales are important wildlife subsistence species at many Native coastal villages throughout the Western Arctic Bering and Chukchi seas because of the whales' predictable migration patterns. Although there is evidence that early hunters fashioned whaling harpoons several centuries prior to contact with 19th century Yankee whalers, we still do not know where or when a whaling tradition began or if it was an outgrowth of scavenging stranded whales. One method to help clarify when active whaling began is to overlay the known physical evidence of species remains at early sites with the chronological stratification of archaeological material. Several hundred whale skulls litter the tundra near abandoned village whaling and sealing sites and middens along the shores of St. Lawrence Island. These whale parts, made up mostly of very young animals, may hold evidence of directed whaling activities. Going backward in time, by investigating current to ancient settlement patterns and faunal assemblages and using novel advances in DNA and forensics analyses, the zooarchaeological trail may help elucidate temporal and spatial scales that connect prehistoric to precontact use of large whales and the formation of a whaling tradition.*

INTRODUCTION

Hunting large whales has played a major role in the ethnography of many Inuit coastal communities in the Western Arctic over the past 200 years. Prior to contact with Yankee whalers in the early to mid-19th century, whaling practices had come and gone at various locations along arctic shores. And while no direct cultural ties seem to connect ancient and recent coastal community practices (Rudenko 1961; Bandi 1969; Giddings 1967; Krupnik 1993a; Harritt 1995), a directed whaling culture has survived into the 21st century. Evidence from the literature suggests that the development of Native whaling practices must have been dependent on a predictable and seasonally abundant resource, because hunters not only targeted certain whale species as part of their subsistence catch of marine mammals (Rainey 1941; Schledermann 1980; Marquette and Braham 1982; McCartney 1984; Krupnik and Bogoslovskaya 1999), but they selected certain sizes and ages of whales that could be readily and safely taken (Chelnov and Krupnik 1984; Krupnik 1987, 1993b; and several reviews in McCartney 1995).

Figure 1. *Umiak skin boat shell made from driftwood and covered with walrus skins. The vertical support poles are rib bones from bowhead whales. (Photography by W. Marquette, National Marine Mammal Laboratory.)*

Archaeological evidence collected in the 20th century suggests that coastal inhabitants of the Western Arctic have used the bones and other products from large whales on a regular to intermittent basis since at least 2000 BP (Fig 1) (see reviews by Stoker and Krupnik 1993 and McCartney 1995). What is not clear is whether ancient hunters fashioned whaling 'technology' as an integral part of their exploitation of marine mammals or simply made the best of good opportunities to scavenge stranded animals. Several species of large whales, in particular bowhead (*Balaena mysticetus*) and gray (*Eschrichtius robustus*) routinely occur near coastal villages today and, in all likelihood, have done so since at least the end of the last permanent glacial ice some 8,000 to 10,000 years ago. Not surprisingly, the habitat use patterns of Native whalers over the past few centuries have been linked to the timing and seasonal occurrence of large whales, along with several other species of marine mammals. Thus far, most of the identified 'pre-Eskimo' coastal groups show some evidence of using whale products (McCartney 1984). Additionally, the movements and distribution of bowhead and gray whales have likely long been known to occur near shore (*cf.* Townsend 1935; Tomilin 1957; Berzin and Bandi 1984; Braham 1984; Braham *et al.* 1980a, b, 1984; Moore and Reeves 1993), making them convenient targets for hunters to set up seasonal hunting camps that might later develop into permanent settlements

(Ackerman 1988; McCartney 1984, 1995). Because the movements and behavior of bowhead and gray whales is predictable, important *a priori* conclusions are possible about where and when to expect them and when just the right sizes will be available for the taking, and thus a predictable food source.

This paper presents the biogeography of bowhead and gray whales and is one of several studies being carried out by a joint Russian-American program sponsored by the National Science Foundation and the University of Alaska on the biological, climatic, and social changes as factors influencing the settlement and development of native coastal societies in the Western Arctic. Here, I discuss why the movements and behavior of bowhead and gray whales have been critical in the formation of seasonal hunting camps that later could became permanent coastal settlements. In future studies, other researchers should look for evidence of when ancient hunters switched from scavenging for whale parts to fashioning implements that led to today's subsistence whaling practices and a whaling tradition in the Western Arctic.

The information presented in this paper comes from early Yankee and Native whaling records from the 19th and 20th centuries as well as aerial, vessel, and shore-based research data collected by national, State of Alaska, and private (mostly oil) companies from the 1970s into the 1990s and from oral conversations during my visits to Alaskan Eskimo villages from 1975 to 1997.

BIOGEOGRAPHY OF LARGE WHALES

Bowhead and gray whales routinely occur near human settlements where the activities of Native hunters have been linked to the seasonal occurrence of these whales, along with other important marine mammal species including walrus (*Odobenus rosmarus*), various seals (*Phoca* spp. and *Erignathus barbatus*) and beluga or white whale (*Delphinapterus leucus*). All of these species are important subsistence resources at most coastal Native villagers throughout the Western Arctic.

Sightings and strandings of other species of cetaceans (collectively whales, dolphins and porpoises) occur in arctic waters south of the Chukchi Sea, including the minke whale (*Balaenoptera acuterostrata*), humpback whale (*Megaptera novaeangliae*), adult male sperm whale (*Physter macrocephlus*), right whale (*Balaena glacialis*), fin whale (*Balaenoptera physalus*), killer whale (*Orcinus orca*), Dall porpoise (*Phocaenoides dalli*), and harbor porpoise (*Phocaena phocaena*). While humpback, right, and sperm whales are also considered 'large' whales, bowhead and gray whales are the only large whale species routinely found in the archaeological record or taken by Native whalers and thus they are the focus of this work.

Bowhead Whale, *Balaena mysticetus*

Bowhead whales in the Western Arctic spend the winter months (December to March) in open water south of the Bering Sea pack ice or actually in the pack ice, which in 'average' years ranges across the central Bering Sea at approximately 62°N (Niebahuer and Schell 1993). The behavior and distribution of the whales in winter are not specifically known, nor are there any data on the historical connection between the Bering Sea population (sometimes called a 'stock') of bowhead whales and those residing in the Okhotsk Sea to the south and west of the Bering Sea. These two groups probably have been isolated from one another since at least the last glacial retreat and certainly since the very large take of bowhead whales by Yankee whalers from both stocks during the second half of the 19th century (see Bockstoce 1986).

As the spring pack ice begins to recede, bowheads begin their annual migration north into High Arctic waters to feed and rear their young (Braham *et al.* 1980a, 1984; see Moore and Reeves 1993 for a review). Calves are born during the migration, with births generally peaking from March to June (Nerini *et al.* 1984). The calves spend the productive summer and remainder of the year with their mothers before their first migration with the rest of the herd the following year. Between late March and well into June, the main herd of the bowhead population (variously called the Bering Sea or Western Arctic stock) migrates from the Bering Sea into the Chukchi and Beaufort seas, generally taking a coastal route that corresponds to the breakup of 'shore-fast' ice near shore. These routes take the whales around both the eastern and western shores of St. Lawrence Island in the Bering Sea (now principally around the western end of the island since at least the mid-19th century), along the eastern offshore waters of Chukotka (far eastern Russian territory of the Chukchi Peninsula), through Bering Strait (a component moves along the north shore of Chukotka during favorable open-water years), and the main part of the herd moves north along the western shores of the Seward Peninsula and the northwest arctic coast of North America.

The distance from shore that the migration occurs generally depends on the extent of sea ice; in some years the pack ice is extensive and thick while in other years it is scattered with considerable open water. Wind also plays an important role, especially along the leeward sides of islands and promontories where, near open-water areas (called polynyas), wildlife tends to congregate to avoid the ice and rough seas. In either case, the whales generally migrate along the coast in relatively confined corridors that can vary from less than a km wide to more than 50 km wide (pers. observation). Depending on the nature and extent of the ice in the Arctic Ocean, most bowheads migrate eastward into Canadian waters toward Banks Island. This migration pattern is further offshore since the pack ice concentrates to many miles offshore along the north coast of the Beaufort Sea. When more open water is available, some late-arriving animals move toward the western Chukchi Sea and east Siberian Sea or toward Wrangel Island and Herald Shoal (into Russia across the International Date Line), where they meet up with bowheads migrating back (west and south) in autumn (Miller *et al.* 1986; Moore *et al.* 1995). Undoubtedly, the specific patterns of movement vary among seasons depending on a host of environmental and biological conditions, including ice, wind, and food availability.

The autumn migration takes from August/September to December, again depending on the extent of the pack ice and availability of prey (e.g., copepods), and generally follows the spring route in reverse in the western Beaufort Sea. Bowheads moving among the permanent pack ice in the High Arctic move west to Wrangel Island and south to the north coast of Chukotka where they feed extensively, often coming close to shore (Miller *et al.* 1986). This area along the north coast of Chukotka appears to be a major feeding or staging area in fall (Moore *et al.* 1995). Their route from the north coast of Chukotka is then south along the east coast of Chukotka, through Bering Strait into the Bering Sea, generally in advance of the southern-moving pack ice. An additional component of the population in some years also moves south along the west coast of Alaska, south from Barrow (*cf.* Moore and Reeves 1993). The definitive winter distribution and behavior of bowhead whales in the Bering Sea is generally unknown, although sightings and anecdotal records suggest that they can occur in open water from the south-central Bering Sea to Anadyr Gulf off Chukotka (for a review, see Moore and Reeves 1993).

Prior to the period of extensive Yankee whaling for bowheads during the mid-19th century (e.g., see Marquette and Bockstoce 1980; Bockstoce and Botkin 1983), when the population of whales in the Western Arctic perhaps exceeded 20,000 (Woodby and Botkin 1993), bowheads were distributed throughout the Bering Sea and Arctic Ocean during the short open-water, summer feeding season (about June to October; Townsend 1935; Dahlheim *et al.* 1980; Bockstoce and Botkin 1983). If the whales stayed within a few km of shore, any number of bowheads were available to coastal Native hunters for perhaps eight months of the year (April-November), during both the period of ice breakup and open water and especially in autumn (Krupnik 1993b). Given the amount of food and other resources provided by even a small bowhead whale, it is no wonder that the bowhead has historically been an attractive target for Native hunting from probably very early in the development of human settlements in the region.

Gray Whale, *Eschrichtius robustus*

Gray whales from the eastern North Pacific population spend the winter months along the west coast of North America, from approximately British Columbia to the Gulf of California and throughout the outer coast of Baja California (Rice *et al.* 1984). Beginning in November, their migration route is both coastal and offshore south from the Arctic, into the Gulf of Alaska, where the route taken is very close to shore, and again northbound close to shore on the return trip that begins in March in Alaskan waters. In some locations (Fig. 2), the animals can be found in the surf zone while migrating (Braham 1984). The peak occurrence off Baja California is in January and February, where both mating and calving occur, but the whales also mate and give birth during the southward migration all along the west coast of the United States from December to January (Rice and Wolman 1971).

Figure 2. *An adult gray whale (arrows) exhales, just a few meters outside of the surf-zone, during the northward migration in southeastern Bering Sea. (Photograph by the author.)*

Adult and juvenile males and nonparturient females begin entering the Bering Sea along the Alaska Peninsula in April and move along and into the ice front (pers. observation). There, they feed while waiting for the breakup of the pack ice before heading north into open-water areas in their prime feeding grounds in the northern Bering to southern Chukchi seas. Although the migration route in the northern Bering Sea is largely dependent on the extent and location of the pack ice, the migration is decidedly coastal up to the central Bering Sea (Braham 1984). The migration is diffuse thereafter because the eastern Bering Sea is both shallow and generally biologically unproductive for gray whales, and thus the whales tend to migrate into the central and western Bering Sea where the waters are rich and productive with benthic prey (Nerini 1984; Moore *et al.* 2003). The inter-seasonal selection of coastal and shallow habitats in the Bering and Chukchi seas does not vary greatly and their distribution over these feeding domains principally occurs in open water and light ice conditions (Moore 2000; Moore *et al.* 1999).

The autumn migration out of the Bering Sea is usually completed by late December, but as with most species of migrating animals, some exceptions occur. Near the conclusion of the main feeding period in the Arctic, which is approximately May to November during 'normal' ice years, adult female gray whales begin migrating south out of the Chukchi and Bering seas both along the coast and at sea. This portion of the migration has not been studied in Alaskan waters to any extent, but the evidence that gray whales have historically been taken at coastal sites along the east and north coasts of Chukotka (Marquette and Braham 1982; Krupnik 1993b; Mitchell and Reeves n.d.) and St. Lawrence Island (where the Yupik-speaking inhabitants are closely related, ancestrally, to those of coastal Chukotka) suggests that gray whales may remain close to shore in autumn as well (also see Tomilin 1957; Berzin and Bandi 1984; Blokhin 1989). Rugh and Schulman-Janiger (1999) and Rugh *et al.* (2001) have recently reviewed the gray whale migration patterns in northern waters.

WHALE BEHAVIOR, SEA ICE, AND WHALING TRADITION

Long before humans set foot on Western Arctic shores, bowhead and gray whales had developed predictable life history strategies that maximized the survival of their offspring. As a consequence of these whales returning to the same places year after year, and likely century after century, early hunters developed the means to exploit the whales for food and shelter. The places humans built shelters and later settled along the coast are a product of their own success at utilizing these whales and other marine mammals and birds in a sustainable manner. What are these environmental and oceanographic features, and the behavior and ecology of large whales that bring them close to coastal areas that helped establish a whaling tradition?

Nutrient-rich cold water moves up the Bering Sea slope, north and east along the coast of Chukotka and east along the western and southern shores of St. Lawrence Island. These areas are particularly productive for mid-water and benthic zooplankton, the predictable food source used by large whales during much of the year, and for seasonal migrants such as seabirds, walrus, and various species of fish. Less nutrient-rich waters flow along the American coast in the eastern Bering Sea and pass along the eastern shores of St. Lawrence Island en route to Bering Strait. Conversely, cool upwelling water moves through the Strait of Anadyr between St. Lawrence Island and Chukotka (see Neibauer and Schell 1993 and Loughlin and Ohtani 1999 for an overview), resulting in high prey productivity. Some of the most productive zooplankton feeding grounds for large whales are found just north of St. Lawrence Island west to Chukotka in an area called the 'whale kitchen' and north of Bering Strait, known to Yankee whalers as the 'cow yard' for its large abundance of bowhead and gray whales during the open-water whaling seasons (*cf.* Nerini 1984; Nerini and Oliver 1983; Bockstoce 1986).

Figure 3. *Drifting pack ice during spring in the northern Bering Sea. Small to large open water 'potholes' and long stretches of open water 'leads' are used by numerous species of birds, seals, and whales during northward migration to feeding grounds in the Arctic. (Photograph by the author).*

The second element relating to the use of large whales by Native hunters is the predictability of open water areas. As wind and water currents begin to move north and west in spring, ice that had been held fast to the shore and just offshore in shallow waters begins to break up and form nearshore polynyas (Fig 3.). Also, extensive cracks in the ice can form both nearshore and offshore, facilitating the movements of migrating animals from areas of less productivity to the south (as in the central Bering Sea) to more productive areas further north (e.g., Burns *et al.* 1981). The cracks and open-water areas (called leads) predictably occur at areas where currents move along the coast of Chukotka, St. Lawrence Island, and numerous places along the route of the bowhead migration from Bering Strait north into the Beaufort Sea (Braham *et al.* 1980a). The length of time the animals stay in the polynyas or leads depends upon the prevailing wind and the strength and direction of the current. These are the same places undoubtedly used by early hunters to take whales, seals, and walrus, much as do modern hunters.

Figure 4. *Diomede Islands, in Bering Srait, northern Bering Sea, an area where the bowhead whale migration can experience a slowdown, due to a bottleneck effect, during periods of severe ice coverage. The Yupik-speaking villagers living on the island to the right are citizens of the United States, while those on the island to the left are Russian. (Photograph by the author.)*

The movement of nutrient-rich currents along the coast of Chukotka and St. Lawrence Island and along the western U.S. shores north of Bering Strait provides a predictable food source for whales and other animals during their migration and feeding. The extent of ice and availability of suitable habitat for migration to occur in spring can, of course, vary within seasons and from year to year. Some years, ice in the Bering Sea can extend as far south as the Pribilof Islands, often resulting in a slightly late or protracted migration of bowhead and beluga whales (Johnson *et al.* 1981; Brueggeman 1982; Braham *et al.* 1984; Ljungblad *et al.* 1986). This is especially true if the ice is extensive or if wind compacts the ice, thus reducing the number of open-water areas and leads available for migration corridors. Often the absence or lateness of whales and other marine mammals are noted (especially in more recent records) at various villages along the migration corridor. In mild years, whales may never leave the northern Bering Sea or may arrive very early in the Gulf of Anadyr (see Bogoslovskaya *et al.* 1982). In severe ice years, the pack ice may form an ice plug at Bering Strait, preventing or delaying the migration well beyond that of average years (Fig. 4); (e.g., Johnson *et al.* 1981).

Figure 5. *Inuit hunters from Barrow, Alaska, pursuing a young bowhead whale whose wake can be seen a few meters in front of the umiak skin boat. (Photograph by D. Withrow, National Marine Mammal Laboratory.)*

Traditional whaling by definition requires that the whale species hunted be seasonally predictable in their occurrence and abundance to allow the hunting effort to sustain the subsistence economy, and that the risk of injury or death to the hunters be minimal. Both bowhead and gray whales fit this model. Behaviorally, both species are approachable by small boats under enough circumstances to be taken close to shore (Fig 5.). Since these animals are often 'resting' during migration (both species), mating (bowheads), feeding (both), or caring for their calves (both), they spend significant amounts of time in relatively calm waters not exposed to high winds or in ice fields often close to shore (i.e., within a few miles) and approachable by small boats.

Generally, bowheads migrate in spring when the Arctic is choked with pack ice, whereas in autumn they migrate through large, open-water areas, as do gray whales in spring and autumn. The distribution of bowheads appears to be more dependent on ice conditions in autumn than in spring, in that the whales are observed over inner-shelf waters during open water periods but move into deeper waters during moderate to heavy ice conditions (Moore 2000). These differences in migration patterns take the whales close to shore in some areas in spring and to different areas in autumn. For coastal whalers, hunting these large animals is made easier because the animals usually travel through open-water areas, sometimes very nearshore, when weather conditions are less extreme and ice floes are moved by the currents and not so much by wind. Water conditions, thus, can be very calm, and the animals often congregate to interact, 'sleep,' or feed. If calves are present, as happens during the spring for both species, the whales can be approached and the calves taken either to attract the cow for taking (19th century Yankee whalers often used this tactic, but Native hunters have not taken calves in recent years) or as an easy prey to kill (as precontact native hunters apparently did). Young-of-the-year calves are generally one-fifth to one-third the size of adult females yet may supply a substantial amount of meat and oil. The 'economy of scale,' then, for a village is to take the smallest whale that produces the most food, oil, or other products, with the least effort by the whalers.

A review of the literature, as well as personal experience at most currently active whaling villages in the United States and visits to several abandoned village sites on St. Lawrence Island, strongly suggests to me that a whaling tradition was established in the Western Arctic based on taking very young whales (certainly subadult and perhaps mostly young-of-the-year calves), except in recent years (*cf.* Braham 1995), and indeed the establishment of such a tradition was as much 'fortuitous ecology' (Krupnik 1993b) as actual 'need.' A larger share of the subsistence needs of most native villages, in at least the recent past, was from walrus, seals, fish, and caribou, and less than 50% (the proportion varies by village and year) of the subsistence take was from large whales (*cf.* Foote 1965 in Stoker and Krupnik 1993). The take of small whales fits the model for the take of bowhead and gray whales because of the whales' nearshore behavior patterns in spring and early summer (e.g., at St. Lawrence Island, Siriniki [and adjacent locations along the Russian coast], the Diomede Islands, Wales, Point Hope, the NW coast, and Barrow). Slightly older animals (subadults and adults) were, and are, taken in late summer and autumn along the U.S. NW coast and Chukotka—evidenced by dozens of known historical sites and a few current whaling villages along the Russian coast (see Fig. 15.1 in Stoker and Krupnik 1993 for locations of most villages current and past).

Although it is clear that bowheads can break through ice of considerable thickness (perhaps 60 cm; George *et al.* 1989), and certainly they can negotiate through large expanses of ice-covered sea during migration, bowheads tend to congregate at particular places each year at certain times of the year. Gray whales also congregate in the same general areas each year, and in the northern Bering Sea many of these places are the same for bowheads. For example, St. Lawrence Island, the coast of Chukotka, Bering Strait, outer Kotzabue Sound, and along the NW coast toward Barrow are all locations where traditional whaling practices have been carried out for at least the past 200 years, and most likely in some of these areas (St. Lawrence Island, Wales, and NE Cape in Russia) well before the precontact period.

DISCUSSION

The predictable nature of whale migrations and the historical take of small (younger) whales, in the absence of definitive archaeological data, provide important *a priori* information for understanding the development of early whaling (see Braham *et al.* 1980a; Marquette and Braham 1982; Bostoce 1986 Krupnik 1987, 1993b; McCartney 1995; Savelle and McCartney 1999). Unfortunately, there have been no studies to locate and systematically describe the whale remains and artifacts in the archaeological sites across the Western Arctic covering at least the past few hundred years. The evidence to support a direct link between ancient settlements (e.g., Choris and 'Old Whaling') and precontact whaling tradition at or near those sites is at best equivocal (Mason and Gerlach 1995). A lack of information about any connection between current whaling settlements and precontact whaling also hampers our understanding of how ancient whaling traditions developed and what influence they had on indigenous societies today (Hall 1990; Sheehan 1995; Savelle and McCartney, this volume).

Archaeologists working in high latitudes historically did not take full advantage of information on the ecology of wildlife species for describing events leading to early human settlement patterns. Local hunters and other knowledgeable indigenous people who might have had such knowledge were often not included in the process of exploration and discovery. Another reason was the near absence of reliable information from the scientific community on the natural history of most wildlife species. During the first half of the 20th century, at a time when archaeological fieldwork in the Bering and Chukchi seas was in its infancy, few published accounts existed on when and where subsistence wildlife species could be found, much less their behavior, information which native hunters possessed. This was especially true for the large whales upon which coastal villages had depended far back into this millennium (*cf.* Ackerman 1988,

1998). Fewer than a handful of authoritative publications existed on the principal species taken by precontact coastal hunters prior to the mid-20th century, and only anecdotal 19th century Yankee whaling records have survived that describe the geographic locations and behavior of whales taken.

Figure 6. *Cranium and nasal bones of a young gray whale located near the current shoreline at Kialegak, St. Lawrence Island. (Photograph by M. Nerini, Natonal Marine Mammal Laboratory).*

During my visits to St. Lawrence Island in the mid-1970s, Mary Nerini, Bud Fay, and I, along with Mathew Iya from Savoonga, located more than 100 subadult (mostly calf) bowhead and over 200 subadult gray whale skulls at one or more locations on the island: Kialegak, a pre-19th century abandoned hunting settlement near Southeast Cape (Fig. 6); Puwowlak (Pawooliak), along the south coast near the existing whaling camp used today by hunters from Savoonga (Fig. 7); and inland and south of the current village of Gambell *(Sivokak)* on the western end of the island. These are some of the identified sites that need to be explored during future visits *(cf.* note 1 in Braham 1995). The specimens there will help make up the samples needed to investigate the age and size distribution of whales taken by local hunters and also help to find forensic evidence linking the actual taking of animals with the development of whaling practices.

Figure 7. *Deteriorating cranium of a young bowhead whale located inshore at the pre-20th century village site east of Gambell, Alaska, St. Lawrence Island. (Photograph courtesy of the National Marine Mammal Laboratory.)*

The existing evidence for active precontact whaling (before about 1848) comes from the identification of whaling artifacts and the remains of whales, mostly calves and subadults, found near settlements or among archaeological midden remains. Since the occurrence and behavior of most whales is generally predictable on a seasonal scale (using current data; e.g., Dyke *et al.* 1995), a meaningful relationship appears to exist—and probably has existed for centuries—between the occurrence of certain whale species, and their younger age-classes, and the introduction of directed whaling practices. A consistent pattern of midden remains of these animals and associated tool usage may suggest a 'switchover' from passive to active whale hunting. However, only a fraction of the known sites in the Western Arctic have been investigated, and most require accurate dating of the evidence and assessing the stratigraphic relationships of the artifact assemblages and bone samples (to species and age-classes).

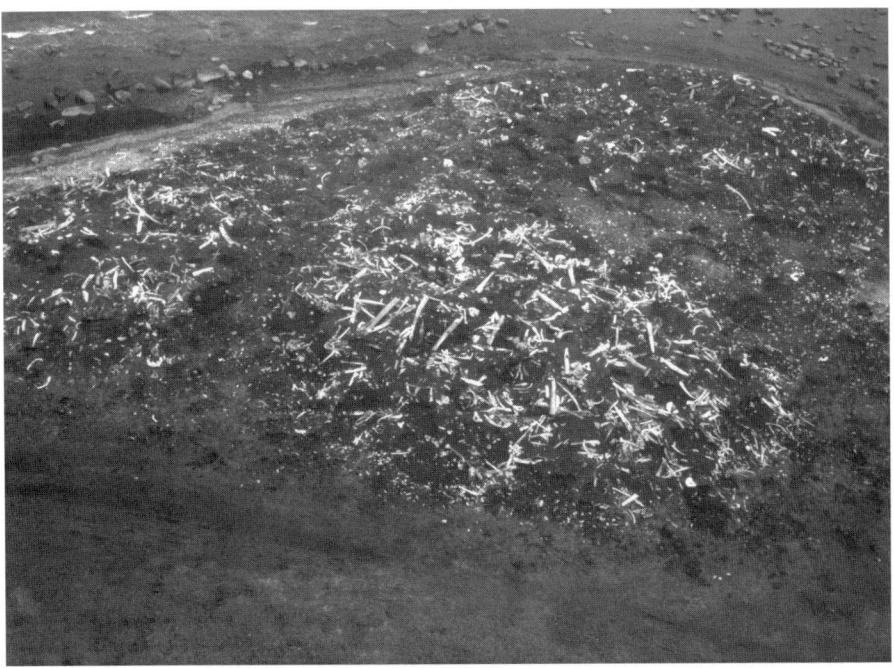

Figure 8. *Partial view of the miden mounds near the pre-20th century Inuit village of Kialegak, north of SE Pt., St. Lawrence Island, Alaska. (Photograph by the author.)*

Although the definitive data from archaeological sites have yet to be collected, the seasonal occurrence of bowhead and gray whales (along with other marine mammals) played a major role in the establishment of seasonal hunting camps, which in all likelihood in turn led to transitional villages and eventually permanent settlements. Some of these settlements have lasted into modern times or remain to be investigated at abandoned archaeological sites, such as Kialegak and other coastal sites on St. Lawrence Island (Fig. 8). Indirect evidence suggests that whaling may have been an active enterprise on or near St. Lawrence Island for many centuries (Geist and Rainey 1936; Collins 1937; Rainey 1941; Giddings 1960), but the work there did not focus on the identification or use of whales. Unfortunately, very little specific work has been done linking the use of whaling implements (harpoon or flensing tips) to the actual taking of whales. Nor is there evidence that early hunters pursued the large whales in skin boats, as has been done for the past few centuries. Because it appears that early directed whaling may have started in the Bering Sea (Yesner 1995), and St. Lawrence Island has a rich archaeological history, this is a good place to begin to locate and (re)investigate archaeological sites for the presence of whale and other marine mammal subsistence fauna.

Whether early inhabitants of the Western Arctic were truly 'whale hunters' is not fully evident from the existing archaeological record. It seems unlikely that a sustained culture could have existed in the absence of dependable whaling vessels such as *umiaks* (Inuit skin boats) or strong harpoon tips (Mason and Gerlach 1995). It is not clear how long Native hunters have used these instruments at a level of utility that would have turned whale hunting into a successful activity leading eventually to a whaling tradition. With the help of local community leaders and hunters, future work should explore site-specific evidence for when and where directed whaling began, how the record relates to other circumpolar archaeological sites, and the historical significance of panarctic whaling in the development of early human societies.

RECOMMENDATIONS

The following areas of investigation are recommended if this work is to continue:

1. Selected archaeological sites along the Russian and American coasts (such as the rich faunal assemblages on St. Lawrence Island and eastern Chukotka) need to be located and mapped for their general topography and faunal content, in some cases based on previous excavations (e.g., Geist and Rainey 1936; Collins 1937; Rainey 1941; Larsen and Rainey 1948; Bandi 1967; Chlenov and Krupnik 1984). Previously undescribed sites throughout the area also should be identified for future excavations (e.g., on St. Lawrence Island). The pattern of whale usage and approximate antiquity of each site should be determined to assess settlement patterns and the relationship between the fauna present and the biology of the species identified. Such work may also help us understand more about when and why native hunters switched from bowhead to gray whales, as suggested in the record at about the 5th to the 9th century A.D. (see Krupnik 1987; Mitchell and Reeves n.d.).

2. A comparison of the life history and ecology of large whales with the physical evidence (e.g., whale bones) from the chronological (aging) stratification of identified archaeological sites needs to be made. This will provide direct evidence that large whale species were used by ancient hunters, presumably during prehistoric and precontact periods.

3. Other studies should include identifying body fluid stains found on ancient harpoon tips (*cf.* Mason and Gerlach 1995) to identify what was killed. Also, samples should be taken of oil from the burning of fat and blubber of birds, seals, walrus, and whales collected from ancient building sites, interiors, cooking tools, and the like, which could be identifiable to species and tested with the ages of settlement artifacts.

4. Detailed measurements of the whale remains in particular, using standard faunal site analyses (Grayson 1984; Davis 1987), should be taken. The identification of whale parts, and specifically, preliminary identification of whale species will be crucial using a set of standard measurements developed by Edward Mitchell (Los Angeles County Museum) and generously applied by Allen McCartney and James Savelle (*cf.* McCartney 1995) and species identifications located in True (1904), Andrews (1914), and others.

5. Bone samples need to be collected and analyzed for DNA (see Pääbo 1989; Baker and Palumbi 1996) to identify individuals to species and to compare whaling patterns such as size/age distribution among sites. Also, comparison of samples from museums and those collected at various sites throughout the Western Arctic, Okhotsk Sea, and Eastern Arctic needs to be done to assess local 'extinctions' (e.g., Roy *et al.* 1994) and the impact of whaling activities. Bone fragments should be analyzed for stable isotope ratios (Lee-Thorp *et al.* 1989; Lajtha and Michener 1994) as a means to determine large-scale changes in diet within the species' geographic range over long periods of (zooarchaeological) time. Standard radiocarbon dating of suitable sites (*cf.* Stuckenrath and Anderson, n.d.) also should be done. Artifacts should be collected and analyzed for blood and other proteins (Gerlach *et al.* 1996) to determine which species of fauna were being utilized at the site. To my knowledge, none of this work on ancient DNA, stable isotopes, and blood proteins has been carried out on gray or bowhead whales.

Acknowledgments: Support for this work was provided by a National Science Foundation grant (OPP-9807051) to the University of Arkansas, administered by Dr. Allen McCartney, whom I particularly thank for his support and encouragement. Marcia Muto of the National Marine Mammal Laboratory, AFSC, NOAA, Seattle, Washington, was especially helpful in the early stages of this work with library archives at the University of Washington. Donna J. Braham provided invaluable editing. During my 20 years of research in arctic Alaska, I was fortunate to have learned firsthand about the natural history of bowhead and gray whales and about whaling practices, from dozens of village whalers and from scientific colleagues who generously helped educate me to life in the North. A few in particular who made my career rich in more ways that I can say were: Conrad Oozeva, Tommy Awtinawmi, Roger Silook, and Vernon Slwooko from Gambell; Matthew Iya and Jerry Wongittilin, Sr., from Savoonga; Patrick Omiak from Little Diomede; Arnold Brower, Sr., and Kenny Toovak from Barrow; Bud Fay, University of Alaska, deceased; John Burns, Alaska Department of Fish and Game; Gennadii Fedoseev, TINRO, former Soviet Union; and Cliff Fiscus, U.S. Fish and Wildlife Service, and National Marine Fisheries Service, Seattle, who started me on this wild adventure!

REFERENCES

Ackerman, R.
> 1988 Settlements and Sea Mammal Hunting in the Bering-Chukchi Sea Region. *Arctic Anthropology* 25(1): 52-79.
> 1998 Early Maritime Traditions in the Bering, Chukchi, and East Siberian Seas. *Arctic Anthropology* 35(1): 247-262.

Andrews, R.C.
> 1914 'Monographs of the Pacific Cetacea,' Volume 1: *The California Gray Whale.* New York: New Series, Memoirs of the American Museum of Natural History.

Baker, C.S. and S.R. Palumbi
> 1996 'Population Structure, Molecular Systematics and Forensic Identification of Whales and Dolphins,' in J.C. Avise and J.L. Hamrick, eds., *Conservation Genetics: Case Histories from Nature.* New York: Chapman and Hall.

Bandi, Hans-Georg
> 1967 'The Archaeology of St. Lawrence Island, Alaska,' pp. 485-491 in V.L. Kontrimavichus, ed., *Beriangiia v Kainozoe.* Vladivostok: Academy of Science.
> 1969 *Eskimo Prehistory.* Fairbanks: University of Alaska Press.

Berzin, A.A. and Bandi, Hans-Georg
> 1984 'Soviet Studies of the Distribution and Numbers of Gray Whale in the Bering and Chukchi Seas,' from 1968 to 1982, pp. 409-419 in M.L. Jones, S.L. Swartz, and S. Leatherwood, eds., *The Gray Whale, Eschrichtius robustus.* Orlando, FL: Academic Press.

Blokhin, S.A.
> 1989 A Note on the Spatial Distribution of Gray Whales off Chukotka. *Reports of the International Whaling Commission* 39: 309-311.

Bockstoce, J.R.
> 1986 *Whales, Men and Ice: The History of Whaling in the Western Arctic.* Seattle: University of Washington Press.

Bockstoce, J.R. and D.B. Botkin
> 1983 The Historical Status and Reduction of the Western Arctic Bowhead Whale (*Balaena mysticetus*) Population by the Pelagic Whaling Industry, 1848-1914. *Report of the International Whaling Commission,* Special Issue 5: 107-141.

Bogoslovskaya, L.S., L.S. Votrogov, and I.I. Krupnik
    1982 The Bowhead Whale off Chukotka: Migrations and Aboriginal
        Whaling. *Reports of the International Whaling Commission* 32:
        391-399.

Braham, H.W.
    1984 'Distribution and Migration of Gray Whales in Alaska,' pp. 249-
        266 in M.L. Jones, S.L. Swartz and S. Letherwood, eds., *The
        Gray Whale, Eschrichtius robustus*. Orlando, FL: Academic Press.
    1995 'Sex and Size Composition of Bowhead Whales Landed by
        Alaskan Eskimo Whalers,' pp. 281-313 in Allen P. McCartney,
        ed., *Hunting the Largest Animals. Native Whaling in the Western
        Arctic and Subarctic*. Edmonton: Canadian Circumpolar Institute
        (CCI) Press, University of Alberta, Occasional Publication No. 36
        Studies in Whaling No. 3.

Braham, H.W., M.A. Fraker, and B.D. Krogman Braham, H.W.
    1980a Spring Migration of the Western Arctic Population of Bowhead
        Whales. *Marine Fisheries Review* 42(9-10): 36-46.

Braham, H.W., B.D. Krogman, and G.M. Carroll
    1984 Bowhead and White Whale Migration, Distribution, and
        Abundance in the Bering, Chukchi, and Beaufort Seas, 1975-78.
        NOAA Technical Report NMFS SSRF-778.

Braham, H.W., W.M. Marquette, T.W. Bray, and J.S. Leatherwood
    1980b The Bowhead Whale: Whaling and Biological Research. *Marine
        Fisheries Review* 42(9-10): 1-96.

Breiwick, J.M. and H.W. Braham
    1984 The Status of Endangered Whales. *Marine Fisheries Review*
        46(4): 1-64.

Brueggeman, J.J.
    1982 Early Spring Distribution of Bowhead Whales in the Bering Sea.
        Journal of Wildlife Management 46: 1036-1044.

Burns, J.J., L.H. Shapiro, and F.H. Fay
    1981 'Ice as a Marine Mammal Habitat in the Bering Sea,' pp. 781-797
        in D.W. Hood and H.A. Calder, eds., *The Eastern Bering Sea
        Shelf: Oceanography and Resources*, Volume 2. Seattle:
        University of Washington Press.

Chlenov, M.D. and I. Krupnik
    1984 Whale Alley: A Site on the Chukchi Peninsula, Siberia.
        Expedition 26(2): 6-15.

Collins, H.B. Jr.
    1937 *Archaeology of St. Lawrence Island, Alaska*. Washington, D.C.:
        Smithsonian Institution, Miscellaneous Collections 96(1).

Dahlheim, M.E., T.W. Bray, and H.W. Braham
    1980 Vessel Survey for Bowhead Whales in the Bering and Chukchi
        Seas, June-July 1978. *Marine Fisheries Review* 42(9-10): 51-57.

Davis, S.J.M.
    1987 *The Archaeology of Animals*. New Haven: Yale University Press.

Dyke, A.S., J. Hooper, and J.M. Savelle
    1995 A History of Sea Ice in the Canadian Arctic Archipelago Based on
        Postglacial Remains of the Bowhead Whale (*Balaena mysticetus*).
        Arctic 49(3): 235-255.

Geist, O.W. and F.G. Rainey
1936 *Archaeological Excavations at Kukulik, St. Lawrence Island, Alaska: Preliminary Report.* University of Alaska Miscellaneous Publication 2. Washington, D.C.: U.S. Government Printing Office.

George, J.C., C. Clark, G.M. Carroll, and W.T. Elison
1989 Observations on the Ice-Breaking and Ice Navigation Behavior of Bowhead Whales (*Balaena mysticetus*) near Point Barrow, Alaska, Spring 1985. *Arctic* 42(1): 24-30.

Gerlach, S.C., M. Newman, E.J. Knell, and E.S. Hall, Jr.
1996 Blood Protein Residues on Lithic Artifacts from Two Archaeological Sites in the De Long Mountains, Northwestern Alaska. *Arctic* 49(1): 1-10.

Giddings, J.L.
1960 The Archaeology of the Bering Sea. *Current Anthropology* 1(2): 121-138.

1967 *Ancient Men of the Arctic.* New York: Alfred A. Knopf.

Grayson, D.
1984 *Quantitative Zooarchaeology.* Orlando, FL: Academic Press.

Hall, E.S., Jr.
1990 *The Utqiagvik Expedition.* Barrow, AK: North Slope Borough Commission on Iñupiat History, Language, and Culture.

Harritt, R.K.
1995 'The Development and Spread of the Whale Hunting Complex in Bering Strait: Retrospective and Prospects,' pp. 33-50 in A.P. McCartney, ed., *Hunting the Largest Animals. Native Whaling in the Western Arctic and Subarctic.* Edmonton: Canadian Circumpolar Institute (CCI) Press, University of Alberta, Occasional Publication No. 36, Studies in Whaling No. 3.

Johnson, J.H., H.W. Braham, B.D. Krogman, W.M. Marquette, R.M. Sonntag, and D.J. Rugh
1981 Bowhead Whale Research: June 1979 to June 1980. *Reports of the International Whaling Commission* 31: 461-475.

Krupnik, I.I.
1987 Bowhead vs. Gray Whale in Chukotka Aboriginal Whaling. *Arctic* 40(1): 16-32.

Krupnik, I.I.
1993a *Arctic Adaptations: Native Whalers and Reindeer Herders of Northern Eurasia.* Hanover and London: University Press of New England.

1993b Prehistoric Eskimo Whaling in the Arctic: Slaughter of Calves or Fortuitous Ecology? *Arctic Anthropology* 30(1): 1-12.

Krupnik, I.I. and L.S. Bogoslovskaya
1999 Old Records, New Stories: Ecosystem Variability and Subsistence Hunting in the Bering Strait Area. *Arctic Research of the United States* 13: 15-24.

Lajtha, K. and R.H. Michener
1994 *Stable Isotope Ratios in Ecology and Environmental Science.* Oxford: Blackwell Scientific Publications.

Larsen, H. and F. Rainey
1948 Ipiutak and the Arctic Whale Hunting Culture. *Anthropological Papers of the American Museum of Natural History* 42.

Lee-Thorpe, J.A., J.C. Sealy, and N.J. Van Der Merwe
    1989 Stable Carbon Isotope Ratio Differences Between Bone Collagen
        and Bone Appetite, and their Relationship to Diet. *Journal of
        Archaeological Science* 16: 585-599.
Ljungblad, D.K., S.E. Moore, and J.T. Clarke
    1986 Assessment of Bowhead Whale (*Balaena mysticetus*) Feeding
        Patterns in the Alaskan Beaufort and Northeastern Chukchi
        Seas via Aerial Surveys, Fall 1979-84. *Report of the International
        Whaling Commission* 36: 265-272.
Loughlin, T.R. and K. Ohtani
    1999 *Dynamics of the Bering Sea.* Fairbanks: University of Alaska Sea
        Grant.
Marquette, W.M. and J.R. Bockstoce
    1980 Historical Shore-based Catch of Bowhead Whales in the Bering,
        Chukchi, and Beaufort Seas. *Marine Fisheries Review* 42(9-10):
        5-19.
Marquette, W.M. and H.W. Braham
    1982 Gray Whale Distribution and Catch by Alaskan Eskimos: A
        Replacement for the Bowhead Whale? *Arctic* 35(3): 386-394.
Mason, O.K. and S.C. Gerlach
    1995 'The Archaeological Imagination, Zooarchaeological Data, the
        Origins of Whaling in the Western Arctic, and 'Old Whaling' and
        Choris Cultures,' pp. 1-32 in A.P. McCartney, ed., *Hunting the
        Largest Animals. Native Whaling in the Western Arctic and
        Subarctic.* Edmonton: Canadian Circumpolar Institute (CCI)
        Press, University of Alberta, Occasional Publication No. 36,
        Studies in Whaling No. 3.
McCartney, A.P.
    1984 'History of Native Whaling in the Arctic and Subarctic,' pp. 79-
        111 in H.K. s'Jacob, K. Snoeijing, and R. Vaughan, eds., *Arctic
        Whaling: Proceedings of the International Symposium on Arctic
        Whaling.* Gröningen: Arctic Centre, University of Gröningen.
    (ed.) 1995. *Hunting the Largest Animals. Native Whaling in the Western
        Arctic and Subarctic.* Edmonton: Canadian Circumpolar Institute
        (CCI) Press, University of Alberta, Occasional Publication No. 36,
        Studies in Whaling No. 3.
Miller, R.V., D.J. Rugh, and J.H. Johnson
    1986 The Distribution of Bowhead Whales, *Balaena mysticetus*, in the
        Chukchi Sea. *Marine Mammal Science* 2: 214-222.
Mitchell, E.D. and R.R. Reeves
    n.d. Aboriginal Whaling for Gray Whales in the Eastern North Pacific
        Ocean., in H.W. Braham, G.P. Donovan, and M.M. Muto (eds.),
        Gray Whales, Eschrichtius robustus. Cambridge, UK:
        International Whaling Commission.
Moore, S.E.
    2000 Variability of Cetacean Distribution and Habitat Selection in the
        Alaskan Arctic, Autumn 1982-91. *Arctic* 53(4): 448-460.
Moore, S.E., D.P. DeMaster, and P.K. Dayton
    1999 Cetacean habitat Selection in the Alaskan Arctic during Summer
        and Autumn. *Arctic* 53(4): 432-447.

Moore, S.E., J.C. George, K.O. Coyle, and T.J. Weingartner
    1995 Bowhead Whales Along the Chukotka Coast in Autumn. *Arctic*
        48(2): 155-160.
Moore, S.E. and R.R. Reeves
    1993 'Distribution and Movement,' pp. 313-386 in J.J. Burns,
        J.J.Montague, and C.J. Cowles, eds., *The Bowhead Whale.*
        Lawrence, KS: Allen Press: Society of Marine Mammalogy,
        Special Publication No. 2.
Moore, S.E., J.M. Grebmeier and J.R. Davies
    2003 Gray whale distribution relative to forage habitat in the northern
        Bering Sea: Current conditions and retrospective summary.
        Canadian Journal of Zoology 81(4): 734-742.
Neibauer, H.J. and D.M. Schell
    1993 'Physical Environment of the Bering Sea Population,' pp. 23-43
        in J.J. Burns, J.J. Montague and C.J. Cowles, eds., *The*
        Bowhead Whale. Lawrence, KS: Allen Press: Society of Marine
        Mammalogy, Special Publication No. 2.
Nerini, M.K.
    1984 'A Review of Gray Whale Feeding Ecology,' pp. 423-450 in M.L.
        Jones, S.L. Swartz and S. Leatherwood, eds., *The Gray Whale,*
        Eschrichtius robustus. Orlando: Academic Press.
Nerini, M.K., H.W. Braham, W.M. Marquette, and D.J. Rugh
    1984 Life History of the Bowhead Whale, *Balaena mysticetus. Journal*
        of Zoology (London) 204: 443-468.
Nerini, M.K. and J.S. Oliver
    1983 Gray Whales and the Structure of the Bering Sea Benthos.
        Oecologia (Berlin) 59(2-3): 224-225.
Pääbo, S.
    1989 Ancient DNA: Extraction, Characterization, Molecular Cloning
        and Enzymatic Amplification. *Proceedings of the National*
        Academy of Sciences 86(6): 1939-1943.
Rainey, F.G.
    1941 Eskimo Prehistory: The Ovik Site on the Punuk Islands.
        Anthropological Papers of the American Museum of Natural
        History 37(4).
Rice, D.W. and A.A. Wolman
    1971 The Life History and Ecology of the Gray Whale (*Eshrichtius*
        robustus). *American Society of Mammalogists, Special Publication*
        No. 3.
Rice, D.W., A.A. Wolman, and H.W. Braham
    1984 The Gray Whale, *Eschrichtius robustus. Marine Fisheries Review*
        46(4): 7-14.
Ross, W.G.
    1977 Whaling and the Decline of Native Populations. *Arctic*
        Anthropology 14(2): 1-8.
Roy, M.S., D.J. German, A.C. Taylor, and R.K. Wayne
    1994 The Use of Museum Specimens to Reconstruct the Genetic
        Variability and Relationships of Extinct Populations. *Experiential*
        50: 551-557.

Rudenko, S.I.
    1961 *The Ancient Culture of the Bering Sea and the Eskimo Problem*
        Arctic Institute of North America, Anthropology of the North,
        Translations from Russian Sources No. 1. Toronto: University of
        Toronto Press.
Rugh, D. and A. Schulman-Janiger
    1999 The Gray Whales Great Migration. *Alaska Geographic* 26(3): 38-
        45.
Rugh, D.J., K.E.W. Shelden, and A. Shulman-Janiger
    2001 Timing of the Gray Whale Southbound Migration. *Journal of
        Cetacean Research and Management* 3(1): 31-39.
Savelle, J.M. and A.P. McCartney
    1999 Thule Eskimo Bowhead Whale Interception Strategies. *World
        Archaeology* 30(3): 437-451.
Schledermann, P.
    1980 Polynyas and Prehistoric Settlement Patterns. *Arctic* 33(2): 292-
        302.
Sheehan, G.W.
    1995 'Whaling Surplus, Trade, War, and the Integration of Prehistoric
        Northern and Northwestern Alaska Economies, AD 1200-1826,'
        pp. 185-206 in A.P. McCartney, ed., *Hunting the Largest Animals.
        Native Whaling in the Western Arctic and Subarctic.* Edmonton:
        Canadian Circumpolar Institute Press, Occasional Publication
        No. 36, Studies in Whaling No. 3.
Stoker, S.W. and I.I. Krupnik
    1993 'Subsistence Whaling,' pp. 579-629 in J.J. Burns, J.J.
        Montague, and C.J. Cowles, eds., *The Bowhead Whale.*
        Lawrence, KS: Allen Press: Society of Marine Mammalogy,
        Special Publication No. 2.
Stuckenrath, R. and D.D. Anderson
    n.d. Radiocarbon Dates from the American Arctic. *American Antiquity.*
Tomilin, A.G.
    1957 Cetaceans, *in* V.G. Heptner, ed., *Mammals of the USSR and
        Adjacent Countries* Volume IX. Moscow, USSR: Izdatl' sto
        Akademi. Nauk SSSR.
Townsend, C.
    1935 The Distribution of Certain Whales as Shown by Logbook
        Records of American Whaleships. *Zoologica* 19(1): 1-50.
True, F.W.
    1904 The Whalebone Whales of the Western North Atlantic.
        Smithsonian Contribution to Knowledge No. 33.
Woodby, D.A. and D.B. Botkin
    1993 'Stock Sizes Prior to Commercial Whaling,' pp. 479-629 in J.J.
        Burns, J.J. Montague and C.J. Cowles, eds., *The Bowhead
        Whale.* Lawrence, KS: Allen Press: Society of Marine Mammalogy,
        Special Publication No. 2.
Yesner, D.R.
    1995 'Whales, Mammoths, and Other Big Beasts: Assessing Their Role
        in Prehistoric Economies,' pp. 185-206 in A.P. McCartney, ed.,
        *Hunting the Largest Animals. Native Whaling in the Western Arctic
        and Subarctic.* Edmonton: Canadian Circumpolar Institute Press,
        Occasional Publication No. 36, Studies in Whaling No. 3.

The Bowhead Whale Off Chukotka: Integration of Scientific and Traditional Knowledge

Lyudmila S. Bogoslovskaya
Translated by Petr Aleinikov, Edited by Igor Krupnik

Abstract. *This paper addresses recent contributions by Native sea mammal hunters of Chukotka, Russia, to modern scientific studies of the bowhead whale (Balaena mysticetus) of the Bering-Chukchi stock. Thanks to hunters' knowledge and observations, an additional western route for the bowhead spring migration was identified and the summering of whales off the Chukchi Peninsula has been discovered. A new and highly specialized type of bowhead whale, takyshkak, has been described. The significance of bowhead whales and whaling for ancient and present-day residents of Chukotka is emphasized. The prospects for cooperation between scientists and indigenous people in the studying of arctic marine species that are particularly difficult to observe and monitor, such as bowhead whale, are discussed.*

INTRODUCTION

Bowhead whaling is a key traditional occupation of indigenous peoples of Chukotka—the Yupik Eskimo and the Maritime (coastal) Chukchi. Over the course of two millennia, Native hunters have successfully pursued these gigantic marine mammals, using large walrus-skin boats and sophisticated toggle-head harpoons. There is abundant evidence to corroborate that this harvest was once highly successful. Large quantities of bowhead bones at ancient village sites, numerous archaeological finds of whaling tools (harpoons, lances, etc.), scenes of whale hunting depicted in prehistoric rock art, recorded folklore texts, and stories told by Native elders all demonstrate the importance of traditional whaling.

Over the long history of indigenous whaling, local people have learned the peculiarities of whale behavior and their migration routes. It is not an accident that all of the large coastal communities of Eastern Chukotka were located at sites where whales approach the shore in the course of their annual spring and fall migrations.

Most of the present-day Native hunters are not only good and ardent observers but also skillful narrators. They describe colorfully, and in detail, the habits and the behavior of whales. In many Yupik and Maritime Chukchi families, the passion for hunting as well as the profession of a marine hunter has been passed from father to son, together with the knowledge and practical skills accumulated from generation to generation. Hence, the present study is based upon the author's findings, data from many colleagues in the field, and literary sources, in addition to information obtained by and from the Native residents, largely from marine hunters of the Chukotka Autonomous Area (*Okrug*). The long-term cooperation of scientists and indigenous people discussed below has proved highly fruitful in another matter. In 1997–1998, thanks to their joint effort, Native shore-based hunting for bowhead whales, which was banned by the USSR government in the 1970s, was resumed in Chukotka.

The loss of subsistence whaling for almost 30 years (and for a much longer time in many Native communities) was a frustrating experience to the Yupik and Chukchi people of Chukotka. The ban on bowhead whaling was painful, not only because the meat of the bowhead whale is far superior to that of the gray whale in terms of its taste and nutritional properties, but also because it can be stored successfully in family meat caches (ice cellars). What was even more important was the key role that the bowhead whale played in the social, cultural, and spiritual life of the people of Chukotka, primarily in the Yupik (Eskimo) communities. Even today, bowhead whaling has retained its importance in maintaining ethnic pride and cultural awareness, and it supports cultural and language traditions of indigenous people. All the residents of Native communities in the Bering Strait Region, both Yupik (Eskimo) and Chukchi, are fully aware of that role.

The resumption of traditional bowhead whaling off Chukotka by Native hunters in 1997–1998, will undoubtedly increase our knowledge of that species' biology. Local hunters continue to contribute new information about the timing and routes of seasonal migrations, and the distribution, population dynamics, and specific behavioral features of the bowhead whales. Such information is of tremendous value for marine scientists; eventually it will make it easier to develop specific measures to ensure the conservation and overall survival of the bowhead whale stock. Furthermore, integration of traditional and scientific knowledge about bowhead whales based on long-term local monitoring would be highly useful in designing similar programs for other unique and endangered large mammal species.

ON THE HISTORY OF SCIENTISTS AND NATIVE HUNTERS' COOPERATION IN CHUKOTKA[1]

Scientific research about the bowhead whale population off Chukotka does not have a long history. Its most successful efforts have been linked to several projects and programs that eventually resulted in

---

[1] This section is co-authored with Igor Krupnik.

more than 20 years of cooperative work by scientists and local marine hunters from coastal communities in Eastern Chukotka.

The outset of that cooperation is strongly associated with Leonard Votrogov (1929–1982), the captain of the state whaling boat *Zvezdnyi*. From 1969, his vessel, a common Russian whaler, harvested gray whales in the Bering Strait area 'on behalf of the indigenous peoples of Chukotka' (as stated in many Soviet documents of the time). Captain Votrogov was truly an outstanding person, a successor in the line of old whaling captains, such as the Englishman, William Scoresby and the American, Charles Scammon, who were at the same time devoted naturalists and researchers. Those 'Great Whaling Captains' of the old days made tremendous contributions to the scientific studies of marine mammals and about bowhead whales, in particular.

In the course of his entire service as a Russian whaling captain, Captain Votrogov monitored whales and other marine mammals and he documented his observations in a truly scientific way. He also interviewed Native hunters in the coastal communities where his boat transported and delivered the harvested whales. In response, several hunters started writing him letters to report their observation of whales, particularly during the winter period when Votrogov's boat was harbored in the south, at its main wintering base in Vladivostok. Votrogov also wrote personal letters to some villages along the arctic coast of Chukotka that were not in the regular zone of his operations. These letters included special questions about sightings of whales, and in this way Votrogov received first-hand information from local residents. As a result he collected an invaluable set of records about the sightings and seasonal presence of bowhead whales along the entire coast of Chukotka, from the Gulf of Anadyr in the south to Bering Strait to Chaun Bay far northwest along the Arctic Coast. In addition, he interviewed some 500 captains and navigators of Russian arctic cargo ships and of the State Hydrographic service as well as several old whaling captains regarding the distribution of gray and right whales along the entire Pacific coast of Russia (for details, see Bogoslovskaya *et al.* 1984).

Votrogov was also the first to propose that local hunters should undertake systematic long-term observations of whales and other marine mammals and record those data on their own. In 1977–1978, upon Votrogov's personal initiative, Andrei Ankalin (*Angqalen*[2]), a Yupik hunter from the village of Sireniki, conducted spring monitoring of the bowhead whale migration from his hunting camp at Anytykuk (*Angetequq*), east of Sireniki. Excerpts from Ankalin's

---

[2] Most of the Native names and place-names from Chukotka are cited with a parallel St. Lawrence Island Yupik spelling given in parenthesis. The Yupik spelling was added by Igor Krupnik and it was checked by Willis Walunga, a Yupik elder from Gambell, St. Lawrence Island—ed.

logbook of 1977–1978 were published in 1982 (Bogoslovskaya *et al.* 1982). His observations offered the first evidence to challenge the accuracy of the American census of the Chukotka-Bering sea bowhead stock that was conducted at the same time (May 1978) off Barrow. During the next few years, several other local hunters made similar observations and kept their personal records of bowhead whales and other marine mammals at different sections of the Chukotkan coast. Those data were subsequently processed by Votrogov and Bogoslovskaya (e.g., Bogoslovskaya and Votrogov 1981; Votrogov and Bogoslovskaya 1980; Bogoslovskaya *et al.* 1982; 1984), who launched a coordinated study of the distribution and population dynamics of bowheads off Chukotka.

During the same period (1975–1979), Igor Krupnik interviewed Native elders and made extensive recording of stories by senior hunters, who were born during the first decades of the 20th century. These interviews (Krupnik 1987, 2001) documented the extensive level of local knowledge on bowhead whaling traditions in Chukotka. Additional interviews of elders and searches of archival and literary sources were undertaken between 1977–1981, under the study of *Whale Alley*, a prehistoric memorial site built of dozens of bowhead skulls and mandibles on a small island off the Chukchi Peninsula (Arutyunov *et al.* 1982; Chlenov and Krupnik 1984). Thus, the present-day system of knowledge about traditional indigenous whaling and biology of marine mammals off Chukotka was developed due to close cooperation among biologists, anthropologists, and local people. Above all, this included knowledge of bowhead whales and Native whaling. This pattern of research and the ways the data were collected in the 1970s differed radically from the common practices of marine mammal scientists who usually make their observation from whaling boats and/or scientific vessels.

Between 1977 and 1992, bowhead whale and whaling studies in Chukotka were organized primarily around annual surveys carried out by teams of marine zoologists from the Severtsov Institute of Evolutionary Morphology and Animal Ecology in Moscow (under Lyudmila Bogoslovskaya's leadership). These were supplemented through field surveys undertaken by anthropologists Mikhail Chlenov and Igor Krupnik from the then Miklukho-Maklay Institute of Ethnography, USSR Academy of Sciences in Moscow (leader: M. Chlenov). As a result of these surveys, several maps were produced, showing the distribution of the principal biological marine resources off the Chukchi Peninsula, including major cetacean species, walrus haul-outs, concentrations of seals, and sea-bird colonies. These maps cover the entire Chukchi Peninsula, from Cape Serdtse-Kamen on the Arctic Coast to Cape Bering on the Gulf of Anadyr coast. Additional maps and databases cover the sites of all the coastal villages and hunting camps of indigenous people who existed since prehistoric times to the middle of the 20th century (Krupnik 1983;

Bogoslovskaya 1993; Fig. 1). Major village sites have been described, and annual subsistence cycles and patterns of marine resources used have been documented, based upon elders' recollections.

Probably the most important outcome of such collaboration was the establishment of a network of trained observers recruited among Native marine hunters from various communities along the coast of Chukotka. Since 1992, this network of Native observers has become the foundation for a joint Russian-American program in marine mammal and subsistence research in Chukotka, with special emphasis upon bowhead whales. The American side was represented by the Department of Wildlife Management, North Slope Borough, under the overall supervision of Dr. Thomas Albert from Barrow, Alaska. Originally, the chief Russian partner in this project was the (Native) *Naukan* Association of Marine Hunters based in the town of Lavrentiya in the northern Chukotsky District (Chairman: Mr. Mikhail Zelensky; Biological Supervisor: Dr. Vladimir Melnikov). Beginning in 1994, observers from the Providensky District recruited by the Eskimo Society of Chukotka, *Yupik*, under the leadership of Lyudmila Ainana (*Aynganga*), and biologists Nikolay Mymrin and Lyudmila Bogoslovskaya joined the Russian-American monitoring project.

The initial bowhead whale monitoring project of 1992–1997 was followed in 1998 by another program that involved collaboration between Native hunters of Chukotka and the U.S. National Park Service. That program also focused on the collection of data concerning traditional subsistence among Chukotka indigenous peoples, with special reference to the use of marine mammal resources (see Table 6).

These long-term programs have demonstrated the success of joint studies by scientists and indigenous people and the need for integration of traditional and scientific knowledge. Of course, Votrogov, Bogoslovskaya and Krupnik were far from being the first Russian scientists to interview local people about behavior and migrations of marine mammals. Many references to the importance and value of traditional environmental knowledge are to be found in the earlier studies by Russian biologists and ethnographers of the 1930s–1950s, including Avenir G. Tomilin, P.N. Nikulin, Boris A. Zenkevich, Vladimir A. Arsenyev, Ekaterina S. Rubtsova, and others. However, the Chukotka bowhead whale studies over the last 25 years contributed an absolutely new insight into traditional knowledge and the role of local hunters in long-term research. We can now claim with confidence the equality of traditional and scientific knowledge about marine mammal populations, in terms of its systemic approach, profundity, and scientific accuracy. The sections below offer a review of the current data about the bowhead whale off Chukotka obtained as a result of such joint effort by scientists and marine hunters.

NEW DATA ON THE BIOLOGY OF BOWHEAD WHALES OFF CHUKOTKA

The following summary is based upon my personal observations of 1978–1998 (Table 1, Figs. 2–4), surveys of the available literary sources, and information obtained from Native marine hunters from several local communities (Tables. 2–5, Figs. 5–14). It provides an insight into seasonal distribution, population dynamics, and behavior of bowhead whales off Chukotka. Since the main body of scientific data on the biology of that species was presented in earlier studies (Bogoslovskaya 1995; Bogoslovskaya *et al.* 1982, 1984, etc.), the main objective of this review is to organize the information contributed by the indigenous people of Chukotka in a more systematic and generalized manner.

Spring Migration

Numerous authors have noted the persistence of bowhead whales in relation to preferred habitat and seasonal migration routes. In fact, Karl Bogdanovich (1901:20) wrote almost 100 years ago that despite the detrimental impact of American commercial whaling on the Chukotka bowhead stock, '[...] It still sticks to Cape Chaplin in its spring northward migration.'

According to the records from several years of scientific observation in northern Alaska, the bowhead whales first arrive in March and this first appearance is subsequently followed by three waves ('pulses') of spring migration (Moore and Reeves 1993:336-38):

- the first wave ('pulse') in early April– mid-May;

- the second wave ('pulse') in mid–May;

- the third (and less defined) wave ('pulse') in late May-mid-June.

American scientists emphasize the fact that although the ice conditions are highly variable from year to year, the three migration waves ('pulses') are very constant. In addition, these migration waves are highly differentiated in terms of whale age and sex. The first wave consists primarily of calves; the second wave includes animals of different age and sex; whereas the third wave is made up mostly of females with newborn calves and big males.

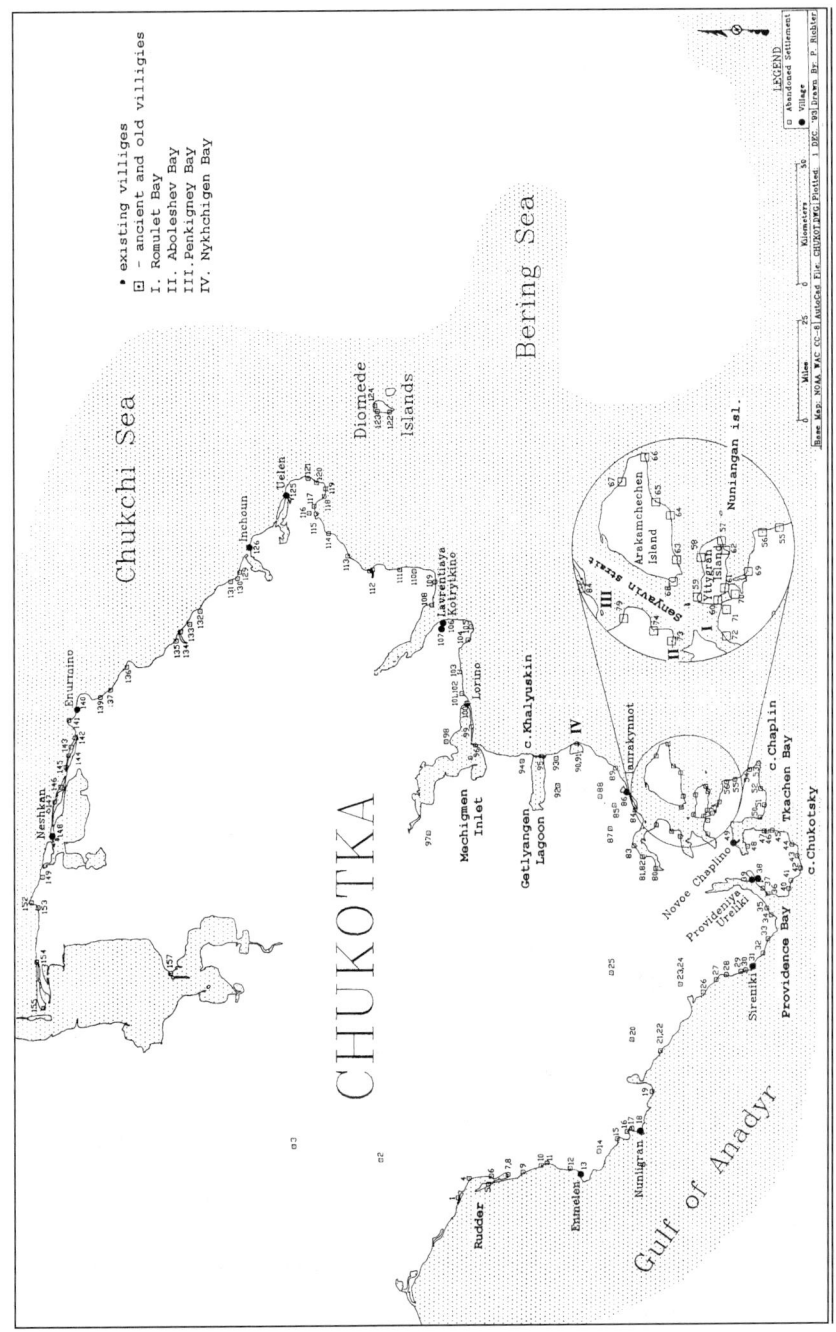

Figure 1: *Map of ancient and modern villages of the Chukotka Peninsula (Bogoslovskaya 1993).*

Table 1. Sightings of 110 bowhead whales in the Seas of the Russian Arctic. *Observer L. S. Bogoslovskaya, aircraft observations during survey and transit flights: May 21–31, June 21–31, June 23–30, 1993*

Data	Site	Number, additional information
05.22	Barents Sea: Franz Josef Land, 150 km NNE of the Graham-Bell Island	7 big whales: 2+3+2 whales in an ice split; 2 , 3, and 7 farther away
05.22	Barents Sea: Franz Josef Land, southern polynya off the archipelago, 50 km S of Northbrook Island	4 whales: 3 big and 1 small whale
05.24	Barents Sea: Novaya Zemlya, 45 km NW of Cape Zhelanya	5 whales: 2 very big single whales and a group of 3 small whales among close fields of drifting floes
05.24	Novaya Zemlya. Entry to Nordenskjold Bay	Group of 4 big whales among drifting floes of 80% ice cover
05.29	Novaya Zemlya. Miller Bay	3 whales: 2 big and 1 small
05.31	Kara Sea: Uedineniya Island, 130 km WSW	4 big whales in polynya
06.02	Kara Sea: Severnaya Zemlya, 90 km NW of Oleny Peninsula, Bolshevik Island	5 big whales: 2+3 in ice leads
06.02	Kara Sea: Severnaya Zemlya, 180 km NNE of Arktichesky Cape, Komsomolets Island	4 whales in small polynya
06.05	Laptev Sea: New Siberian Islands, 120 km NNW of Belkovsky Island	4 whales: 3 big and 1 small whale a wide ice lead
06.07	East-Siberian Sea: Cape Shelagsky, 260 km NNW	5 whales: 3 very big and 2 medium-sized in ice leads
06.07	East-Siberian Sea: Bear (Medvezhyi) Island, 20 km N	2 whales in small polynya
06.14	East-Siberian Sea: Edge of Great Siberian Polynya	Tight group of 8 whales
06.14	East-Siberian Sea: De Long Islands, between Zhokhov and Bennet Islands	2 whales
06.25	East-Siberian Sea: Anjou Islands, 150 km SE of New Siberian Islands	5 very big whales in polynya
06.27	Chukchi Sea: off the village of Vankarem, 40 km NW	30 whales: 10 big+14 big+4 big+2 small in large open water areas
06.27	Chukchi Sea: off the village of Vankarem, 90 km NW	18 big whales, following one another closely among drifting ice floes of 70% cover

216

Table 2: Results of observations (June 1994) of marine mammals from the old village Imtuk (Fig. 1, No. 32). Observer: Panaugy'e T. [Legend: ↑= N; ↓=S; ←=E; →=W; •= lingering; 0 = no animals; — = no observations]

Date	Ice (%)	Bowhead	Gray	Minke	Humpback	White whale	Orca	Walrus	Bearded Seal	Ringed Seal	Larga Seal
01/06	0	1↗	0	0	0	0	0	10-12•			0
02/06	0	2↗	0	0	0	0	0	5-7→↓	0	4-5→↓	0
03/06	0	1↗	1←	0	0	0	0	0	7-8→	2←	0
04/06	0	7-9→	0	0	0	0	0	5-7→	0	0	0
05/06	0	7-10-12→	0	0	0	0	0	2→	0	0	0
06/06	0	7-10→	0	0	0	0	0	0	1→	0	0
07/06	—	—	—	—	—	—	—	—	—	—	—
08/06	—	—	—	—	—	—	—	—	—	—	—
09/06	—	—	—	—	—	—	—	—	—	—	—
10/06	0	0	3←	0	0	0	0	0	8-10→	0	0
11/06	0, mist	0	1•	0	0	0	0	0	0	0	0
12/06	0, mist	0	1•	0	0	0	0	0	2→	0	0
13/06	0, mist	0	1•	0	0	0	0	0	0	0	0
14/06	0	0	4-6↗	0	0	0	0	0	0	0	0
15/06	0	0	0	0	0	0	0	0	8-10→	0	0
16/06	0	0	0	0	0	0	0	0	5→	0	0
17/06	0	0	1•	0	0	0	0	0	0	0	0
18/06	0	0	1•	0	0	0	0	0	0	0	0
19/06	0	0	1•	0	0	0	0	0	0	0	0
20/06	0, mist	0	0	0	0	0	0	0	0	0	0
21/06	0, mist	0	0	0	0	0	0	0	0	0	0
22/06	0, mist	0	0	0	0	0	0	0	0	0	0
23/06	0, mist	0	0	0	0	0	0	0	0	0	0
24/06	0	0	3←	0	0	0	0	0	0	0	0
25/06	0, mist	0	0	0	0	0	0	0	0	0	0
26/06	0, mist	0	0	0	0	0	0	0	0	0	0
27/06	—	—	—	—	—	—	—	—	—	—	—
28/06	—	—	—	—	—	—	—	—	—	—	—
29/06	—	—	—	—	—	—	—	—	—	—	—
30/06	—	—	—	—	—	—	—	—	—	—	—

Table 3: Results of observations (June 1994) of marine mammals from the Enmelen village and its vicinity (Fig. 1, No. 13). Observer: Gyrgol'naut G. [Legend: ↑= N; ↓=S; ←=E; →=W; •= lingering; 0 = no animals; — = no observations]

Date	Ice (%)	Bow-head	Gray	Minke	Hump-back	White whale	Orca	Walrus	Bearded Seal	Ringed Seal	Larga Seal	Ribbon Seal
01/06	0	1↑←	2	0	0	0	0	5-10↓	10-20↓	10-20↓	5↓	0
02/06	0	0	2↑←	0	0	0	0	0	5-10↓	5-10↓	0	0
03/06	0	2↑←	1↑←	0	0	0	0	1	20-40	10-20	0	0
04/06	0	2↑←	1↑←	0	0	0	0	1000•	0	0	0	0
05/06	0	0	2•	0	0	0	0	1000•	0	0	0	0
06/06	0	2←	2←	0	0	0	0	20-30↓	10-20↓	20-50↓	5-10↓	0
07/06	0	2←	3↑→	0	0	0	0	1↓	5-10↓	10-20↓	0	0
08/06	0	1←	2	0	0	0	0	40-50↓	10-20	10-20↑→	0	0
09/06	0, mist	2←	3↑→	0	0	0	0	5-10↓	50	10-15	0	0
10/06	0, mist	2←	1	3↓	0	0	0	0	10-20↓	10-20↓	0	0
11/06	0	2•←	1•←	0	0	0	0	0	5-10↓	20-40	0	0
12/06	0	3•←	2↑	0	0	0	3-5•↑	0	5-10	10-20	10-20	0
13/06	0	1↑	3	0	0	0	0	0	20-40↓	20-40↓	0	8-10↓
14/06	0	4↓	5↓	0	0	0	0	5	0	0	0	0
15/06	0	2•←	3•↓	0	0	0	5↓	10-15↓	1-5↓	15-20↓	0	0
16/06	0	2/•??	0	0	0	0	0	0	0	0	0	0
17/06	0	2↓	1•↓+3↑	0	0	0	0	0	0	0	0	0
18/06	0	2←+1↓	1↓	0	0	0	0	0	0	0	0	0
19/06	0	0	5•+6+3•→	0	0	0	0	0	10-20↓	5-10↓	0	0
20/06	0	0	1•←+2↑	0	0	0	0	0	40-50↓	10-20↓	0	0
21/06	0	2	3+4+5	0	0	0	0	0	10-20↓	10-20	0	0
22/06	0	2•←	2←+3↑+1•↑	0	0	0	0	2↑	0	80-100↓	0	0
23/06	0	2•←+2↑	2•←+1↓	0	0	0	0	0	20-40	20-40	0	0
24/06	0	0	3+2→	0	0	0	0	2	80-100↓	80-100↓	0	0
25/06	0	0	2+2→	0	0	0	0	0	10-20↓	10-20↓	0	0
26/06	0	3↓	2•+2↓	0	0	0	0	0	10-20↓	10-20↓	0	0
27/06	0	6↓	1•+2+1•	0	0	0	0	1	20-40↓	10-20	0	0
28/06	0	5↓+2•→	1↓	2↓	0	0	5↑	1↓	10-20↓	10-20↓	0	0
29/06	0	2•→	2→	0	0	0	0	1↓	>40↓	100↓	0	0
30/06	0	0	2→	0	0	0	0	0	10-20↓	10-20↓	0	0

Table 4: Results of observations (July 1994) of marine mammals from the Enmelen village and its vicinity (Fig. 1, No. 13). Observer: Gyrgol'naut G. [Legend: ↑= N; ↓=S; ←=E; →=W; •= lingering; 0 = no animals; — = no observations]

Date	Ice (%)	Bow-head	Gray	Minke	Hump-back	White whale	Orca	Walrus	Bearded Seal	Ringed Seal	Larga Seal	Ribbon Seal
01/07	0	2	2	0	0	0	0	10-20	40-50	0	0	0
02/07	0	1	2	0	0	0	0	10-20	0	100-150	0	0
03/07	0	0	2	2	0	0	5	5	100-150	50-100	0	0
04/07	0	3	2	0	0	0	0	10-20	200-250	100-150	0	0
05/07	0	0	2	0	0	0	0	1	10-20	10-20	0	0
06/07	0	0	2	0	0	0	0	0	0	0	0	0
07/07	0	2	1	0	0	0	0	1-5	1-5	1-5	0	0
08/07	0	3	1	0	0	0	0	10-20	1-5	1-5	0	0
09/07	0	1	2	0	0	0	0	10-20	1-5	1-5	0	0
10/07	0	0	1	0	0	0	0	10-20	0	1-5	0	0
11/07	0	0	3	0	0	0	0	10-20	0	1-5	0	0
12/07	0	6	3;1	0	0	0	5	10-20	1-5	1-5	0	0
13/07	0	1	2	0	0	0	0	20-40	0	0	0	0
14/07	0	0	2	0	0	0	0	10-20	0	0	0	0
15/07	0	2	3	0	0	0	0	10-20	0	0	0	0
16/07	0	1	3	0	0	0	5	40-50	1-3	0	0	0
17/07	0	3	8;6	0	0	0	0	10-20	1-3	1-5	0	0
18/07	0	0	3;4	3	0	0	0	0	1-3	5-10	0	0
19/07	0	3;2	3;5	0	0	0	5	200-400	2-5	2-5	0	0
20/07	0	0	3	0	0	0	0	0	0	0	0	0
21/07	0	2	2	0	0	0	0	80-100	1-5	1-5;1-3	0	0
22/07	0	0	0	0	0	0	0	0	1-3	0	0	0
23/07	0	0	1	0	0	0	0	0	0	0	0	0
24/07	0	1;4;3	0	0	0	0	0	0	0	0	0	0
25/07	0	3	1	0	0	0	0	0	0	0	0	—
26/07	—	—	—	—	0	—	—	—	—	—	—	—
27/07	—	—	—	—	0	0	0	—	—	—	—	—
28/07	0	4	2	1	0	0	0	0	0	0	0	0
29/07	0	0	2	0	0	0	0	10-20	0	0	0	0
30/07	0	0	1	0	0	0	2	5-10	2	2	0	0
31/07	0	0	1	0	0	0	0	0	0	0	0	0

These same migration waves occur roughly at the same time in southeastern Chukotka, although the first wave there begins earlier than in Alaska, usually in late March. Unlike northern Alaska, however, their direction and extent vary at different sites along the Chukotkan coast (Ainana *et al.* 1995, 1997; Table 2–4; Fig. 5-8):

> *Near the Village of Enmelen at Cape Bering* (Fig. 1, No 13): The spring migration is mostly directed southward.

> *Near the Village of Nunligran at Preobrazhenya* Bay (Fig. 1, No 18): In March-June the stock movement actually splits, with several whales migrating in two opposite directions: one group moves westward, toward the village of Enmelen while the other goes southeastward, towards the village of Sireniki.

> *Near the Village of* Sireniki (Fig. 1, No 31): From late March and early April, all whales are moving along the coast in a southeastern direction, towards Cape Chukotsky, which constitutes the southernmost edge of the Chukchi Peninsula.

In the course of spring migration, the largest concentration of bowhead whales is usually recorded in the vicinity of the old abandoned village of *Imtuk* (Fig. 1, No 32) and the lagoon bearing the same name. There are numerous reasons to believe that in the area from Imtuk Lagoon to Cape Chaplin, the whales from the Sireniki polynya meet with whales that winter in the more southerly sections of the Bering Sea. Although the Yupik villages of Sireniki *(Sighinek)* and Imtuk *(Imtuk)* were only 8 km apart, that small distance makes the boundary between two regions. The one extends further west and covers the area where migration of the 'Sireniki whales' is taking place, and the other spreads to the east-southeast, where 'Sireniki' whales usually merge with those from other groupings.

This is indicated by the sharp differences in the number of migrating whales off the village of Sireniki and near the *Imtuk* lagoon, to which local hunters pointed repeatedly. For instance, on May 25, 1978, only a single whale was recorded off the village of Sireniki (according to Andrey Ankalin's observations). At the same time, '[...] there were so many whales [near the Imtuk Lagoon] that the hunters detoured around them in their walrus-skin boats in order to avoid collision' (Bogoslovskaya *et al.* 1984:198-199). Again, on May 25–29, 1994, two experienced local observers sighted only a single whale off the village of Sireniki. On the same day, near the Imtuk Lagoon, Timofei Panaugye *(Pangawyi)* observed two large groups of bowhead whales "[that were] so numerous that it was impossible to count them." The whales were moving toward the east-southeast and Provideniya Bay (Ainana *et al.* 1995:58).

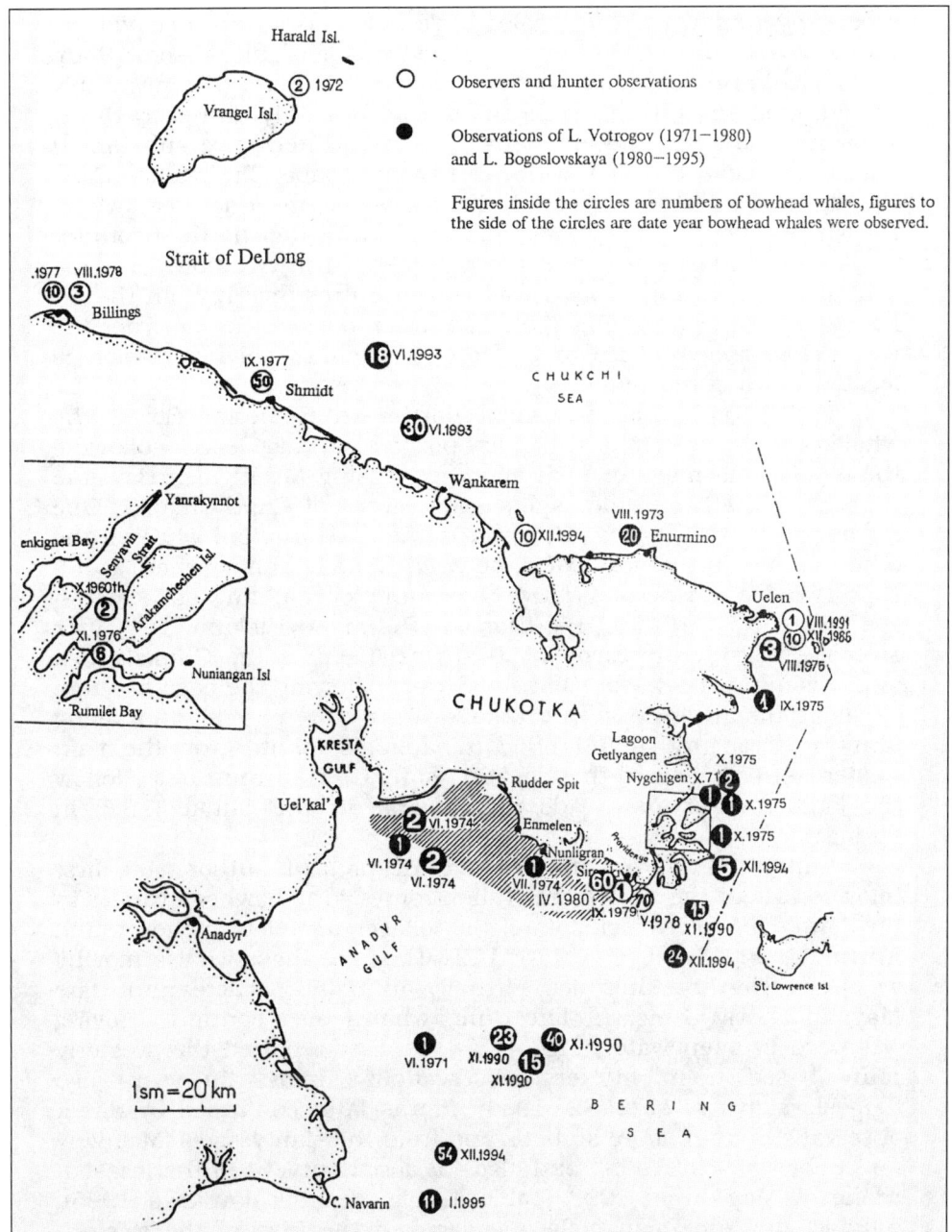

Figure 2: *Observations of bowhead whales in the area around the Bering Strait and the Strait of DeLong*

Numerous elderly hunters from the village of Sireniki reported repeatedly that when the whales were not passing by their village, hunters used to travel to the abandoned village of Imtuk (some 8 km to the east) and farther on to Provideniya Bay. There, they invariably encountered several big bowheads during late May and early June. Thus, our data and observations by local hunters give substantial reasons to believe that the Imtuk Lagoon—Cape Chukotsky—Cape Chaplin Region (Fig. 1, Nos. 32, 42, 53) is the 'meeting area,' where the whales coming from the Sireniki polynya further north encounter whales coming from the Khatyrka–Navarin Region in the southwestern Bering Sea, as well as those presumably from the St. Lawrence Island area (Fig. 15). The latter normally come from the vast fields of sparse ice and a large stationary polynya, which is located south of the island.

For their further movement northward to Bering Strait, the whales use the system of stationary polynyas and ice leads. Following the continuous mass of solid shore-ice closing Mechigmen Bay, the route of migrating whales splits again north of Cape Chaplin. One stream is directed eastward, towards the Alaska shore whereas the other stream moves northward, between Cape Dezhnev (where half a century ago a prosperous Native community of Naukan was located— see Fig. 1, No. 119) and Big Diomede (Ratmanov) Island. The latter stream subsequently turns westward as it enters the Chukchi Sea and it continues towards De Long Strait, following the edge of shore ice along the arctic coast of Chukotka. It is quite possible that along that way some groups and individual animals splinter from the main westerly stream toward the Alaskan shore. On that route, they follow the long, arch-shaped leads that emerge in the Chukchi Sea in spring.

Captain Leonard Votrogov and the present author were first informed about the western migration route of the bowhead whales in the Chukchi Sea by residents of the villages of Uelen, Inchoun, and Enurmino (Fig. 1, No 125, 126, 140). They see those whales moving westward from the shore ice, virtually annually, particularly in late May and early June. In late June whales also continue moving westward in open water (Figs. 8, 9). We first reported this westerly route based upon hunters' observations almost 20 years ago (Bogoslovskaya *et al.* 1982, 1984). It was later confirmed by direct observations from shore stations and from Russian vessels (Melnikov and Bobkov 1991, 1993), and also via aerial surveys in the Russian Arctic in May–June, 1993 (Table 1; Fig. 3; Bogoslovskaya 1995). Initially, our American colleagues greeted the idea of the western route for the bowhead whale spring migration in the Chukchi Sea with great skepticism. However, later observations by Native hunters under the joint Russian–American whale monitoring project put an end to those doubts (Ainana *et al.* 1997).

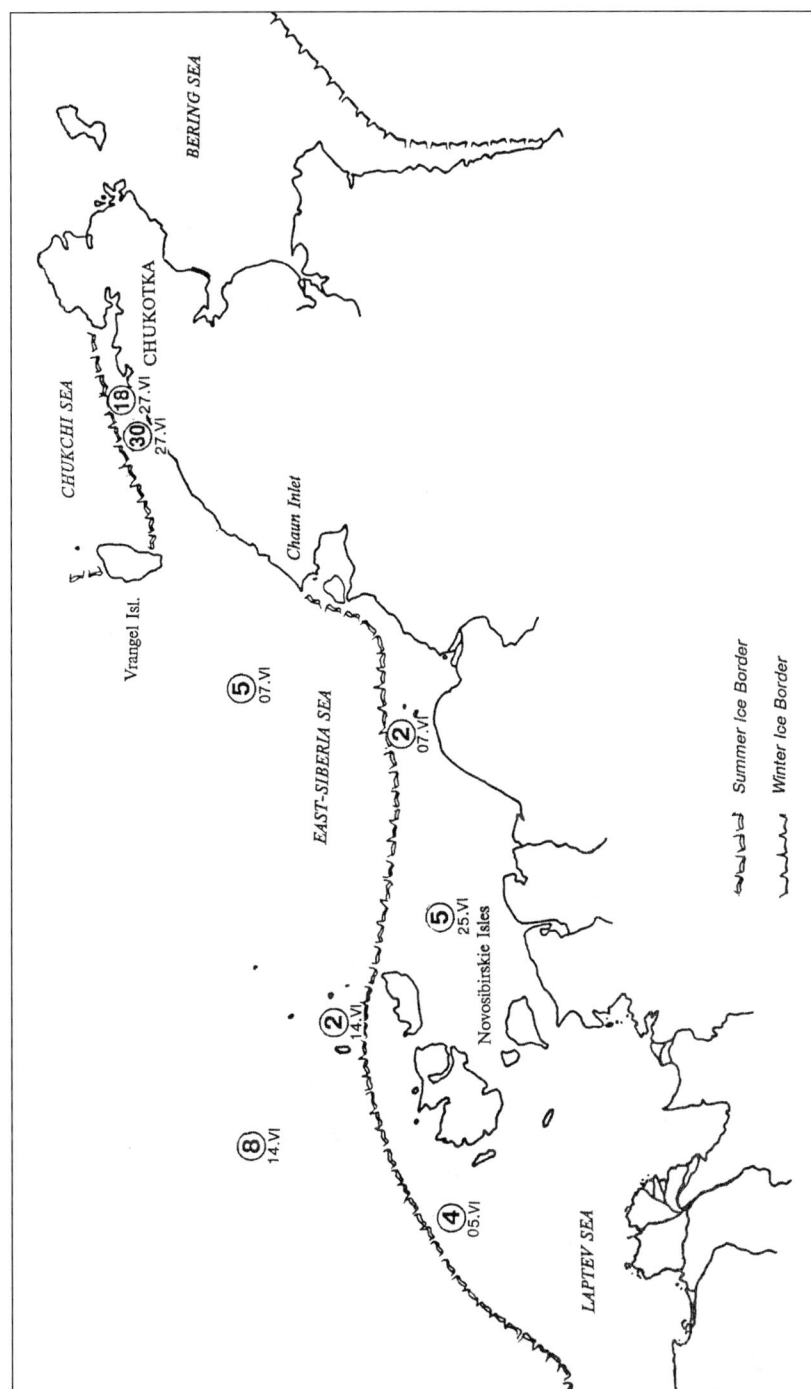

Figure 3. Sightings of bowhead whales in the eastern sector of the Russian Arctic, 1993 (See Table 1).

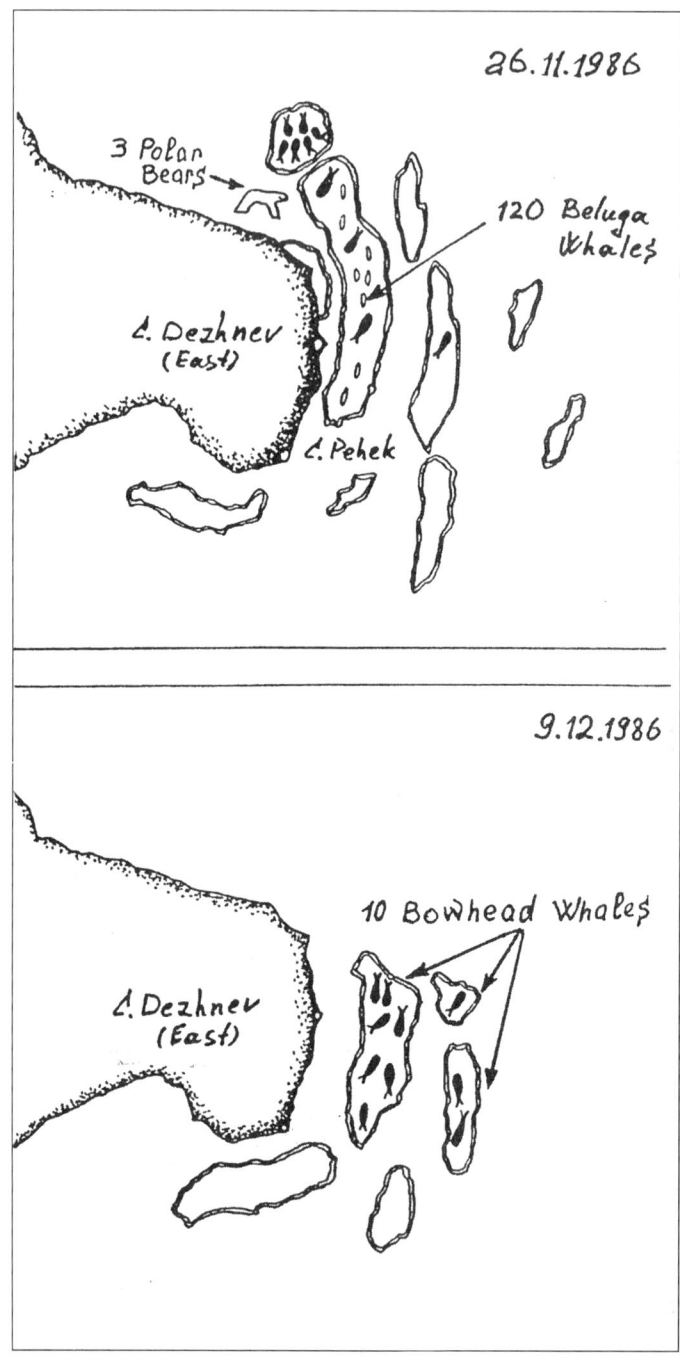

Figure 4. *Bowhead whales and white whales in polynyas near Cape Dezhnev. Sketch by V. Shcrobot, MI-8 Helicopter commander.*

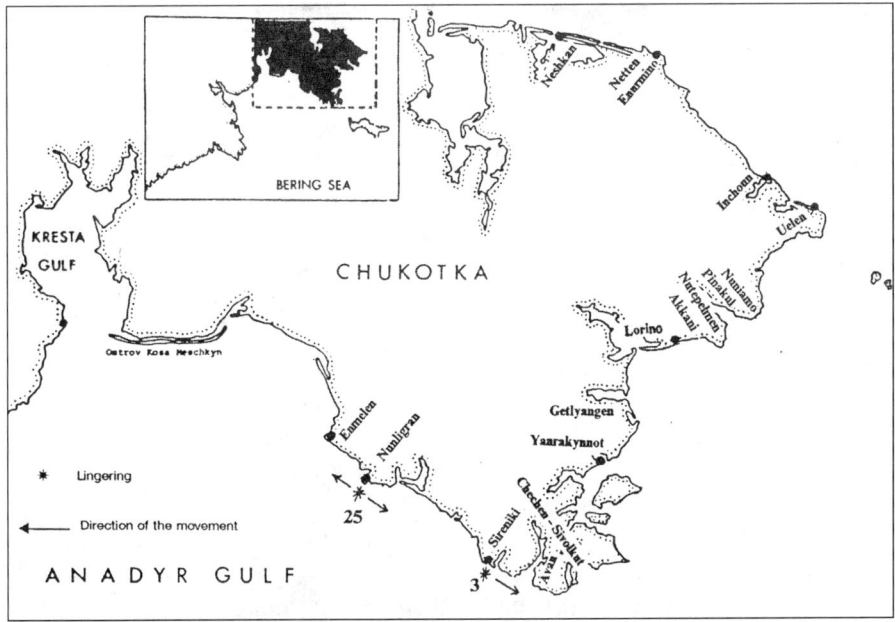

Figure 5. *Number of bowhead whales observed off Chukotka Peninsula: March 1995—28 whales.*

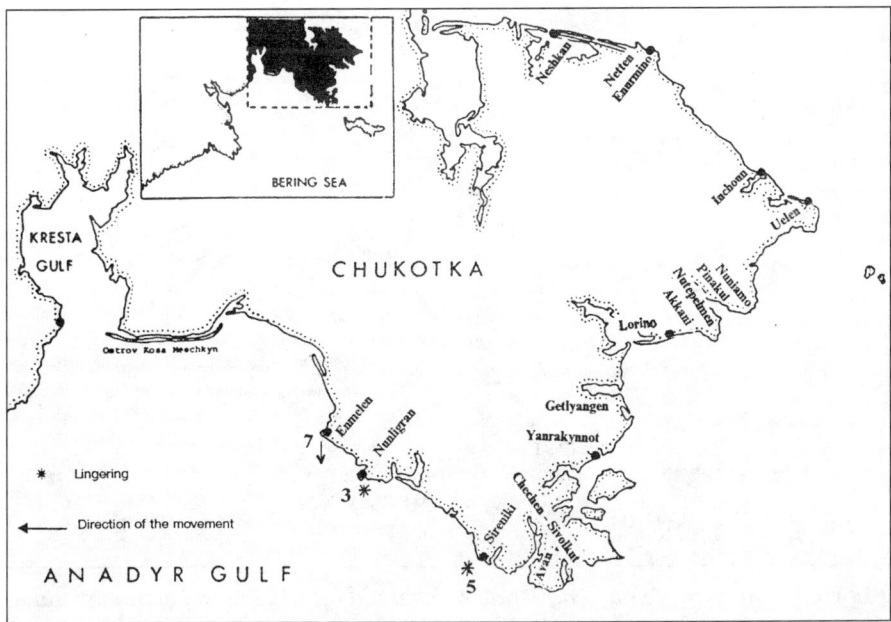

Figure 6. *Number of bowhead whales observed off Chukotka Peninsula: April 1995—15 whales.*

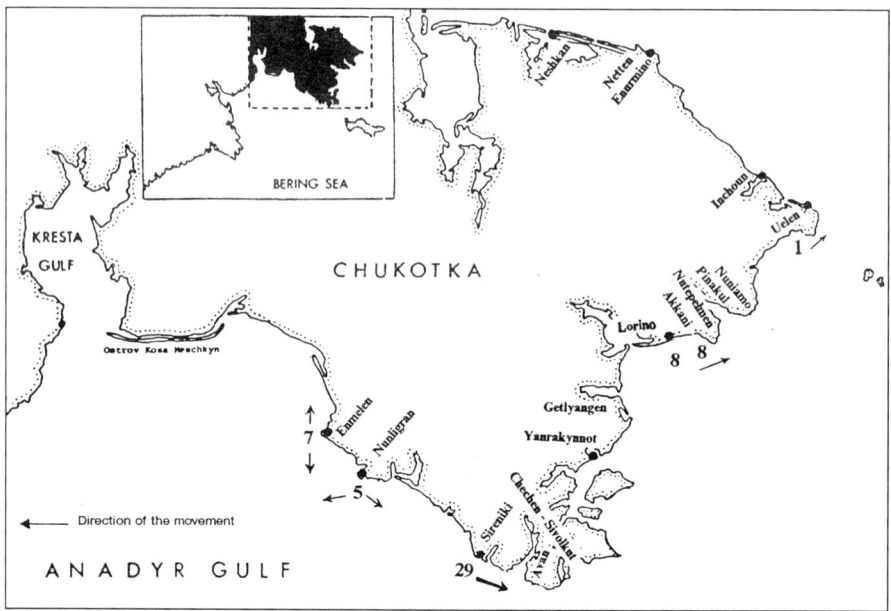

Figure 7. *Number of bowhead whales observed off Chukotka Peninsula: May 1995—58 whales.*

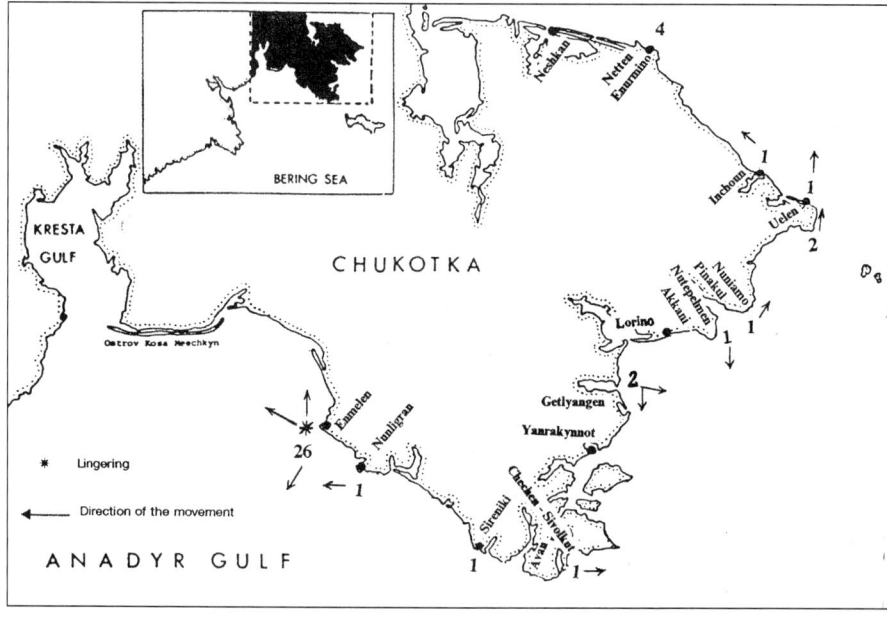

Figure 8. *Number of bowhead whales observed off Chukotka Peninsula: June 1995—41 whales.*

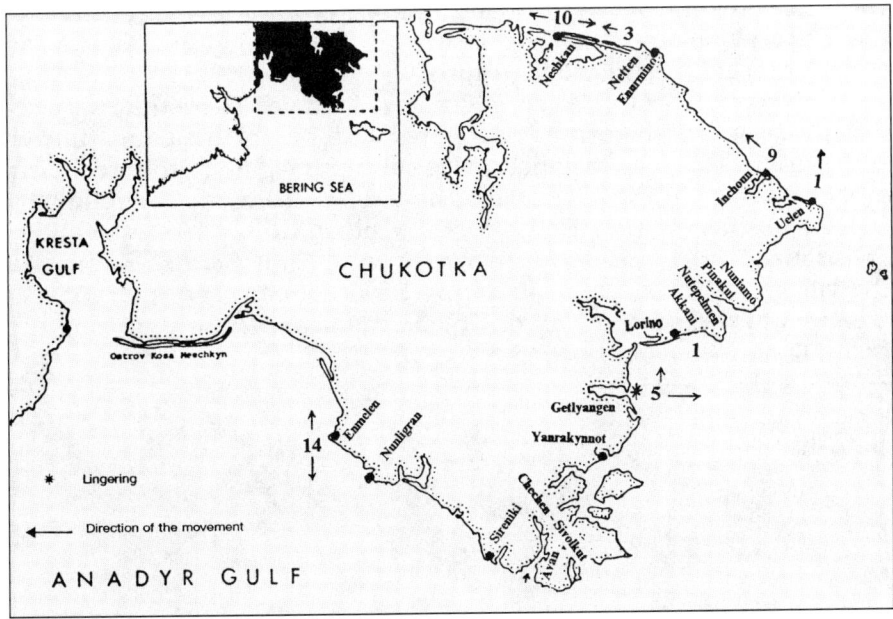

Figure 9. *Number of bowhead whales observed off Chukotka Peninsula: July 1995—43 whales.*

The comparison of the dates for the spring migration of the bowhead whales off Chukotka and Alaska offer substantial support to the following scenario. In the western (Siberian) sector, the main flow of migrating whales enter the Arctic Ocean roughly a month later than in the eastern (Alaskan) sector (Bogoslovskaya *et al.* 1982, 1984).

Fall Migration

According to hunters' observations, supported by statements of other local residents, the westernmost boundary of the bowhead distribution off Chukotka in late summer is probably positioned near Ayon Island and Chaun Bay of the East-Siberian Sea (Fig. 3). Whales can be seen there in August–September but not every year. At this time, they usually approach an area off the village of Billings (near Cape Shmidt) in the western section of the De Long Strait. Most frequently they move in groups of ten or more animals. Local residents repeatedly reported whales 'playing,' that is, jumping out of the water. Off the settlement of Shmidt of the De Long Strait, whales are sighted regularly in late August–September. First, scores of individual whales may be seen and then groups of 60 and more animals arrive. They appear in conjunction with the formation of young ice or the arrival of old drifting ice floes (Fig. 2).

In October, local hunters and scientists have observed bowhead whales in the area between Wrangel and Gerald islands in the

Western Chukchi Sea (see Bogoslovskaya *et al.* 1984). Presumably, these are whales on their westward fall migration from Alaska, as stated several times in the literature.

However, a well defined migration of whales *eastward*, towards Bering Strait is also seen in September–October, at the same time or even earlier. Large aggregations of whales are sighted quite regularly off the village of Vankarem (on the Arctic Coast of Chukotka) by coastal observers (e.g., Bogoslovskaya *et al.* 1982; 1984; Melnikov and Bobkov 1991, 1993). Farther east, from Cape Serdtse-Kamen' to the village of Uelen, right at Bering Strait, a bowhead migration is observed in the second half of October and, particularly, during the first half of November.

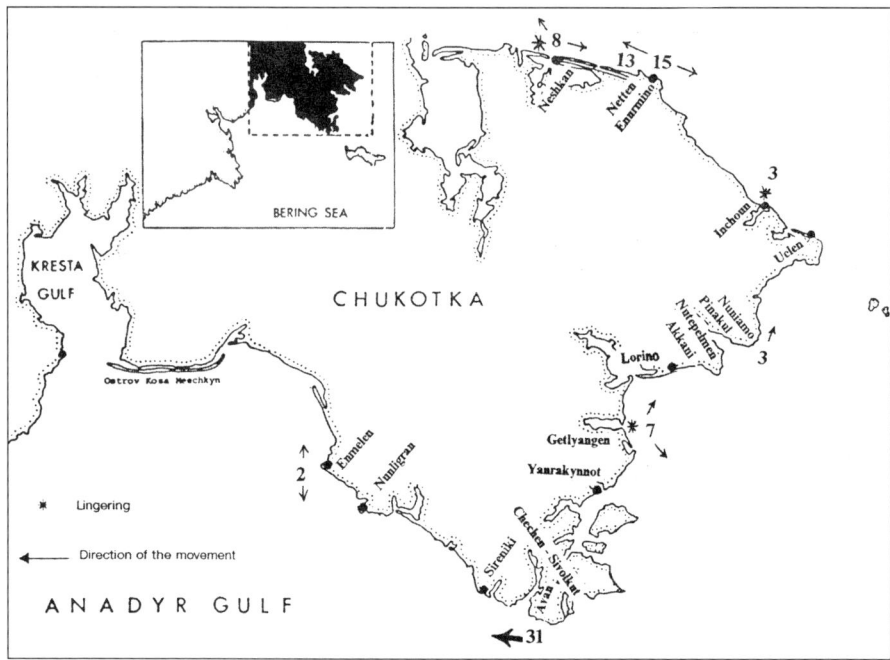

Figure 10. *Number of bowhead whales observed off Chukotka Peninsula: August 1995—82 whales*

During the same weeks (from the second half of October until the first half of November), the mass migration of bowheads is also observed along the eastern and southern shores of the Chukchi Peninsula, from Cape Dezhnev in Bering Strait to Kresta Bay (in the Gulf of Anadyr; Fig. 2). This fall migration is probably best documented off the Yupik village of Sireniki, on the southern shore. Here, the greatest number of whales usually passes by during the first half of November. Due to heavy storms during late fall, hunters rarely go out hunting in boats, but they observe the whales from shore. In the fall, until the arrival of new ice, whales occur off the islands in the Senyavin Strait, and they enter the narrow channels separating the islands and the mainland (Figs. 12–14). In November, 1976, Eugene Paulin watched six or seven whales breaking young ice with their backs in the Senyavin Strait, south of the Kynkai Island (Fig. 2).

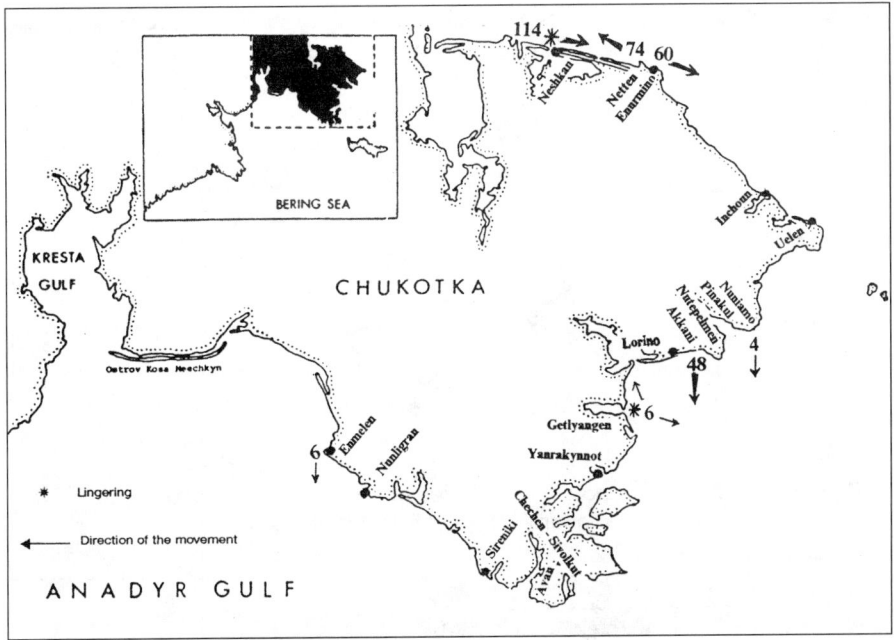

Figure 11. *Number of bowhead whales observed off Chukotka Peninsula: September 1995—312 whales*

Thus, in the Asiatic (Siberian) sector of the Chukchi Sea, the fall migration of bowhead starts very early, probably beginning in late August. There are still too few documented observations to discuss the status of whales that appear off Ayon Island and near the village of Billings in the East Siberian Sea. It is unclear whether these are migrating or summering whales or whether they even belong to the Bering-Chukchi stock. With regard to other more easterly stations along the Arctic Coast of Chukotka, such as Vankarem, whales quite frequently appear there in late summer and during the fall. Further east, off Cape Serdtse-Kamen, their eastward migration towards Bering Strait becomes obvious, particularly off the village of Inchoun and at Uelen at the northern entrance to the Bering Strait. South of Uelen, the fall migration takes place within a more limited period of time compared to the spring migration. It depends little upon the strength of wind, snowfall, or temperature, but it is closely associated with the arrival of sea ice. As such, it can shift by ten or more days to an earlier or later date.

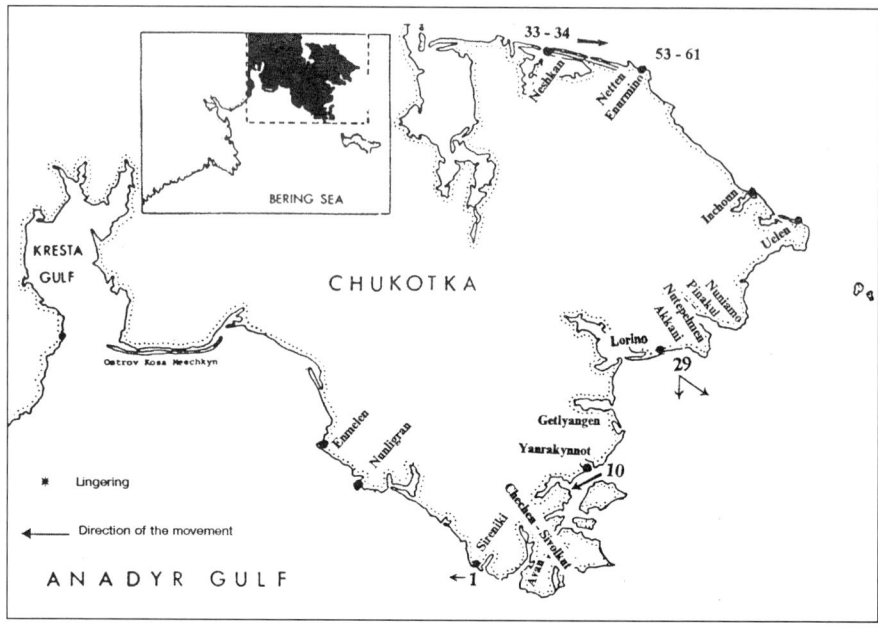

Figure 12. *Number of bowhead whales observed off Chukotka Peninsula: October 1995—115-124 whales*

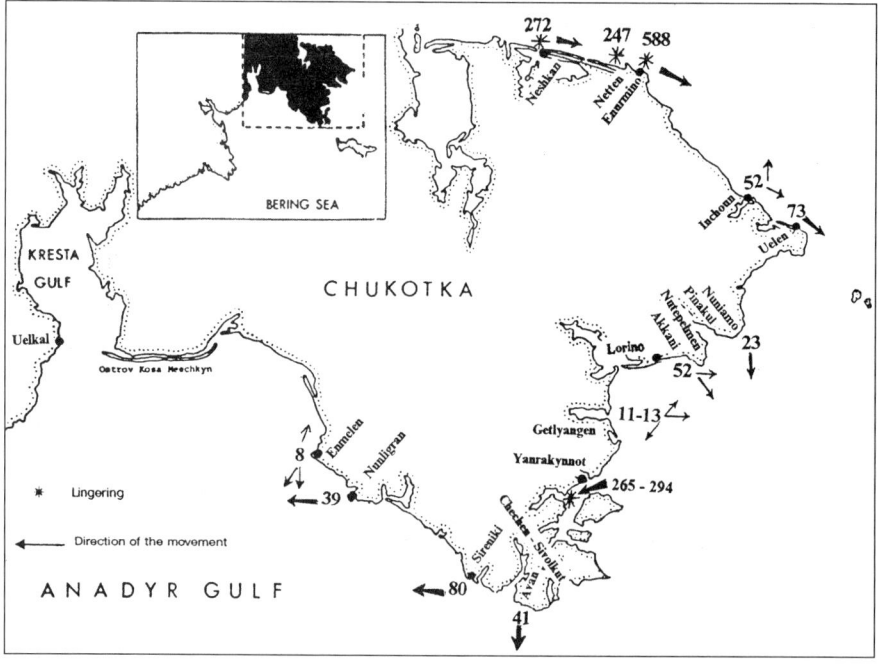

Figure 13. *Number of bowhead whales observed off Chukotka Peninsula: November 1995—1751-1782 whales*

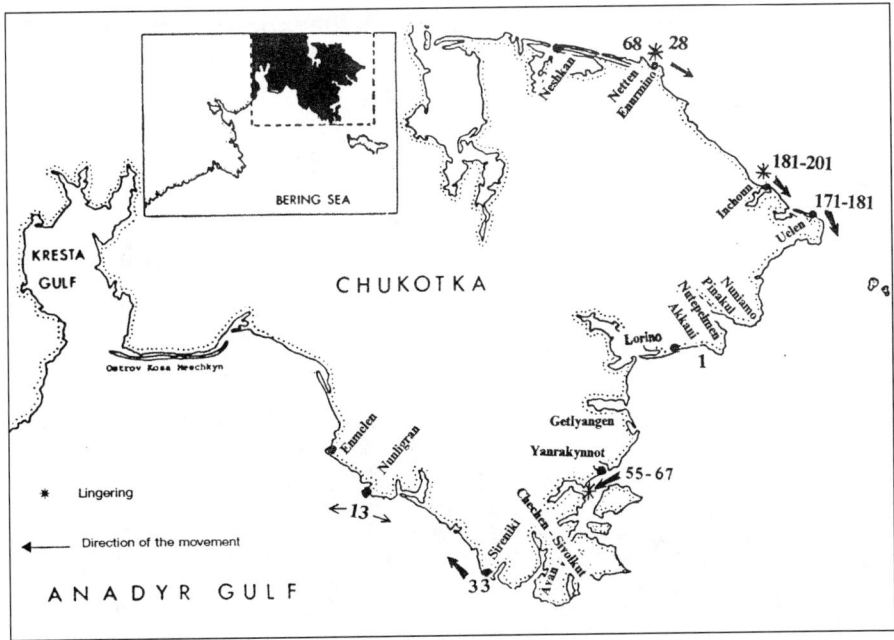

Figure 14. *Number of bowhead whales observed off Chukotka Peninsula: December 1995—550-592 whales.*

It should be noted that in some years many whales are observed concurrently off Sireniki (on the southern shore) and off Uelen (to the north of Bering Strait). For instance, in 1977, 1980, 1985, 1987, the peak of the fall migration took place between November 4 and 12 in Uelen, and during the first half of November in Sireniki. At the latter station there were 'very many bowheads' observed on November 7 and 10. The reported speed of the migration is such that during these few days only a small portion of the 'Uelen' whales (if any?) could reach the area off Sireniki. Subsequently, in Sireniki the second migration wave was observed in late November. Those might have been the 'Uelen' whales.

To summarize, in the Asiatic/Siberian sector the first wave of fall migration is usually observed in September. It starts simultaneously near several stations in Chukotka and proceeds concurrently with the movement of summering whales. The most common span of active fall migration is from late October until the first half of December (Table 5; Figs. 12–14), with the peak in early mid-November. In November, 1995, some very large aggregations of feeding whales were reported at several stations off the coast: 272 whales off the village of Neshkan, 247 at Cape Netten, 585 at Cape Serdtse-Kamen, and 276–307 whales in Penkigney Bay, at the entrance of the Senyavin Strait (Fig. 13). The routes of fall and spring migrations off the coast of Chukotka mostly overlap (Figs. 15-16). However, it is unclear whether the same groups of whales move along

231

the same routes in spring and fall or whether those routes are used by different segments of the whale population in different seasons.

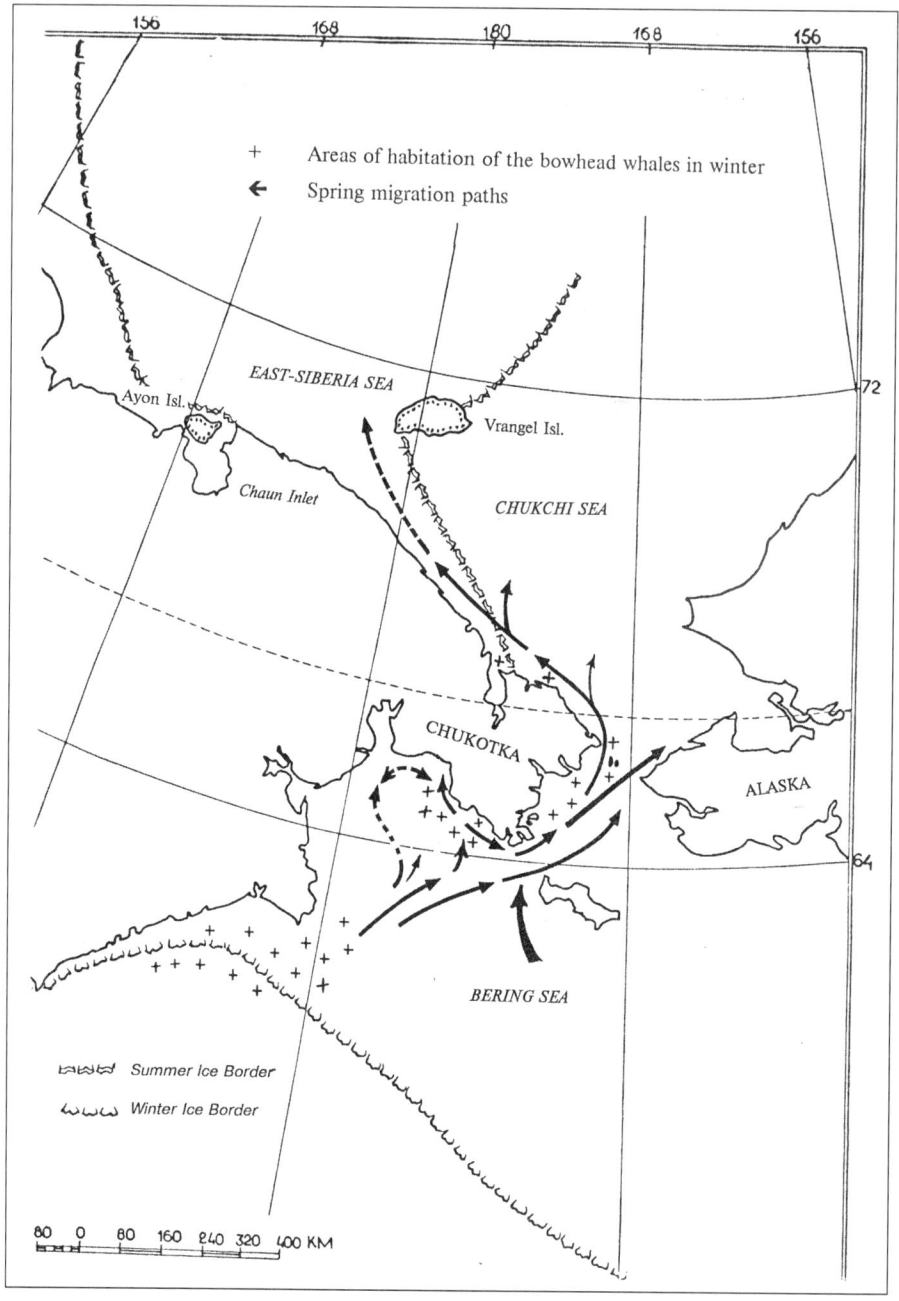

Figure 15. *Movement and location of bowhead whales around the Chukotka Peninsula in winter–spring.*

SHAPE AND SIZE OF WHALES, GROUPINGS, AND ASSOCIATION WITH ICE

According to reports by Yupik hunters from Sireniki, a particular variety ('morph') of the bowhead whale visits their area. They call it 'the other bowhead whale' or *takyshkak (takeshkak)* in Yupik, in contrast to the regular bowhead whale called *aghvepik*. *Takyshkak* is reportedly drop-shaped and it is notably shorter and thicker than the regular bowhead whale well known to hunters and scientists alike. *Takyshkak* can be seen in spring, in the course of the last migration wave (in late May and early June). Such a whale was taken on May 27, 1998, by Petr Typykhkak's (Tepeggkaq) family boat crew, a team of local Yupik hunters from Sireniki.

Interestingly, the Yupik Eskimo from the village of Ungaziq (Old Chaplino—see Fig. 1, No 42) once used the name *takyshkak* to refer to another variety of bowhead whale. This was a long animal with a thin skin and a thin layer of blubber or *mangtak/muktuk* (Rubtsova 1971; author's field data).

During the last 10-15 years, local hunters constantly observe bowhead whales of various sizes; but occasionally some very small ones ('two-walruses long') are sighted near or at Sireniki. This confirms some earlier reports about migration of (adult) whales accompanied by nursing calves towards Provideniya Bay (Bogoslovskaya and Votrogov 1981). The former residents of the village of Avan at the entrance of the bay (Fig. 1, No 40; the village was closed in 1942) remember when whales were also taken there during the summer. Judging from the size of old bones in the prehistoric underground dwellings, most of those whales once taken at Avan were very large animals. By contrast, the old dwellings in the abandoned settlement of Singak (*Singhaq*) to the west of Sireniki (Fig. 1, No 28), were built of skulls and mandibles of very small bowhead whales (see also Krupnik 1993).

Hunters at the Anytykuk (*Angetequq*) camp to the east of Sireniki (Fig. 1, No 33) have been watching bowhead whales in May for many years. They observed whales playing in shallow water, some 150-200 m off the shore. The whales were seen jumping from the water, spinning, and showing pectoral flippers. Andrey Ankalin (*Angqalen*), who was the first to report this behavior, even heard some sound signals of a low-frequency whistle type that accompanied such activity (Bogoslovskaya *et al.* 1982).

A peculiar pattern of whale behavior—also observed by hunters—is the so-called *standing*. A whale may stay at the same site for about a day, directly at or close to the water surface, sinking to a small depth and surfacing again. Some hunters believe that bowhead whales rest and even sleep in that position. During seasonal migration peaks (in May and November), the animals move quietly in very large groups, often without diving deep into the water and obviously in no hurry. Local hunters usually compare them to a 'fleet of barges.' During this time, whales usually approach the shore without being disturbed by humans or boats. Such large groupings are later followed by single whales that seem to be chasing the groups that have already passed. Those individual whales move on fast and they dive frequently (Petr Typykhkak/*Tepeggkaq* and Anatoly Kanikhin, pers. comm.).

Analysis of local observations from several sites in southeastern Chukotka during February–December, 1995, reveal the following patterns of whale behavior:

- In February and September, only a few individual whales were seen at some best positioned observation sites;
- From February through December, the following groupings have been observed:
    o 97 single whales, of which 20 were 'standing at the same site;'
    o 35 pairs of moving whales, of which two pairs were seen 'standing at the same site;'
    o 18 whales in groups of three, of which two groups were seen 'standing at the same site;'
    o 7 groups of four whales, of which two groups were 'standing at the same site.'
- Larger aggregations of whales were always reported during migration peaks. These appeared to be temporary groupings of whales moving in the same direction. The largest aggregations of whales were observed during fall migration in November: near the village of Yanrakynnot (Fig. 1, No 86) on November 25, 1995, 40-45 whales swimming in a westward direction; on November 26, 57-62 whales were migrating westward; and near the village of Sireniki on November 27, 39 whales were moving westward.

Local shore observations reveal that bowhead whales travel in dense ice and open water alike. The first whales are usually seen off the villages of Nunligran and Sireniki in February–March, sometimes in 100% ice cover. Hunters from the village of Yanrakynnot report that the late fall arrival of whales to the Penkigney Bay area (Fig. 1, III) often coincides with the appearance of young ice, which does not deter the whales. The whales raise their backs to break the ice, which is 10–30 cm thick, leaving some characteristic structures that hunters call 'little houses.' However, not infrequently, bowheads get ice-trapped. Figure 4 illustrates the postion of ten large bowhead whales trapped with 120 white whales (*belukha*) in three polynyas off Cape Dezhnev. After two weeks, the white whales left the polynyas, whereas the big whales stayed until Jan. 4, 1987, i.e., for 40 days.

GENERAL PATTERN OF BOWHEAD WHALE DISTRIBUTION OFF CHUKOTKA

Winter Grounds

The distribution of bowhead whales off the Chukchi Peninsula is closely associated with the system of stationary polynyas and leads. In this regard, the large Sireniki polynya is of particular importance. This expanse of open water or thin ice allows whales to remain in the area off the southern coast of the peninsula, even during particularly harsh winters, when the sea-ice cover is about 90–100%. Of great importance too is another big polynya that is formed during December off and along Penkigney Bay. In late fall and early winter, it is visited by hundreds of apparently foraging whales.

The maps (Fig. 2–14) and tables (1–5) provided summarize available data on numbers, distribution, and the direction of movement of bowheads off the Chukchi Peninsula in wintertime. The following pattern is obvious: along the southern shore of the Peninsula (off the villages of Enmelen, Nunligran, and Sireniki), bowheads can be observed every year from February through June, that is, up to the end of the last wave of spring migration.

Summer Grounds

American commercial whalers of the 1800s believed that July to September was the best season to hunt bowheads in the northern Bering Sea and southern Chukchi Sea (Townsend 1935). This information can be matched with recent reports about later dates of spring whale migration along Chukchi Peninsula (compared to the Alaskan side) and also about some whales turning westward from Bering Strait along the arctic coast of Chukotka rather than northeastward, towards the Beaufort Sea. This, together with other evidence points to the former summer grounds once used by a considerable number of whales in the northwestern section of the Bering Sea, the Bering Strait area, and in the western Chukchi Sea. Those whales that summered along the Chukchi Peninsula coast were exterminated by commercial whalers during the second half of the 19th century[3]. Today, however, there remain some individual whales or small groups that spend the summer months in this area; they generally move as individuals or as groups of two or three.

---

[3] See Burns 1993:755, for a different (and more cautious) approach to the issue of the summer bowhead stock in the Bering Strait area that was reportedly exterminated by commercial whaling during the late 1800s—ed.

Table 5. Bowhead whales sighted off the Chukotka Peninsula in 1995, by site (from shore and from boat).

No.	Place	Jan.	Feb.	Mar.	Apr.	May	Jun.	Jul.	Aug.	Sept.	Oct.	Nov.	Dec.	Total
1	Uel'kal													0
2	Enmelen	2			7	7	26	14	2	6		8		70
3	Nunligran		1	25	3	5	1					39	13	89
4	Sireniki		4	3	5	29	1				1	80	33	156
5	Avan								31			41		72
6	Chechen–Sivolkut						1							1
7	Yanrakynnot										10	265-294	55-67	330-371
8	Getlyangen						2	5	7	6		11-13		31-33
9	Lorino					8								8
10	Akkani					8		1		48	29	52	1	139
11	Nutepenmen						1							1
12	Pinakul						1					23		24
13	Nunyamo				3				3	4				10
14	Uelen					1	3	1				73	171-181	249-259
15	Inchoun						1	9	3			52	181-201	246-266
16	Enurmino						4		15	60	53-61	588	28	748-756
17	Netten							3	13	74		247	68	405
18	Neshkan							10	8	114	33-34	272		437-438
	Total	2	5	28	18	58	41	43	82	312	126-135	1751-1782	550-592	3016-3098

One local 'summer grouping' off the southern Chukchi Peninsula reportedly remains off the village of Enmelen (see Table 3-5; Fig. 6-11). There, whales may be observed from shore all summer, from late June through September–October, before the beginning of the fall migration. These shore observations are in full accordance with reports by Votrogov, who frequently encountered summer aggregations of large baleen whales, including bowheads, in the area between Enmelen and Rudder Bay, some 30 mi. westward (Votrogov, unpublished data of 1963; 1969–1982). Several photographs of bowhead whales taken from the catcher boat *Zvezdny* in July–August are preserved in Votrogov's personal archives. There is a good photograph of a female bowhead, 14 m long, taken by Votrogov's boat and delivered to Enmelen in July, 1970.

Another small local grouping frequents the area off Getlyangen Lagoon, where whales may be observed almost every year from June through August (Table 5; Fig. 8–10). In summer and fall of 1971 and 1973–1975, Votrogov's catcher boat, *Zvezdny*, encountered such 'summer' bowhead whales at various sites along the southern and eastern shore of the Chukchi Peninsula, from the Gulf of Anadyr up to Bering Strait. Local people report summer sightings of bowhead whales along the entire shoreline of the Chukchi Peninsula, from Kresta Bay in the south to Cape Vankarem in the Arctic Ocean and further westward, up to Cape Billings and Chaun Bay in the East Siberian Sea (Fig. 2). The year 1995, for which we have the most detailed record, was no exception. Individual animals and groups of whales were reported by local observers at several sites, from Sireniki on the southern shore to the village of Neshkan on the Arctic Coast (Table 5; Fig. 8–10). Altogether, 11 whales were seen in June, 36 in July, and 80 whales in August, that is, well before the beginning of fall migration. Thus, we may assume that bowhead whales stay off the Eastern Chukotka shores all year round, forming large aggregations during the time of spring and fall migrations.

The area off Cape Navarin (at the southern boundary of the Gulf of Anadyr) is a critical region for the Bering-Chukchi bowhead whale stock or, at the least, for its Chukotka component. While observing whales from an airplane in November 1990, we sighted four groups (with a total number of 94 animals) that were moving across the Gulf of Anadyr, from Cape Chaplin toward Cape Navarin. Two additional sightings of 54 and 11 whales, respectively, were reported off Cape Navarin on the same day (Fig. 2). In November 1995, an observer at the abandoned village of Avan sighted a group of 41 whales (Fig. 13) that was, again, moving almost directly south, towards Cape Navarin (Ainana *et al.* 1997). These observations indicate that the region off Cape Navarin is the most obvious wintering ground for a considerable number of bowheads of the Bering-Chukchi stock.

THE STRUCTURE OF THE BERING-CHUKCHI STOCK

The issue of geographical (sub-population) structure of the Bering-Chukchi bowhead stock[4] has never been considered in detail. Today, we may rely upon both observations by scientists (e.g., Bogoslovskaya *et al.* 1982; 1984; Bogoslovskaya 1995; Melnikov and Bobkov 1991, 1993) and on the data collected in the course of a joint whale monitoring program by the Society of the Yupik Eskimo of Chukotka in 1994–1996 and the Association *Naukan* in 1992–1996. Matching of various observation records brings us to the following conclusions:

(1)   There are two routes of spring migrations of bowhead whales through Bering Strait, from their winter grounds in southern and central Bering Sea region:
– Northeastward, along the coast of Alaska. This route is well known and is well described in the literature; it is used for regular counting of migrating bowhead whales at ice camps off Barrow;
– Westward, along the northern coast of the Chukchi Peninsula, towards De Long Strait and the East Siberian Sea. The whales follow the system of polynyas and ice leads formed in the western section of the Chukchi Sea. This route has been known to Bering Strait area Native hunters for generations but it was not reported in the scientific literature until recently (Bogoslovskaya *et al.* 1982).

(2)   Some bowhead whales do not move northward to the Chukchi Sea and stay for the summer off the Chukchi Peninsula. In addition to individual whales and small groups observed at numerous sites (see above), two regularly occurring local groupings have been documented. One uses the summer grounds off the village of Enmelen at Cape Bering and the other frequents the Getlyangen Lagoon area (Tables 3–5).

(3)   The presence of bowhead whales has been documented for spring, summer, and fall in the De Long Strait, the East Siberian Sea, and the Laptev Sea areas (Fig. 3) as well as in some other areas in the Russian High Arctic (Bogoslovskaya 1995).

These data, I believe, indicate that the Bering–Chukchi bowhead whale stock consists of at least three distinct groupings that are observed during the summer months:

---

[4] The 'Bering-Chukchi stock' is currently the most common name for this population of bowhead whale that inhabits the northern section of the Bering Sea, the Chukchi, and the Beaufort Sea. However, several other names are often used, such as 'Western Arctic,' 'Bering Sea,' 'Bering–Chukchi–Beaufort' stock, and others (Moore and Reeves 1993:314).

Grouping A: Uses the summer grounds off the southeastern and eastern Chukchi Peninsula, in the Gulf of Anadyr, Chirikov Basin, and Bering Strait;

Grouping B: Uses the summer grounds off the Arctic Coast of the Chukchi Peninsula, in the western Chukchi Sea and the eastern section of the East Siberian Sea; and,

Grouping C: Uses summer areas off the northern coast of Alaska and western Canada, in the eastern Chukchi Sea and the Beaufort Sea.

Whales from these areas meet and interact regularly during their seasonal migrations and in winter. The most intriguing issue is the possibility of genetic exchange between these three (summer) groupings of bowhead whales; within the Bering–Chukchi stock as a whole; and between the Bering–Chukchi whales and other large geographical groupings of bowheads (for instance, those that inhabit the more westerly sectors of the Arctic Ocean). This crucial issue requires additional research.

All three groupings listed above compose the Bering–Chukchi stock of bowhead whales. They are very much comparable to the three breeding groupings of the Pacific walrus identified by scientists that are located (in wintertime) in outer Bristol Bay, in the Cape Navarin area, and in the Sireniki polynya area. These three groupings form the Pacific walrus stock.

THE STATUS AND NUMBERS OF BOWHEAD WHALES OFF CHUKOTKA

The total number of bowheads off Chukotka, as well as that of the share of 'Chukotka whales' in the combined Bering-Chukchi bowhead stock, can now be assessed on the basis of various new data. These include: recent publications, scientific surveys, and files of observation logbooks by Native observers from the *Yupik* Eskimo Society of Chukotka and the Association *Naukan*, who worked in 1994-1996 under a project funded by the Department of Wildlife Management, North Slope Borough (Ainana *et al.* 1995, 1997).

As already stated, the spring whale migration along the southeastern shore of the Chukchi Peninsula takes place approximately at the same time as that off Point Barrow in North Alaska. Therefore, a better total population estimate could presumably be obtained by counting whales simultaneously at several sites in Chukotka and using the same technique(s) at the spring whale counts off Barrow.

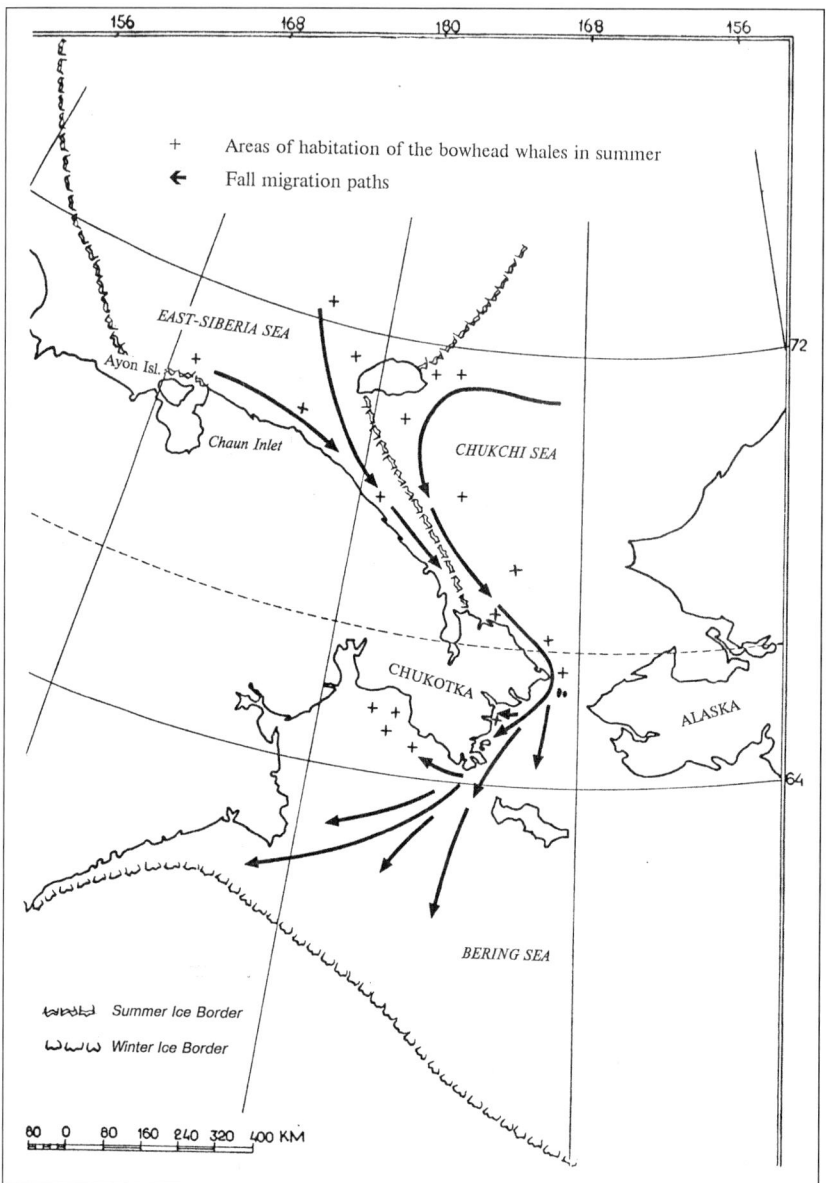

Figure 16. *Movement and location of bowhead whales around the Chukotka Peninsula in summer–fall.*

The most convenient sites to observe and count whales migrating from the Sireniki polynya are located at or near the villages of Kenlegaq and Sireniki (Fig. 1, Nos. 30-31) and at Cape Ulyakhpen, between the village of Sireniki and Imtuk Lagoon. The counting of whales belonging to this group, along with the animals arriving from more southerly areas such as the Khatyrka-Navarin region and St.

Lawrence Island, could best be accomplished from the abandoned village sites of Imtuk, Avan, and Chaplino (Ungaziq). Several capes at the entry to Provideniya Bay, such as Cape Stoletiya, Cape Lesovsky, and Cape Lysaya Golova, may also be considered among the best potential observation/census sites on the southeastern Chukchi Peninsula.

Our hypotheses about the whales' subsequent whereabouts, after they pass this area on their annual spring migration, are as follows:

1. Some 'Siberian' bowheads stay off the southern and eastern coast of the Chukchi Peninsula during the summer months. For instance, some bowheads are regularly observed off Enmelen (Cape Bering) in the northern section of the Gulf of Anadyr, Getlyangen Lagoon, and other places.

2. Some whales evidently migrate slowly along the eastern and northeastern coast of the Chukchi Peninsula and then proceed northeastward towards Alaska. These may be late migrants that pass Barrow, when local spring observation and whale censuses are already completed, because of the disintegration of shore-fast ice.

3. Some whales after they pass through Bering Strait remain in the western sector of the Chukchi Sea during the summer months. Berzin (1981) noted that the total number of those whales might be as high as 1000–1500; and,

4. Finally, some whales move further westward, through De Long Strait, into the East Siberian Sea (Bogoslovskaya 1995).

Finally, some whales would not be counted along the southeastern section of the Chukchi Peninsula, from Enmelen-Sireniki to the Getlyangen Lagoon. Such uncounted whales may well remain in the northern sector of the Gulf of Anadyr, in between the villages of Uellkal and Sireniki. During summer months, they may keep crisscrossing the western section of the Bering Sea, between the Gulf of Anadyr and the Khatyrka-Navarin region (Ainana *et al.* 1995, 1997).

Based upon the wide spectrum of recent data from shore, vessel, and aerial observations during 1978–1998, and hunters' statements, all the above scenarios are quite likely to occur. Thus, one of the main objectives for future surveys and conservation efforts should be to build a working network for thorough and reliable annual counts of all the groupings of bowheads off Chukotka.

Earlier data (i.e., Bogoslovskaya *et al.* 1982, 1984) and more recent on-shore observations by Native hunters revealed that beyond the almost year-round presence of many bowheads off the southern and eastern Chukchi Peninsula, the stock peaks markedly twice a year. It happens during spring migration (May through June), and in the fall and early winter (September through December). The fall migration commonly brings large numbers of whales close to the Chukchi Peninsula. In November 1995, for example, over 3000 whales were counted at one time in the Siberian sector (Table 5).

To summarize, we believe that there is convincing evidence for an annual migration along the Chukchi Peninsula that involves at least 3000–5000 bowhead whales. The majority of them belong to the three groupings A, B, and C, indicated above.

WHERE DID ANCIENT HUNTERS TAKE THE WHALES?

The importance of bowhead whaling in the traditional subsistence economy of Chukotka Native people is amply demonstrated by the sheer number of whale bones preserved in abandoned underground dwellings and storage structures. In addition, numerous ancient ritual structures built of bowhead skulls, mandibles, and, occasionally, scapulae and ribs are reported at many sites along the coast. Such monumental structures abound among the ruins of ancient and old villages, with the most extensive of all, the so-called *Whale Alley*, constructed on Yttygran Island, in Senyavin Strait (Arutyunov *et al.* 1982).

Not long ago, one could also encounter numerous vertical poles made of whale mandibles erected at many sites along the coast (Arutyunov *et al.* 1982). According to a Native hunter, Leonid Kutylin, from the village of Yanrakynnot, these poles were put up after every successful whale hunt '...so that the ancestors could see us hunting and observing the traditions.' A few of the formerly numerous whale bone poles are still standing at some abandoned sites. They are often accompanied by bowhead whales skulls with traces of sacrifices well preserved into the modern days. These are mostly symbolic food offerings—bones of bearded seals, ringed seals, and reindeer, inserted into the bone tissue.

The first detailed examination of whale bone remains, ancient settlement ruins, and ritual structures along the Chukchi Peninsula was conducted in 1977–1981 by the author and Igor Krupnik, with associates (see reports in: Arutyunov *et al.* 1982; Bogoslovskaya *et al.* 1979, 1982, 1984; Chlenov and Krupnik 1984; Krupnik 1983, 1987, 1993; Krupnik *et al.* 1983). In 1985, 1987 and 1988, Russian marine zoologists, working independently, completed coastal surveys of the surface bones at ancient sites along the southern, eastern, and northern shoreline of the Chukchi Peninsula, from Rudder Bay to Cape Serdtse-Kamen in the Arctic Ocean, including on Big Diomede (Ratmanov) Island.

Table 6. Some Statistical Data on the Shore-Based Marine Hunting in Native Villages of Chukotka, 1998.

Districts	Chukotsky		Providensky		Iultinsky	
Number of villages with marine mammal hunting	6		5		4	
Total Native population of the communities with active marine hunting	3700		2086		750	
Number of crews/total number of hunters	20/159		11/68		7/66	
Reported catch by species and region	Animals taken	Meat, kg per person per year	Animals taken	Meat, kg, per person per year	Animals taken	Meat, kg, per person per year
Gray whale	91	47.2	19	17.3	0	0
Bowhead whale	0	0	1	2.3	0	0
Walrus	604	36.9	238	26	65	8.4
Bearded seal	593	6.7	244	4.6	41	0.95
Ringed seal	1728	5.9	92	0.5	487	2.3
Beluga (White whale)	8	1.04	2	0.2	1	0.1
Total		97.74		50.9		11.75

Note: The table is based on the data of the Department of Agriculture, Food, and Fisheries, Chukotka Autonomous Area (*Okrug*) for 1998.

In the course of these surveys, anthropologists and zoologists identified the species of the whale bone and also collected relevant documentary data, including early hunting records for the 20th century, and interviews with the older and the most experienced hunters. The results of these studies were summarized in a series of publications that documented the complex structure of traditional Native whaling in Chukotka (Arutyunov *et al.* 1982; Bogoslovskaya *et al.* 1982, 1984; Krupnik 1984, 1987, 1993; Stocker and Krupnik 1993; Krupnik *et al.* 1983). Bowhead and gray whales were by far the main cetacean species hunted, with occasional and/or small harvests of humpback and beluga whales.

The bowhead whale was the main target of Native whaling in all traditional and ancient communities along the southern and southeastern shore of the Chukchi Peninsula, up to Cape Chaplin and including the small villages on Ittygran Island (Fig. 1, Nos. 13-63). Farther north, in the Mechigmen Bay area, far fewer bowhead whales were taken and the most important prey were young gray whales (Fig. 1, Nos. 66-103).

A considerable number of bowhead bones are preserved in the ruins of two traditional Chukchi villages, Yandogai and Nunyamo (Fig. 1, Nos. 105, 109) located at the two entry capes of Lavrentiya Bay. Ancient and historical bones are quite numerous farther north, particularly at the abandoned village of Naukan (Fig. 1, No. 119), located at the Bering Strait narrows, and also along the Arctic Shore of the Chukchi Peninsula, between the present-day villages of Uelen and Neshkan (Fig. 1, Nos. 125-148). West of Cape Serdtse-Kamen, the large bones are encountered primarily, if not exclusively, in the remains of ancient coastal villages that extend up to Cape Schmidt (North Cape). Judging from numerous scenes of whaling depicted in ancient rock paintings along the Pegtymel River valley (Dikov 1971), ancient Chukotka residents presumably practiced some bowhead whaling even farther west, in the easternmost sector of the Eastern Siberian Sea, around Cape Billings (Bogoslovsksaya *et al.* 1982; Krupnik 1987).

All traditional communities that hunted the largest numbers of bowhead whales were located at the promontories and 'turn-around' capes that projected into the seasonal migration routes of the whales, such as Chaplino, Yandogai, Nunyamo, and Naukan. Other important whaling centers were located at stationary winter polynyas and ice leads, such as Sireniki, Imtuk, Avan, Uelen, Enurmino, Neshkan, and others. Hunters from those communities located at sites with substantial winter shore ice could pursue bowhead whales, during the summer time or in the course of fall migration only (such as the village of Enmelen and smaller villages in the Mechigmen Bay; Fig. 1, Nos. 90-104). Despite that seasonal limitation, bowhead whale bones are encountered in the ruins of almost all ancient coastal villages in Chukotka. Without the 10–20 ft. whale mandibles used as roof beams to support the bone and sod roof cover, it would have been impossible to construct large semi-underground houses (*nenglu*), that used to be the most common type of Native winter dwelling before the mid– or late–19th century.

CHUKOTKA NATIVE PEOPLE SPEAK ABOUT THE BOWHEAD WHALE

The following statements about the bowhead whale were recorded among the Yupik people of Chukotka in 1994–1995, in the course of the survey supervised by the author and conducted under the Inuit Circumpolar Conference (ICC)–sponsored project on the role of whaling in Inuit (Eskimo) communities (Freeman *et al.* 1997). Earlier field interviews and personal letters from Native hunters were attached to the 1994–1995 survey. The opinions expressed below clearly represent people's desire for restoration of Native bowhead whaling in Chukotka that was unilaterally banned by the Soviet Government in 1976, without any consultations with indigenous people. However, in 1997, after several years of efforts and intensive lobbying, Native whaling was restored in a few Native communities on the Chukchi Peninsula under a modest IWC quota.

AINANA (Aynganga)[5], Lyudmila, born 1934, Chair, the 'Yupik' Society of Eskimos of Chukotka

> Our main whale is the bowhead, although our Eskimo[6] from (the village of) Chaplino would take gray whales too. The meat and *mantak* (skin with blubber) of the bowhead whale are of special importance to us. This is the best and the most favorite food of ours. Now, since the authorities deprived us of the possibility of taking that whale, we are just dreaming of it. I, personally, could occasionally eat some meat and *mantak* in Alaska and on the Island [St. Lawrence Island (USA), where many descendants of the former immigrants from Chaplino (Ungaziq) live today—L.B.], when I am visiting someone there or travel on 'Yupik' Society business. But the majority of our people can only dream of this food.
>
> Scientists learned from our hunters for the first time that in spring the route of bowhead whales to the north forks: some whales migrate towards the Alaskan coast as is well known to scientists but others turn west from Bering Strait to pass off the Chukotka coasts. This finding of the second route is an important contribution of our hunters to the science of the bowhead whale. Hence, the goal of the 'Yupik' Society is to ensure the restoration of the quota for our people to hunt bowhead whale.

---

[5] Proper Yupik name spelling is added in parenthesis, according to St. Lawrence Island Yupik transliteration system—ed.

[6] The Yupik people of Chukotka still call themselves 'Eskimo' in Russian. They are also labeled as 'Eskimo' in today's Russian media, scientific and local documentation, and personal papers—ed.

ANKALIN (Angqalen) Andrei, 1941—1994. A boat-captain, outstanding navigator, and experienced hunter from the village of Sireniki. He was the first among indigenous hunters of Chukotka to participate in scientific studies of bowhead whale migrations off his Native village. Died tragically in the tundra (froze to death) together with his son in 1994.

...When I am at sea, I always think of bowhead whales. I expect to see them and to hear them breathe. When I kill a walrus or a bearded seal I am thinking of the bowhead whale. This is our greatest animal. Apparently, in Africa this is the elephant, and with us it is the bowhead whale. Why did we learn about elephants and tigers at school and there is nothing in our textbooks about the bowhead whale?

Today, young folks observe little, and they don't like to merely watch the sea. When in spring I trap for arctic foxes in the Angetekuq camp (ten miles off Sireniki), I always watch bowhead whales playing. I become a whale myself and I am whirling in the water together with them. You will laugh—once I fell from the shore: I forgot where I was and moved on to meet them.

...It is difficult for me to explain what the bowhead whale means to us, the Sireniki Yupik folk. In fact, we were the last people (in Chukotka) to cease hunting (for bowhead whale from skin boats). When hunting was banned it seemed that our soul was killed altogether. I don't know how we are going to live. And without the bowhead we won't be together—each of us has a temper, just look at me, you know my character!

If the bowhead whale continues to be hunted we shall remain. Otherwise, we shall be lost—you will see! So make sure we have a license (quota). If there is no hunt we shall go down quickly, one after another. My brother *Yuggaq* was the first to go and now he is calling to me. [*Yuggaq*, Ankalin's elder brother and a retired boat-captain, shot himself in the fall of 1979—L.B.]

PANAUGYE (Pangawyi) Timofei, 1939—1995. A marine hunter for 44 years and respected boat captain from Sireniki. Perished at sea with his crew in 1995.

All our whale ceremonies are dedicated to the bowhead whale. This is what unites us, the Yupik people from Chukotka and St. Lawrence Island. The hunting, the whale festivals, and all the rites and rules we follow are almost the same, both here and there. Bowhead whale is the only true game for us, for all Sireniki Eskimos. Many residents here would not taste the gray whale meat, even if food is scarce, although we all eat dried intestines, and particularly, *mantak* (skin) of this whale. But the bowhead *mantak* is so much more tasty.

I believe that whaling is very important to us. Today, it is important that our traditional bowhead whaling should be restored. It is loved by Chukchi as well; but to them it is of no special difference what whale it is, gray or bowhead. But it is only the bowhead whale that counts to us.

We, the Eskimos from Sireniki, can hunt the bowhead whales the same way our forefathers once did it: by lancing and harpooning from skin-boats. Maybe things are different in other villages, but we only want the bowhead whale to be taken by hunters themselves.

In Sireniki, we dedicate the greatest festivals to this whale. Reindeer herders from other villages used to come. And they always were together, both when hunting and at the festival.

TYPYKHKAK (Tepeggkaq) Petr, 1933-2000. A renowned boat captain, respected elder from the village of Sireniki. A very experienced observer, with profound knowledge of whale behavior and hunting traditions of the Yupik people.

If the authorities delay issuing licenses (permits) for the hunt, our elders would pass away without tasting the bowhead meat again. That would be so bad for them. And that would be a shame to us as well, as we, the hunters, have not been able to provide for them.

You are asking me (questions) about whaling, and I am telling you stories and I feel as if I went out to sea and I am looking round in search for the animals. It is so nice even to speak about whales, since whaling is the main joy of our life. These days I can go to the [St. Lawrence] Island to participate in bowhead whale hunt (with local crews). The time of whaling is the best time of life. You respect yourself and you respect nature. My dream is to live until I can see bowhead whaling again right here, off Sireniki (fortunately, he lived to see this happening—ed). The meat and *mantak* of this whale is the tastiest and the most useful of our Native food.

I keep thinking about whales and want bowhead whaling to be restored. It is a good food and traditions, and rituals, which every family has (to perform). The (bowhead) whale unites hunters and their families, it takes away all minor things, all garbage from our life. When we hunt, we become better ourselves.

RENTIN Sergei, born 1960, and YATA Yuri, born 1961, hunters from the village of Novoe (New) Chaplino.

In the past, many people knew how to hunt whales, particularly the bowhead whale. They knew how to do it in open water, in the channels among pack ice, and in the young thin ice. We have heard only scraps of that knowledge but we have not seen it ourselves and so we do not know how to kill a whale.

GALGAUGYE (Galgawyi) Nikolai, 1935—1995. Elder brother of Andrei Ankalin, experienced hunter and a born naturalist/researcher by inclination. Perished at sea in 1995.

There are many wonderful animals, but whales are the best of all. As they pass by near your skin boat, great and quiet, you immediately come to understand your place on the earth. You become warm inside as if you had a good drink.

I think that God or, perhaps, somebody up there created the bowhead whale specially for us. It was impossible to survive without the whale in olden times. It provided home and food, and supplies to feed the dogs, and it gave warmth to people. Nowadays, it is very interesting just to look at whales. But most importantly, it is helpful and you need it in order to become a full-fledged person and a good hunter. Whales give us quietness and the beauty of life.

TRADITIONAL NATIVE WHALE FESTIVALS IN CHUKOTKA

The Whale Festival has been traditionally one of the major community events in the annual life of the Yupik and coastal Chukchi people of Chukotka. However, for the Chukchi communities the species of the whale taken by local hunters (gray or bowhead) was not of that great importance, whereas for the Yupik the bowhead whale had a special status and was considered a truly sacred animal.

The following description of traditional whale ceremonies is based upon elders' accounts documented by Yupik ethnographer, Dr. Tasyan Tein (1984, 1992), who was born in the Yupik village of Naukan at East Cape (Cape Dezhnev). The Naukan Yupik community probably once had the most elaborate and developed cycle of annual whale ceremonies, although several other Siberian Eskimo communities also prepared for their Whale Festival for a long time and very thoroughly. Usually, all the villagers took part in the Whale Festival that took place in the year when a whale was taken. Also invited were relatives and acquaintances from other communities as well as Chukchi reindeer herders from the nearby camps, with whom the Yupik exchanged food, skins, and other goods.

According to Tein's description, at Naukan, during the year when the whale was taken, the Whale Festival was always organized by the clan, whose crew took the whale. Every clan in Naukan reportedly had festival traditions of its own as well as special customs and rites associated with the bowhead whale. Many of these rites were secret; they were observed and transmitted within the clan. In order to participate in the Festival, every clan in Naukan nominated the best dancers and the most agile and strongest hunters to take part in races and other sporting contests. In addition, every clan nominated a weather forecaster to contribute to the organization of the Whale Festival. This was a very important person who was responsible for the rites of exorcising. In order to ask for fine weather, it was necessary to change one's profane or otherwise unclear 'dirty mouth' to a 'clean' one. This substitute—after some special exorcism—was offered by the Heavenly Old Woman to the one who pleaded for it (Rubtsova 1954).

The Naukan Yupik community that once lived at the big cliff village between Cape Paek and Cape Dezhnev also had the longest and the most complicated Whale Festival, named *Pol'a* (*polla*). The Naukan *Pol'a* lasted for a whole month usually in November and early December. That was the last community fall celebration, the festival of thanksgiving to the whale, which offered its life to the village for the long and cold winter.

Before the Whale Festival, all the members of the successful crew would come before dawn to the elder of the Festival for a meeting. There they chose heralds to announce every day, with a solemn call uttered at every dwelling that the Festival has begun. They also announced the news to the wardens of public premises for reception of guests and also to the famous village storytellers. Every day of the Festival was reportedly characterized by a particular rite.

The Whale Festival in Naukan, as recalled by the elders interviewed by Tein during the 1960s and 1970s, was usually started by the heralds. They would come around the village with a solemn call '*Ou-ou-ou, k'amai, saiikh'si!*' This was a request to play the drums, and the beginning of the festival was marked by ritual dances to the accompaniment of the drum. During the first two days, songs and special dances were performed, including some that were danced in masks. On the third day the host of the Festival gave a feast in his house, which was followed by a sporting contest. Day 12 marked exchange of goods, *umak'ut*, between cousins, close acquaintances and friends. One could ask for anything. Refusal to offer the requested object, even when it was of great value, would mean offending the soul of the bowhead whale, the praised guest of the Festival.

Day 20 was also very important. It was the day of spiritual initiation of boys. They were received by an old storyteller and a woman—a wife, or other female relative of the captain of the crew that took the whale. The boys were treated with pieces of whale flippers that were put into walrus skin sacks beforehand. The youngest boys were treated first, then older ones, and finally the eldest teens had their turn. After the children had eaten, the woman would leave and the old storyteller would start telling legends about every clan, about its merits and special features. According to Tein's account, this elderly storyteller was usually one of the most respected elders in the village; he carried great authority among his tribesmen and was a well-to-do person. On Day 28 some complicated rites took place, usually performed with small wooden figures of a whale, boat, and birds. Those performances were mystic; their sacred meaning was already lost by the mid-1900s and cannot be deciphered (Tein referred to them simply as 'puppet performances.').

On Day 31, men and women communicated spiritually with their ancestors and deceased relatives. Before dawn they would sew toy models of clothing. At dawn, the villagers would go to the coastal bluffs. The members of the crew that took the whale would descend to the shore and make a big fire from driftwood. The toy clothes were thrown into the fire, to the shouts of the people, who were calling their deceased. In that way the relatives send messages to those who were gone forever. The Eskimos believed that things sent through the fire reach the other world faster. After that men and women would stand around the fire, join hands and perform various songs and dances. It was thought that the deceased were merrymaking together with the living.

In the afternoon, the elder of the ceremony would take the members of the boat crew that took the whale to the seashore for a symbolic 'washing off' the sins committed before and during the time of the Festival. Every man would beat his clothing worn during the Festival with a snow-beater and whisper certain prayers, in order to exorcise evil spirits. After that, the members of the crew returned to their dwellings and cut each other's hair. It is not until all the ceremonies were accomplished that the Festival was considered to be over.

The bowhead whale, more exactly its soul, was thought to be present at the Festival all the time. The indigenous people believe that the animal, in this case the whale, visits the hunter as a guest in order to offer itself. Hence, when the killed animals were butchered, their bones were not cut. It was thought that the souls of the animals would re-grow new meat on those bones. Instead of the word 'to kill,' the hunters would say that the whale is 'wakeful' in order not to offend the soul of the dearest guest, the whale. If the whale was taken in spring, a special rite of seeing it off to the sea was performed during the fall, and if the whale was taken during the fall, it was seen off the next spring.

Currently, older and middle-aged hunters are still the carriers of the remnants of such age-old cultural traditions associated with bowhead whales (see accounts of the present-day senior hunters cited above). They consider the whales as their friends and spiritual teachers and never as sheer prey.

CONCLUSIONS

I believe that the materials presented above demonstrate the fruitfulness of the joint efforts of scientists and indigenous people of Chukotka in studying the biology of the bowhead whale, a rare and specially protected species, which is very difficult to watch and to monitor.

Thanks to the observations and traditional knowledge of hunters it was proved that there exist:

- Summer groupings of bowhead whales off the southern and eastern coast of the Chukchi Peninsula (as reported by local observers: G. Gyrgolnaut, from Enmelen; N. Kychi, from Nunlingran; A. Ashkamakin, from Yanrakynnot, and others);

- A specific westward route of the spring bowhead whale migration along the northern shore of the Chukchi Peninsula, in the direction of the De Long Strait and the East Siberian Sea (V. Eorelo, from Uelen; M. Keutegin, from Enurmino; hunters from the community of Inchoun);

- A special type (morph) of bowhead whale with a shorter and thicker trunk, referred to by the Yupik Eskimos of Sireniki as *takyshkak*, (Petr Typykhkak, Timofei Panaugye, and others). A whale of this type was taken in Sireniki in 1998, in the first year that subsistence hunting for bowhead whale in Chukotka was resumed.

The success of long-term cooperation between scientists and marine hunters in Chukotka, over more than two decades, illustrates both a new direction and the immense opportunities in the joint study and conservation of endangered species of arctic fauna, including some of the most rare species, such as the bowhead whale.

REFERENCES

Ainana, Lyudmila, Nikolay Mymrin, Lyudmila Bogoslovskaya, and Igor Zagrebin
   1995 *Role of the Eskimo Society of Chukotka in Encouraging the Traditional Native Use of Wildlife Resources by Chukotka Natives in Conducting Shore-Based Observation on the Distribution of Bowhead Whales,* Balaena mysticetus *in Coastal Waters off the Southeastern Part of the Chukotka Peninsula (Russia) During 1994.* Report to Department of Wildlife Management North Slope Borough, Alaska, 137p.

Ainana, Lyudmila, Nikolay Mymrin, Oleg Veter, and Lyudmila Bogoslovskaya
    1997 *Role of the Eskimo Society of Chukotka in Encouraging the Traditional Native Use of Wildlife Resources by Chukotka Natives in Conducting Shore-Based Observation on the Distribution of Bowhead Whales,* Balaena mysticetus *in Coastal Waters off the Southeastern Part of the Chukotka Peninsula (Russia) During 1995.* Report to Department of Wildlife Management North Slope Borough, Alaska, 110p.

Arutyunov, Sergei A., Igor I. Krupnik, and Mikhail A. Chlenov
    1982 *'Whale Alley.' Antiquities of the Senyavin Strait Islands.* Moscow: Nauka Publication, 176p. (in Russian).

Berzin, A.A.
    1981 The Modern Status of the Populations of the Bowhead Whale. *Priroda* 6 (in Russian).

Bogdanovich, Karl I.
    1901 *Essays of the Chukotka Peninsula.* Saint-Petersburg: A.S. Suvorin Printing Office. (in Russian).

Bogoslovskaya, Lyudmila S.
    1993 List of the Villages of the Chukotka Peninsula (2000 BP to present). *Beringian Notes,* NPS, Alaska Region 2 (2): 2–11.
    1995 Bowhead Whales in the Seas of the Russian Arctic, p. 17 in *Proceedings of the International Conference on the Study of Marine Mammals,* Moscow (in Russian).

Bogoslovskaya, Lyudmila S., Mikhail A. Chlenov, and Boris A. Yurtsev
    1979 The Senyavin Strait Islands—A Unique Nature Complex. *Priroda* 9: 91–97. (in Russian).

Bogoslovskaya, Lyudmila S. and Leonard M. Votrogov
    1981 Massive Winterings of Birds and Whales in the Bering Sea Polynyas. *Priroda* 1: 42–43 (in Russian).

Bogoslovskaya, Lyudmila S., Leonard M. Votrogov, and Igor I. Krupnik
    1982 Bowhead Whale off Chukotka: Migrations and Aboriginal Whaling. *Report of the International Whaling Commission* 32: 391-399.
    1984 The Bowhead Whale in the Waters of Chukotka. The History and the Present-day Status of the Population, pp. 191-212 in *Morskie Mlekopitayushchie* (Marine Mammals). Moscow: Nauka (in Russian).

Burns, John J.
    1993 'Epilogue,' pp. 745-764 in J.J. Burns, J.J. Montague, C.J. Cowles (eds.)., *The Bowhead Whale.* Lawrence, KS: Allen Press, Society for Marine Mammalogy, Special Publication No. 2.

Chlenov, Michael and Igor Krupnik
    1984 *Whale Alley*: A Site on the Chukchi Peninsula, Siberia. *Expedition* 26(3): 6-15.

Dikov, Nikolay N.
    1971 *Rock-Art Mysteries of Ancient Chukotka. Petroglyphs of Pegtymel.* Moscow: Nauka (in Russian).

Freeman, Milton M.R., Lyudmila Bogoslovskaya, Richard A. Caulfield, Ingmar
Egede, Igor I. Krupnik and Marc G. Stevenson
    1997 *Inuit, Whaling and Sustainability.* Walnut Creek, CA and London:
        Alta Mira Press.
Krupnik, Igor I.
    1983 'Ancient and Traditional Settlements of Eskimos in the
        Southeastern Chukotka,' pp. 65-95 in V.P. Alekseev, ed.,
        Crossroads of Chukotka and Alaska. Moscow: Nauka Publication
        (in Russian).
    1984 'Gray Whale and the Aborigines of the Pacific Northwest: The
        History of Aboriginal Whaling,' pp. 103-120 in M.L. Jones, S.L.
        Schwartz and S. Leatherwood, eds., *The Gray Whale,
        Eschrichtius robustus.* New York: Academic Press.
    1987 The Bowhead *vs.* Gray Whale in Chukotkan Aboriginal Whaling.
        Arctic 40(1):16-32
    1993 Prehistoric Eskimo Whaling in the Arctic: Slaughter of Calves or
        Fortuitous Ecology? Arctic Anthropology 30(1): 1-12
    2001 *Let Our Elders Speak. Stories of the Yupik (Asiatic Eskimo Elders),
        1975-1987.* Moscow: Institute of Cultural and Natural Heritage
        (in Russian).
Krupnik, Igor I., Lyudmila S. Bogoslovskaya, and Leonard M. Votrogov
    1983 Gray Whaling off the Chukotka Peninsula: Past and Present
        Status. *Report of the International Whaling* Commission 33: 557–
        562.
Melnikov, Vladimir V. and A.V. Bobkov
    1991 About the Migrations of Bowhead Whales in the Chukchi Sea in
        1991. *Okeanologiia,* 33 (15): 729-734 (in Russian).
Melnikov, Vladimir V. and A.V. Bobkov
    1993 Migrations of Bowhead Whales in the Chukchi Sea. In: *Biologiia
        Moria* (Marine Biology), pp. 60-67.
Moore, Sue E. and Randall R.Reeves
    1993 'Distribution and Movement,' pp. 313-386 in J.J. Burns, J.J.
        Montague, C.J. Cowles, eds., *The Bowhead Whale.* Lawrence,
        KS: Allen Press, Society for Marine Mammalogy, Special
        Publication No. 2.
Rubtsova, Ekaterina S.
    1954 *Materials on the Language and Folklore of Eskimos (Chaplino
        Dialect).* Part I. Moscow—Leningrad : USSR Academy of Science
        Publication (in Russian).
    1971 *Eskimo-Russian Dictionary.* Moscow: Sovetskaya Entsiklopedia.
        (in Russian).
Stoker, Sam and Igor Krupnik
    1993 'Subsistence Whaling,' pp. 579-630 in J.J. Burns, J.J.
        Montague, C.J. Cowles, eds., *The Bowhead Whale.* Lawrence,
        KS: Allen Press, Society for Marine Mammalogy, Special
        Publication No. 2.
Tein, Tasyan S.
    1984 *Festivals of Eskimos.* Maagadan: Magadanskoe knizhnoe
        izdatel'stvo, 32 pp. (in Russian).
    1992 *It was this Way... Essays of Traditional Culture of Asiatic
        Eskimos.* Magadan: Magadanskoe knizhnoe izdatel'stvo, 68 pp.
        (in Russian).

Townsend, C.H.
    1935 The Distribution of Certain Whales as Shown by Logbook
        Records of American Whaleships. *Zoologica* 19:1-50.
Votrogov, Leonard M. and Lyudmila S. Bogoslovskaya
    1980 Gray Whales off the Chukotka Peninsula. *Report of the
        International Whaling Commission* 30: 435-437.

Some Observations on the Influence of Environmental Conditions on the Success of Hunting Bowhead Whales Off Barrow, Alaska

John C. George, Stephen Braund, Harry Brower, Jr.,
Craig Nicolson, and Todd M. O'Hara

Abstract. *Analysis of the bowhead whale hunt at Barrow (1990-1997) suggests that hunting success is greatly influenced by wind direction and speed. During the spring hunt along the Chukchi Sea coast, hunters tell us that open leads, moderate to strong offshore winds (easterly component), and stable landfast ice are required to hunt and land whales successfully. This is mainly because easterly winds open lead systems at Barrow by pushing the pack ice offshore. Said another way, it is the presence or absence of sea ice in the nearshore lead that affects spring bowhead hunting success, and wind direction is a reliable indicator of lead conditions and ice cover within the lead. Bowhead whales are generally harvested in spring when winds are offshore (easterly) and are almost never taken when winds are onshore (westerly component). During the fall bowhead hunt offshore of Point Barrow (Beaufort and Chukchi Seas), calm to moderate winds and relatively ice free waters are required to hunt whales effectively. Wind direction, however, does not appear to affect fall hunting success, whereas wind speed has a significant effect. Selected seasons (i.e., spring 1992, 1993, and 1997 and fall 1997) were examined in detail to illustrate extremes in environmental conditions and to quantitatively explore hunters' observations using 'Western' scientific approaches. This analysis does not consider the sociological aspects (e.g., numbers of active crews, cease-fire periods, festivals) of whale hunting which also affect success. Our findings suggest that the bowhead whale hunt at Barrow is highly affected by environmental conditions and that wind speed in the fall and wind direction and ice cover in the spring are the principal variables affecting whale-hunting success. Furthermore, scientific findings all agree well with the hunters' predictions. Such variability in hunting conditions supports flexible hunting regulations that allow for hunting failures (due to environmental factors) during some seasons.*

INTRODUCTION

Bowhead whales (*Balaena mysticetus*) provide an important cultural and subsistence resource for several Arctic Eskimo communities (Alaska Consultants, Inc. and Braund and Associates 1984). The harvest is locally managed through an agreement between the Alaska Eskimo Whaling Commission (AEWC) and the National Oceanic and

Atmospheric Administration (NOAA). The level of allowable harvest is determined under a quota system in compliance with the International Whaling Commission—IWC; (Gambell 1982). The quota is based on (a) the subsistence and cultural needs of Alaskan Eskimos, Russian Eskimos and Chukchis and (b) estimates of the size and growth of the bowhead whale population (Donovan 1982; Braund 1992; IWC 1997).

The subsistence hunt takes place in spring and fall as whales migrate between the Bering and Beaufort seas. These hunts are subject to considerable environmental interference from weather (e.g., wind speed and direction, fog, and temperature), landfast ice stability, and sea-ice conditions and concentration. The success of the hunt is highly affected by these factors and shows considerable variation by year at Barrow and elsewhere. For example, at Barrow the annual harvest of bowhead whales varied from 11 to 30 between 1990 and 1997, whereas the number of AEWC-registered whaling crews (used as an index of effort) remained relatively constant (Table 1). Thus we looked at influences other than whaling effort to explain the variability in year-to-year bowhead harvests. The objective of this paper is to analyze the effect of specific environmental conditions (wind speed, wind direction, temperature, and indirectly, sea-ice concentration) on Barrow bowhead whale hunting success, based on an analysis of the 1990 through 1997 spring and fall whaling seasons.

METHODS

The method involved two procedures: (1) identification of environmental factors that affect whale hunting success through observation of whale hunting and interviews with whaling captains, and (2) testing these findings by analyzing recorded weather data. Information on the critical environmental factors that affect whale hunting was gathered through interviews with three Barrow whaling captains (including author H. Brower, Jr.) and from our own observations of the whale hunt. Based on these interviews and observations, wind speed, wind direction, and temperature were important variables that affected bowhead whaling success at Barrow. Hence, the null hypothesis we wished to test was: environmental factors (wind speed, wind direction, and temperature) do not affect the whale harvest success by Barrow hunters during the spring and fall hunt (1990–1997). We chose the period for this analysis for several reasons: (a) the number of whales harvested was considerably higher than in the 1980s, yielding a larger sample size,[1] (b) we have direct firsthand observations of the sea ice during

---

[1] For instance, no whales were landed at Barrow during the 1982 spring season (however, four whales were struck and lost in a single day).

bowhead census activities in 1992 and 1993; and (c) this period encompasses two years with apparently extreme conditions that help illustrate their effect on hunting success.

Table 1. Registered Bowhead Whaling Captains and Landed Bowheads, Barrow, Alaska (Source: Alaska Eskimo Whaling Commission)

Year	Number of Registered Whaling Captains (Barrow)	AEWC Allocation for Barrow (Prior to Transfers)[1]	Number of Landed Bowhead Whales (Spring)	Number of Landed Bowhead Whales (Fall)	Number of Landed Bowhead Whales (Total)
1990	46	15	6	5	11
1991	45	15	8	4	12
1992	42	18	2	20	22
1993	45	18	16	7	23
1994	47	18	15	1	16
1995	47	22	10	11	21
1996	42	22	5	19	24
1997	47	22	9	21	30

[1] *The AEWC allocates the IWC quota to individual AEWC whaling communities. During the whaling season, any unused strikes in one community may be transferred to other communities, resulting in landed whales greater than the original allocation. The AEWC allocation is included in this table to discount significant changes in Barrow's quota as an explanation of variation in landed bowheads.*

The seasonal hunting periods were defined as: spring (20 April-1 June) and fall (1 September to 21 October) based on our observations of when whales have been harvested during this period.

Weather data were acquired from the NOAA Climate Monitoring and Diagnostics Laboratory (CMDL) for the environmental analysis section. The number of whales captured by day was compiled from the North Slope Borough Department of Wildlife harvest data and combined with the daily average weather conditions. For the analysis, wind speed and direction were binned as follows: wind direction (four categories): NE = 1–89°, SE = 90–179°, SW = 180–269°, NW = 270–360°; wind speed km/h (five categories): 1 = 0–15 km/h, 2= 16–31 km/h, 3 = 32–47 km/h, 4 = 48–64 km/h, and 5= >64 km/h. In a couple of cases, only the wind easterly–westerly component was used. For instance, any wind from 1–179° has at least some easterly wind component where 90° would have the most and 1° almost none. It was calculated as easterly for winds from 1–179°T (true heading) and westerly for winds from 181–359°T. Wind speed differences were tested using a T-test. Chi square statistics were used to compare harvest data vs. wind conditions for the spring and fall seasons. Statistical analyses were conducted in SPSS-PC 7.0 and MS Excel. Only days when crews were actively hunting before the IWC bowhead quota was reached were used in the analysis. Also, days were omitted in the fall after the ocean froze permanently for the season, effectively ending the bowhead hunt

We encountered a problem as to how to address the fall 1991 and 1994 seasons. In these years, the ocean became frozen by approximately 15 October and 10 October respectively, effectively ending the hunt even though Barrow hunters had not reached the quota. Clearly, environmental conditions prevented hunting but this was not attributable to wind conditions. For this analysis, we removed those fall days when hunting was impossible due to heavy ice coverage.

RESULTS AND DISCUSSION

Observations on the Effects of Environmental Factors on Whale Hunting at Barrow 1990-1997

Alaska subsistence bowhead whale hunting falls broadly into two categories: spring hunting and fall hunting. With some local variation, the spring hunt along the northwest coast of Alaska is typified by ice-edge 'still' hunting in lead systems. Still-hunting refers to the hunters waiting silently at the edge of the landfast ice, with skin boats ready for an immediate launch into the open lead if a bowhead is seen within potentially accessible reach. The hunters then quickly launch their boat and quietly paddle within striking distance to the whale. Hunters often strike whales from the ice edge as well. At St. Lawrence Island, the whalers sail their skin boats in pursuit of bowheads in the large polynya (Braund 1988; Stoker and Krupnik 1993). Spring hunts are currently carried out in the Alaska communities of Gambell, Savoonga, Wales, Kivalina, Little Diomede, Point Hope, Wainwright, and Barrow. Fall whale hunting is an 'open water' hunt, which takes place in late August through October. Hunters use an active pursuit technique in powered skiffs ($\leq$ 8 m in length). Currently only the villages of Kaktovik, Nuiqsut, and Barrow hunt in the fall; thus Barrow is the only village that successfully harvests whales in both spring and fall. The St. Lawrence Islanders, however, occasionally take whales in late fall and winter (as in 1997).

The success of whale hunts at Barrow in the spring and autumn depends on a number of sociological factors (e.g., IWC bowhead quota, number of active whaling crews, official 'cease fires,' ability to coordinate between crews) and environmental factors. Although the analysis in this chapter focuses on Barrow, the environmental factors we consider generally hold true for the other villages as well. In 1992, for instance, Wainwright and Point Hope did not strike any whales due to unusual spring westerly wind conditions and closed leads. Wales, Little Diomede, and Kivalina did not strike whales in the spring of 1997, reportedly due to poor environmental conditions. From interviews with whale hunters we compiled critical environmental factors that, in their opinion, affect hunting success (Table 2).

Table 2. Environmental factors and hunting criteria identified by Barrow whale hunters which affect whale-hunting success during the spring and fall bowhead hunts.

Environmen-tal Factor	Spring hunt	Fall hunt
Lead width[1]	> 1 km in width; lead needs to be <u>wider</u> than the normal bowhead dive length (ca 1 km @ 4 km/h)	Not a significant factor
Air temperature	Not a factor; except at low wind speeds and low temperatures (< -20 C) leads tend to freeze.	Not a significant factor
Current	Generally NE[2] current preferred (Chukchi Sea); low speed preferred	Not a significant factor
Wind direction	Prefer east winds, reluctant to strike if winds and currents are SW or W.	Not a significant factor
Visibility (fog)	Fair to good visibility required; whales taken very near ice edge.	Need good to excellent visibility to detect and follow struck whales
Wind speed	< 40 km/h	≤ ca 25 km/h
Wave height	Less than about 0.7 m; large waves can break off ice edge and greatly affect skin boat handling.	Less than 1 m, but hunting is possible, albeit difficult, in larger seas.
Landfast ice stability	Grounded (multiyear) ice is preferred and safer and more stable.	Not a factor
% ice cover	<20% mature, <40% young ice	Prefer 10-40%; there appears to be a range of acceptable conditions; in some cases sea ice can substantially *reduce* wave height and facilitate hunting.

[1]. *Lead width is the distance from the edge of the landfast ice to the edge of the pack ice.* [2] *NE currents tend to prevent landfast ice from breaking free.*

Description of the Barrow Spring hunt. Success during the Barrow spring whale hunt generally requires open, relatively ice-free, leads (≥ 1 km in width), good visibility, fairly high bowhead whale passage rate,[2] stable landfast ice, and winds less than about 40 km/h (Table 2). Based on harvest data collected over the last 18 years, the spring hunting season spans roughly 20 April to 1 June (Fig. 1). During the spring hunt on the Alaska mainland, hunters wait along the ice edge from where they strike migrating animals usually within about 200 m of the ice edge. Wind direction is a critical factor because 'onshore' (i.e., westerly) winds tend to close the leads rapidly or they become ice-choked, both of which prevent hunting. Wind speed is not as critical as direction because there is little fetch in the narrow leads, and wave height is low near the ice edge, with little effect on boat handling (near the lead edge). If the wind speed drops quite low (i.e., 8 km/h or less), however, leads will often close due to sea currents or freeze over. George *et al.* (1995) describe the lead freezing over with young ice due to low wind speeds beginning on 21 April, 1993, and persisting to the end of April. No whales were struck during that period.

Description of the Barrow fall hunt. The fall hunt at Barrow extends from about 1 September to 21 October (Fig. 1). Typically, it is conducted in open water that occurs when the pack ice is at its maximum retreat (September/October), sometimes well in excess of 200 km. Therefore, the hunters contend with significant seas when wind velocity is high. Because Barrow hunters use relatively small boats and hand-thrown hunting tools (i.e., darting gun and harpoon), it is only during periods of low wind velocity that hunters can effectively hunt whales. While wind velocity is critical, the direction of the wind is of little consequence. Sea ice concentration is another critical factor, which we have not included in this analysis. In some years, the pack ice encroaches upon the coast early in the season and can completely preclude whale hunting. Such was the case during the 1991 and 1994 fall hunt at Barrow.

---

[2] While it may seem obvious that the presence of whales is needed to have a successful hunt, it still bears mentioning for the following reasons. For this analysis we chose periods when it is likely a whale would be harvested based on past harvest history. The migratory passage of bowheads past Barrow follows roughly a normal curve across the season that spans the three schools recognized by the hunters. The harvest rate of whales per day and the passage rates as determined from the whale census are quite similar. Thus, harvest success is highest in the first two weeks of May when whale passage rates are highest.

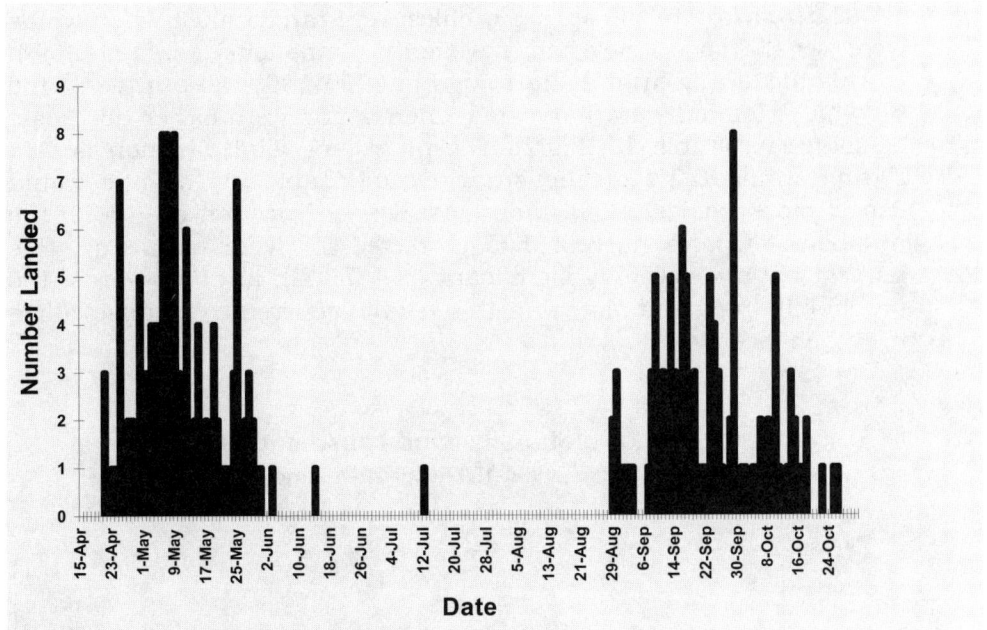

Figure 1. *Number of bowhead whales landed by day by Eskimo hunters at Barrow for the period 1980 to 1997. (For this paper we assumed that the temporal distribution of the harvest (e.g., when whales were landed) defined the whale hunting seasons. Note the suggestion of multiple peaks (25 April, 3-12 May, and 21-25 May) in whale harvest in the spring. This is likely reflecting the pulse-like migratory rates, which the Eskimo hunters refer to as whale 'schools.' See also Bogoslovskaya, this volume).*

ANALYSIS OF SEASONS 1990-1997

Spring

Wind speed. Barrow hunters tolerate surprisingly high wind speeds during the spring bowhead hunt. Wind speed was significantly higher on days when whales were caught (21.4 km/h) than when not caught (17.9 km/h) (p=0.038; Tables 3 and 4a; Figs. 2–3). While not intuitively obvious at first, this makes sense because higher wind speeds (from the east) tend to hold the leads open. Also, because Barrow whalers hunt the lee edge of the lead, which is often protected by pressure ridges, wave height can be relatively low despite high wind speeds. If hunters venture far out into the lead in skin boats during high winds, however, wave height can be a significant problem in that it is hard to maneuver the boat and waves can grow too large to safely handle a small boat. This often occurs after a whale is struck and the crews must pursue the wounded whale far out into the lead.

Wind direction. In the spring, whales were rarely taken when winds were 'west,' that is, they had a 'westerly component' (i.e., 181–359º). During the spring hunt at Barrow from 1990-1997, all but two of the 68 whales landed were harvested when winds were either northeast or southeast (Table 4b; Fig.2). The prevailing wind direction is east during spring (72% during study period, Table 3b) so one would expect more whales taken under east winds. However, even after the number of whales harvested is corrected for wind frequency, this harvest pattern is highly significant (p = 0.002); in other words, the probability (*p*) is quite low that this result occurred by 'chance' alone (i.e., 2 chances in 1000).

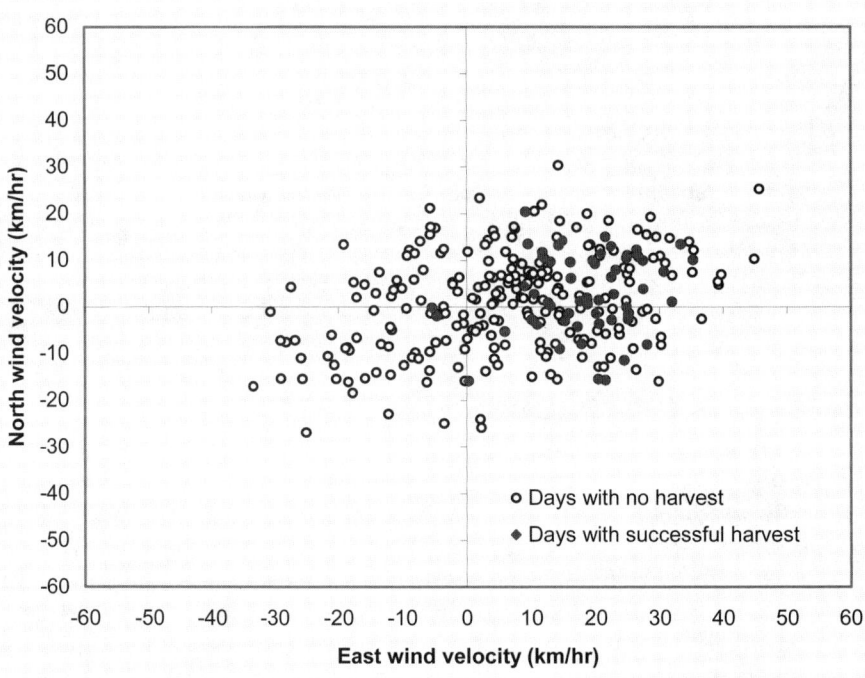

Vector plot: daily wind speed and direction
1990-1997 (Apr 20- 1 June)

Figure 2. *A vector plot showing the distribution of wind speeds and directions during the spring whale-hunting season (1990–1997) indicating when whales were caught (solid dots) and were not (open dots) caught. This type of plot summarizes a considerable amount of data in a fairly intuitive form. Note that the further a wind value is from the origin, the stronger it is. Also, note that nearly all of the whales were taken when the wind had an easterly component (right side of plot), and further, that most whales were taken under fairly strong easterly winds (up to 40 km/h).*

The pattern of successfully striking whales during an easterly wind corroborates the 'predictions' listed in Table 2. This is likely a result of hunters' reluctance to strike whales when the wind is 'on shore' because onshore winds tend to close leads. During lead closure, pack ice can hit and deform (*ivu* or pile up) the landfast ice, which endangers the crews. Furthermore, crews cannot tow whales or haul them on to the ice for butchering during such periods.

Table 3. Wind speed statistics for the spring and fall seasons at Barrow 1990-1997.

Harvested?	Statistic	Season	
		Spring	Fall
Yes	Mean wind speed (km/h)	21.4	13.2
	Number of days	47	59
	Standard deviation	7.0	5.6
	Max. wind speed	36.7	26.6
	Min. wind speed	5.0	2.2
No	Mean wind speed (km/h)	17.9	22.9
	Number of days	250	311
	Standard deviation	9.4	11.4
	Max. wind speed	52.2	60.5
	Min wind speed	1.8	0.4

Data are sorted by days when whales were and were not harvested. Note the marked differences in wind speed when whales were and were not harvested in fall. Note also that wind speeds were actually higher during spring when whales were harvested.

Fall
Wind speed. Wind speed was significantly lower (13.2 km/h) in fall when whales were caught than when they were not caught (22.9 km/h) (p < 0.001; Table 3, Table 5a; Fig. 2 and 3). Note that 79.5% of the whales were harvested when the wind speed was less than 16 km/h (10 mph) while the wind blew at this speed only 37.4% of the time for all seasons (Table 5a). This observation confirms the hunters' criteria that wind speed and wave height must be low during open water conditions in order to hunt whales effectively (Table 1).

Wind Direction. As in spring, the number of harvested whales was tallied within the four wind direction quadrants. However, a chi-square analysis did not detect a statistically significant difference (*p*=0.6) between the numbers of bowheads harvested by wind direction when the wind frequency was considered (Table 5b; Fig. 4). This is not surprising since the fall hunt does not take place within a lead system (as in spring), where pack ice can quickly occlude open water. This analysis corroborates the hunters' criteria in Table 2 that low wind speed is critical for fall hunting success.

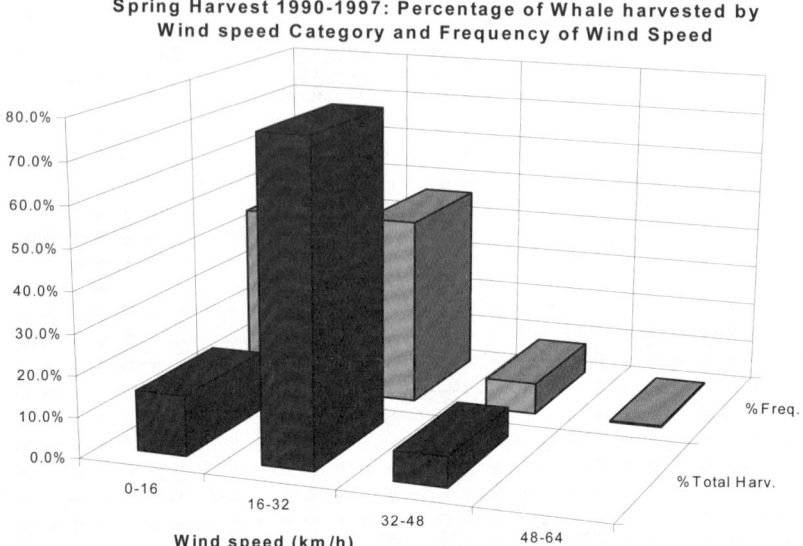

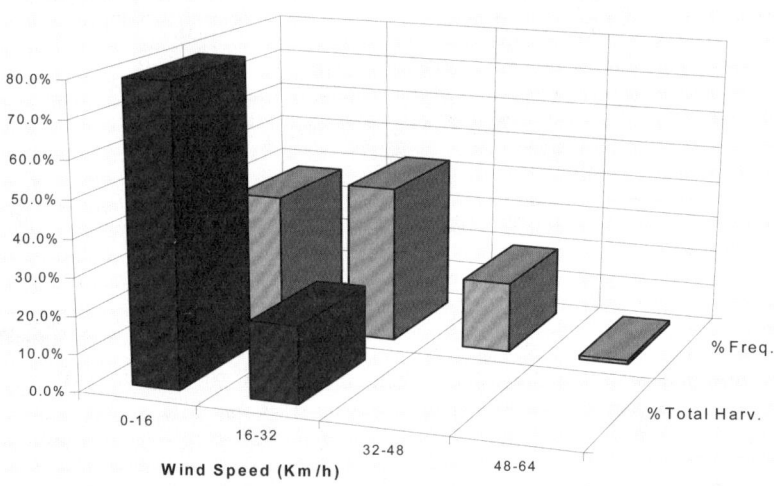

Figure 3. *Comparison of the percentage of total whale harvest by season and wind speed shown with the frequency of wind directions for the period 1990–1997. Note that whales are preferentially taken at moderate to high wind speeds in spring (16-32 km/h) and at low to moderate wind speeds in fall (0-16 km/h). Clearly, the percentage of whales harvested does not follow the occurrence of wind speeds for a given season since none are taken at wind speeds over 32 km/h.*

Table 4. Numbers of whales taken at Barrow in spring under various wind speeds (4a) and wind directions (4b) during the spring hunts (1990–1997).

4a. Wind speed in spring

Wind speed (km/h)	Number harvested	% harvested	% Wind frequency
0-16	10	14.7%	45.8%
16-32	53	77.9%	46.1%
32-48	5	7.4%	7.7%
48-64	0	0%	0.3%
Totals	68		

Percent harvested (% whales harvested) is the percentage of whales harvested under the particular wind speed or direction category. The percent wind frequency (% wind frequency) means the percentage of days of the total that had that particular wind speed or direction.

4b. Wind direction in spring.

Direction	Number harvested	% Harvested	% Wind frequency
NE	42	61.8%	46.1%
SE	24	35.3%	26.3%
SW	2	2.9%	17.9%
NW	0	0%	9.8%
Totals	68		

Table 5. Numbers and percentages of whales taken at Barrow in <u>fall</u> under various wind speeds (5a) and wind directions (5b) during the period 1990-1997.

5a Wind speeds in fall

Wind Speed (km/h)	Number harvested	% Harvested	% Wind Frequency
0-16	70	79.5%	37.4%
16-32	18	20.5%	42.5%
32-48	0	0.0%	19.1%
48-64	0	0.0%	1.0%
Totals	88		

Percent harvested (% harvested) is the percentage of whales harvested under the particular wind speed or direction category. The percent wind frequency (% wind frequency) means the percentage of days of the total that had that particular wind speed or direction.

5b Wind direction in fall.

Direction	Number harvested	Wind Freq.	% Wind Frequency
NE	44	50.0%	43.3%
SE	24	27.3%	28.4%
SW	10	11.4%	15.2%
NW	10	11.4%	13.1%
Total	88		

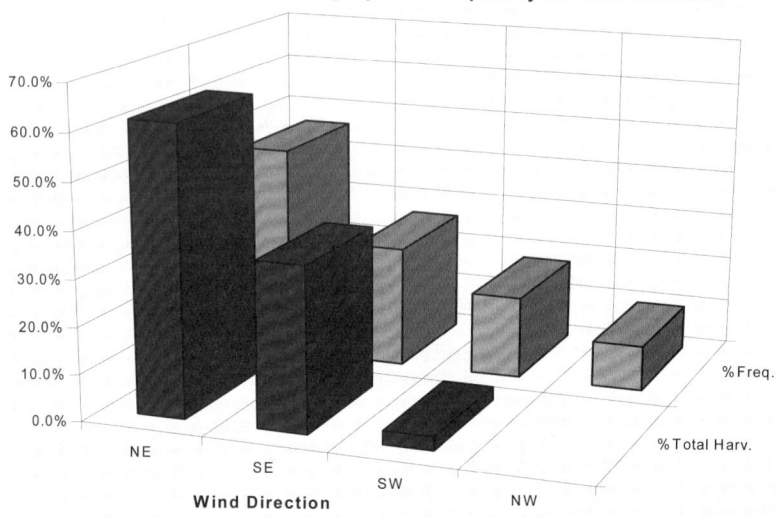

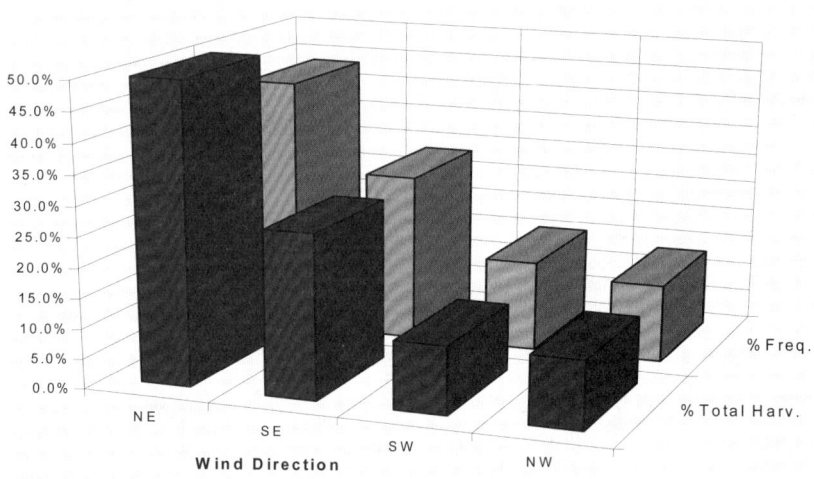

Figure 4. *Comparison of the percentage of total harvest by season (spring and fall) and wind direction shown with the frequency of wind directions for the period 1990-1997. (Note that very few whales are taken in spring when winds are SW or NW. In the fall hunt, however, note that percentages of wind directions (frequency) and the percentage of whales harvested are quite similar, reflecting that wind direction is less important in the fall.)*

COMPARISONS OF SPECIFIC SEASONS

Spring 1992 and 1993

We compared the 1992 and 1993 spring seasons because they exemplify extremes in spring weather and hunting conditions. Hunting success was markedly different between the two seasons.

Spring 1992. The 1992 season was quite unusual as winds persisted from the west during essentially the entire month of May (Fig. 5; Table 6). This west wind kept the lead closed and caused the failure of the 1992 ice-based bowhead whale census resulting in a very poor spring harvest. Wainwright hunters did not strike or land any whales, and Barrow and Point Hope took only two whales each— far below the typical harvest level (Philo *et al.* 1994). The 1992 spring hunt at Wainwright marks the only season, since at least 1978, when hunters failed to strike a bowhead whale. That year, there was considerable multiyear ice along the Alaskan Chukchi Sea coast, driven on shore by westerly winds that completely 'chocked' the nearshore leads. The sea currents were generally from the SW as well. Based on weather records, this was one of the most extreme seasons in terms of persistent west winds during May (when most whales are harvested) since 1922 (Philo *et al.* 1994).

Spring 1993. The 1993 season was quite different from the 1992 season in many respects. In 1993, the winds were consistently east and the leads at Barrow did not close at all during the bowhead census period and hunting season (*ca.* 15 April to 1 June; George *et al.*, 1995; Fig. 5; Table 6). The spring hunt was quite successful, and the Barrow whalers reached their quota by 11 May. Very likely these factors (i.e., east winds and open leads) were also responsible for the excellent visibility and ice conditions during the 1993 bowhead census.

Note that, in both 1992 and 1993, no whales were taken when winds were from the west (Fig. 5, Table 7). In spring, 1993, the mean wind speed was higher when whales were harvested than when they were not taken (Table 6). Although the relationship between wind speed and harvested whales was not statistically significant for the 1993 season, it is when a large sample is considered (1990-1997; Table 3). In 1992, wind speed was also greater when whales were harvested; however, only two whales were taken that season so it was not considered statistically testable (Fig. 5).

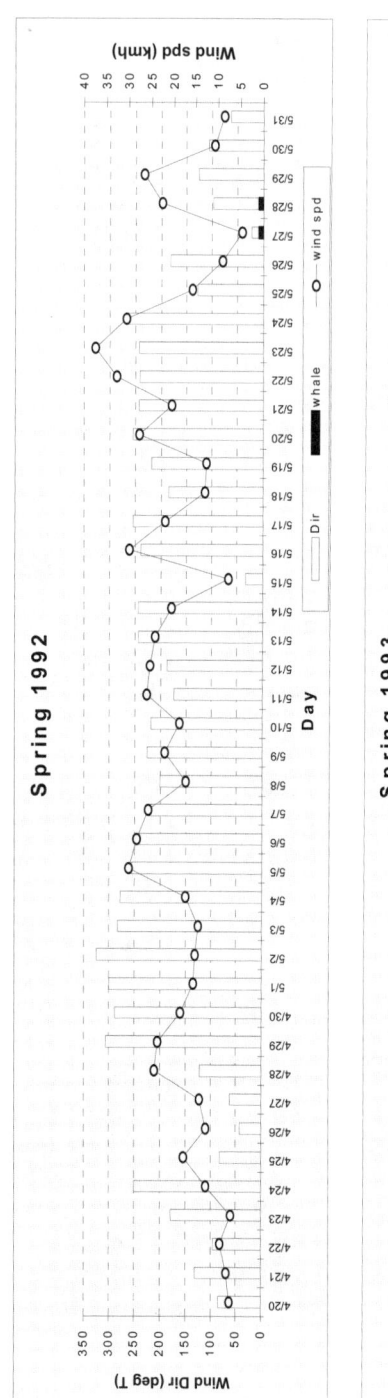
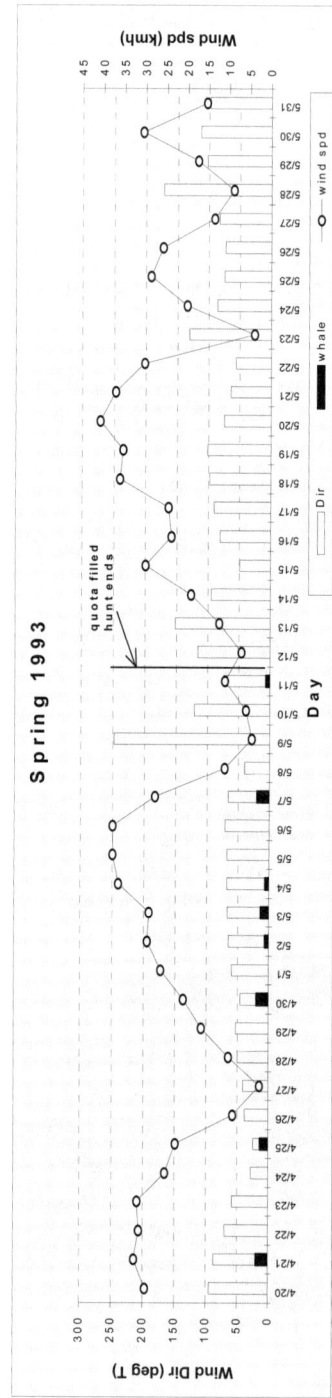

Figure 5. *Relationships between wind speed and direction and numbers of whales taken during Barrow spring hunts. Wind speed (km/h, connected dots), direction (vertical bars, degrees True), and the number of whales landed by day for the spring 1992 and 1993 whaling seasons at Barrow, AK. While this plot appears cluttered with a considerable amount of data, it is easy to see that whales tend to be landed when the wind direction is between 0° and 180° (where the bars are less than 180). Note that whales are taken even when the wind speed is fairly high (up to 35 km/h or about 20 mph). These plots illustrate specific examples of the hunting criteria listed in Table 2.*

Table 6. Meteorological data for spring 1992 and 1993 bowhead hunting seasons at Barrow.

Year	Data	Not Harvested	Harvested	Seasonal Statistics
Spring 1992	Mean wind speed (km/h)	18.4	24.6	18.7
	N (number of hunting days)	40	2	42
	Standard Deviation	8.2	2.8	8.1
	Dominant wind direction			SW
	Mean Temperature C	-9.2	-8	-9.0
Spring 1993	Mean wind speed (km/h)	19.6	25.9	22.4
	N (number of hunting days)	14	8	22
	Standard Deviation	12.7	7.8	10.9
	Dominant wind direction			NE
	Mean Temperature C	-6.8	-12.2	-7.9

Basic meteorological data for the spring 1992 and 1993 bowhead whale hunting seasons at Barrow when bowhead whales were and were not harvested. N (number of hunting days) refers to the length of the whaling season for that year; e.g., in spring 1993, there were only 22 hunting days since the quota was reached by 11 May.

Table 7. Comparison of the number of harvested whales by daily wind direction frequency for the 1992 and 1993 spring seasons.

Wind direction component	1992		1993	
	Number Harvested	Number of Days	Number Harvested	Number of Days
Easterly	2	16	16	21
Westerly	0	26	0	1

Winds with an easterly component were those blowing from 1-179° T and westerly from 181° to 359°T. The 1993 hunting season ended on 11 May when the quota was reached.

1997 SPRING SEASON

Very unstable landfast ice and intermittent west winds characterized the spring 1997 bowhead whale-hunting season at Barrow.[3] According to local hunters, landfast ice formation was delayed (a) due to strong periodic easterly winds, that tore the landfast ice from shore, and (b) the lack of thick, multiyear sea ice to help ground the landfast ice. Thus, the landfast ice consisted of relatively 'young' ice when the 1997 spring hunt began. On 17 May, 1997, about 80% of the landfast ice broke free, setting all the active whaling crews adrift—a total of at least 142 people. Heavy fog set in, but all personnel were rescued owing to the whale hunters' survival skills on sea ice and the skilled helicopter piloting by the North Slope Borough Department of Search and Rescue. Virtually all the conditions that impeded the spring 1997 hunt were also present in spring 1998, when hunters again faced very poor whaling conditions but were extremely careful because of their 1997 experience. As in the 1993 spring season, Barrow hunters tended to land whales in spring 1997 when the wind direction was east (Fig. 6; Tables 8 and 9).

In recent years, the extent and thickness of multiyear ice has declined in the Alaskan Arctic Ocean (Tynan and DeMaster 1997), which could be affecting bowhead hunting success in an indirect manner. The Barrow hunters suggest that multiyear ice helps stabilize the landfast ice through well-grounded terminal pressure ridges. The whale census is equally affected by unstable landfast ice. Large breaks in watch effort due to ice emergency evacuations would increase the variance of the estimates or completely preclude making a population estimate (Krogman *et al* 1985; Philo *et al.* 1994; Raftery and Zeh 1998).

1997 FALL SEASON

The fall hunt in 1997 benefited from very good whale hunting conditions (Fig. 6). Seas were completely ice free and winds were light (<15 km/h) for approximately half the season. As for the 1990–1997 period, the wind speed was significantly lower when whales were captured than when whales were not captured (Table 8). Low wind speed and ice free seas accounted for the very good whale hunting conditions. Again, we mention the great similarity to the 1998 fall season. In fact, the ice retreated at least as far as has ever been documented. Ice free waters extended to 80°N directly north of Barrow.

---

[3] We should also note that very similar ice conditions were present in 1998.

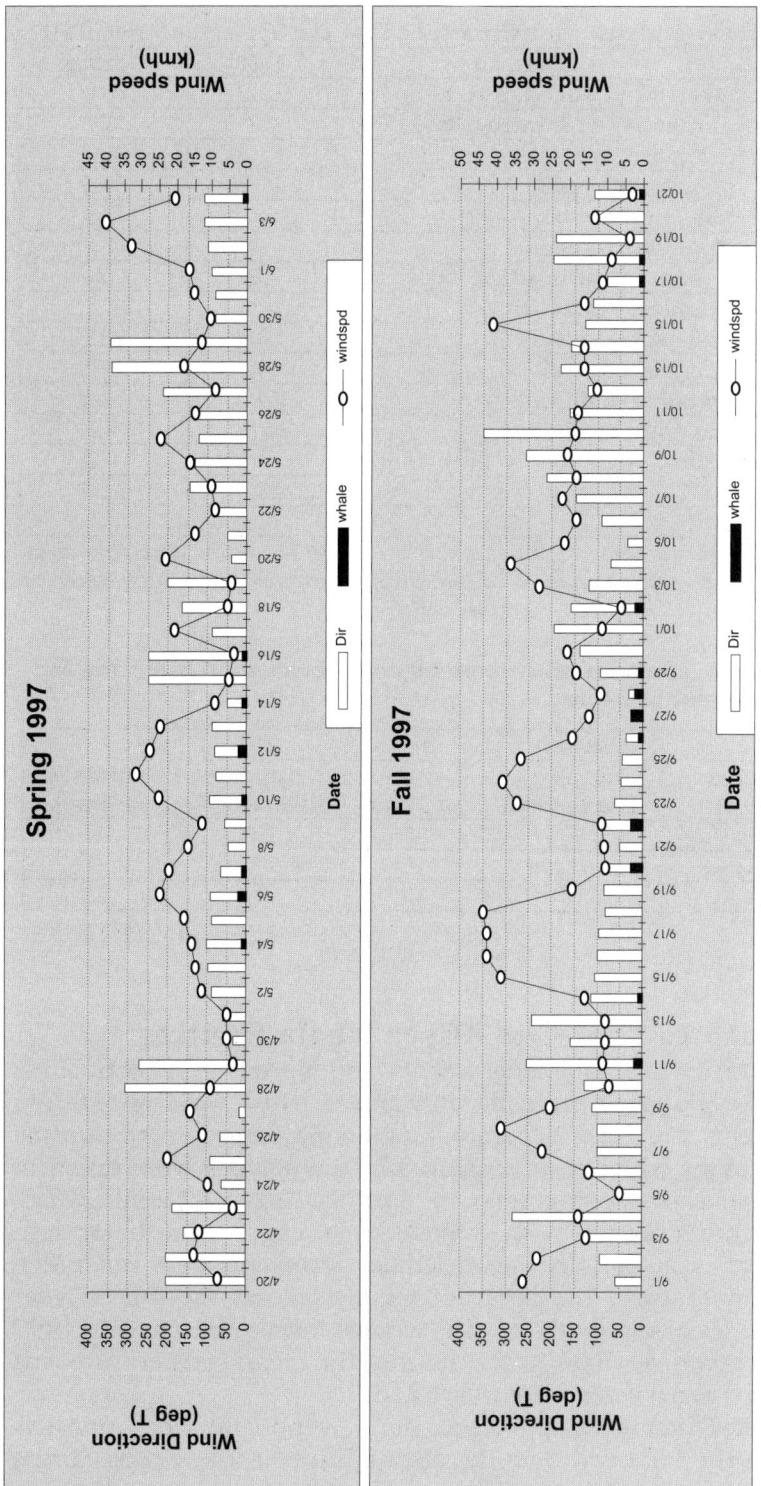

Figure 6. *Wind speed (km/h, connected dots), direction (vertical bars, degrees True), and the number of whales landed by day for the 1997 spring and fall whaling seasons at Barrow, AK. Note that during spring in all but one case, hunters landed whales when the wind was from the east (e.g., < 180°, or sometimes referred to as wind with an 'easterly component'). During the fall season, however, note that whales were taken when wind speed was low or less than about 20 km/h but that some were taken when the wind direction was from the west.*

Table 8. Summary statistics for wind speed for the spring and fall, 1997.

Harvested?	Data	Fall	Spring
No	Mean wind speed (km/h)	23.2	13.3
	n (number of hunting days)	38	35
	Standard deviation	11.3	7.0
	Max wind speed	43.6	31.7
	Min wind speed	4.0	3.6
Yes	Mean wind speed	11.6	18.9
	n (number of hunting days)	13	7
	Standard deviation	4.6	7.5
	Max. wind speed	19.4	27.7
	Min wind speed	3.2	5.0
Seasonal Mean wind speed (km/h)		20.3	14.2
N (number of hunting days)		51	42
Standard deviation		11.2	7.3
Max. wind speed		43.6	31.7
Min wind speed		3.3	3.6

Hunts for days during which whales were and were not harvested. Note that for consistency with other seasons, these statistics exclude the whale taken on 4 June, 1997.

Table 9. Numbers of whales harvested by daily mean wind direction for the 1997 spring and fall hunt.

	Spring		Fall	
Wind direction component	Number Harvested	Number of Days	Number Harvested	Number of Days
Easterly	8	35	18	42
Westerly	1	7	3	9
Totals	9	42	21	51

Easterly winds were those blowing from 1-179ºT and westerly winds were those from 181 to 359ºT.

Other Variables That Can Affect Whale Hunting

Air temperature during spring affects the speed at which leads refreeze; however, leads typically only freeze when wind speeds are low. Obviously, if a lead becomes completely frozen-over, hunting essentially ceases which may explain why few whales were taken at low wind speeds in spring (Fig. 6). For this reason, temperature effects on hunting are probably difficult to assess. As can be seen in Table 6, there is no pattern when whales were and were not taken related to temperature. Temperature mainly reflects the time of year of the hunt. Because both of the whales harvested in the 1992 season were taken very late in the season, the mean temperature was higher when those whales were harvested.

In several instances in recent years, wind and ice conditions have favored the fall hunt over the spring hunt. In the 1980s during spring, temperatures were cold and the landfast ice had much multiyear ice mixed with it and was quite sturdy and generally stable. During the fall hunts, it was not uncommon to have significant ice coverage in October that precluded bowhead hunting. The 1992, 1996, 1997, and 1998 seasons are examples of weather and ice conditions that made for poor spring hunts. The fall weather conditions for these years were generally good (excellent in fall 1992), posing little interference to the hunt. These factors, together with the higher harvest quotas, shifted the bulk of the harvest to the fall in the late 1990s.

CONCLUSIONS

This analysis of hunting seasons appears to document what Barrow whale hunters have long said: that the bowhead whale hunt is heavily influenced by environmental conditions. Hunters indicated that wind direction is a major environmental variable during the spring bowhead hunt, whereby east winds favor a successful hunt and west winds greatly reduce that success. During the fall hunts, wind speed appears to be an important variable while wind direction has less influence on hunting success. Such characterizations should be useful to the AEWC and the IWC aboriginal whaling management program developers in establishing flexible and reasonable hunting guidelines that account for variability in weather and ice conditions. We find it encouraging (and validating) that the statistical results agree with the environmental criteria noted by the hunters in Table 2.

Further analysis should include all seasons for which reliable data exist and account for ice concentration and visibility (fog density) because these factors are known to also affect bowhead-hunting success. Calculating the average number of 'suitable' hunting days, based on weather data from seasons far into the past, is possible. It also seems plausible that hunting seasons dating back to the late 1800s can be characterized based on weather records. Similarly, it may be possible to gain some insights into the *possible future effects* of global change on hunting success based on the modeler's predictions regarding future wind and sea-ice patterns in the Arctic. Finally, we note a trend in the 1990s toward both numerically larger fall harvests and a higher proportion of whales taken during fall.

Acknowledgments. Our sincere thanks go to the AEWC and the subsistence hunters of the AEWC member villages for support of our studies. The AEWC provided data on whales that were landed in villages other than Barrow. Dan Endres and Malcolm Gaylord from the Barrow CMDL observatory assisted in acquiring weather records and helping with interpretation of the data. We thank Dolores Vinas and Lisa Dela Rosa for administrative assistance. Tom Albert, NSB

Senior Scientist, provided valuable advice and support. This project was partially funded by the National Science Foundation (OPP-9909156), for which we are grateful. We also wish to thank Dr. Allen McCartney for his interest in our work and persistence in seeing it completed.

REFERENCES

Alaska Consultants, Inc. and Stephen R. Braund & Associates
  1984 *Subsistence Study of Alaska Eskimo Whaling Villages.* Report
       prepared for the U.S. Department of the Interior.
Braund, S.R.
  1988 *The Skin Boats of Saint Lawrence Island, Alaska.* Seattle:
       University of Washington Press.
  1992 Traditional Alaska Eskimo Whaling and the Bowhead Quota.
       Arctic Research 6(Fall): 37-42.
Donovan, G.P., ed.
  1982 Aboriginal Subsistence Whaling (with special reference to the
       Alaska and Greenland fisheries). *Report of the International
       Whaling Commission, Special Issue 4.*
Gambell, R.
  1982. The Bowhead Whale Problem and the International Whaling
       Commission. *Report International Whaling Commission, Special
       Issue 4:* 1-6.
George, J.C., R.S. Suydam, L.M. Philo, T.F Albert, J.E. Zeh, and G.M. Carroll
  1995 Report of the Spring 1993 Census of Bowhead Whales, *Balaena
       mysticetus,* Off Point Barrow, Alaska with Observations on the
       1993 Subsistence Hunt of Bowhead Whales by Alaska Eskimos.
       Report of the International Whaling Commission 45: 371-386.
International Whaling Commission
  1997 Report of the Scientific Committee. *Report of the International
       Whaling Commission* 47: 55-112.
Krogman, B.D., R.M. Sonntag, J.E. Zeh, and D. Ko
  1985 Environmental Factors Affecting the Results of the 1984 Ice-
       based Visual Census of Bowhead Whales, *Balaena mysticetus,*
       near Point Barrow, Alaska. Report submitted to International
       Whaling Commission (SC/37/PS9).
Philo, L. M., J.C. George, R. Suydam, T.F. Albert, and D. Ramey
  1994 Report of Field Activities of the Spring 1992 Census of Bowhead
       Whales, *Balaena mysticetus,* Off Point Barrow, Alaska with
       Observations on the Subsistence Hunt of Bowhead Whales 1991
       and 1992. *Report of the International Whaling Commission* 44:
       335-342.
Raftery, A.E. and J. E. Zeh
  1998 Estimating Bowhead Whale Population Size and Rate of Increase
       from the 1993 Census. *Journal of the American Statistical
       Association* 93: 1-13.

Stoker, S.W. and I. I. Krupnik
 1993 Subsistence Whaling,' pp. 579-619 in J.J. Burns, J.J Montague, and C.J. Cowles, eds., *The Bowhead Whale.* Lawrence, KS: Society for Marine Mammalogy, Special Publication No. 2.
Tynan, C.Y. and D. Demaster
 1997 Observations and predictions of arctic climate change: potential effects of marine mammals. *Arctic* 50 (4): 308-322.

Fall Whaling in Barrow, Alaska: A Consideration of Strategic Decision–Making

Barbara Bodenhorn

Abstract. *Although it has received relatively little scholarly attention, fall whaling has long been important for Barrow Iñupiat. The environmental conditions in which the fall hunt takes place vary significantly from those affecting the spring whaling season. This paper presents some of the cultural models and social institutions that inform and structure important whaling strategies. Case examples are used to explore some of the implications of a variety of responses to problems experienced during the 1996 and 1997 fall whaling seasons. These responses reveal the degree to which 'environmental conditions' must include social and political factors and invite readers not to confuse shifts in whaling practices with a loss of culture.*

INTRODUCTION

Long time ago, according to an *unipkaaq*[1] related by William Oquilluk (1981:1), 'people did not have to work in their minds to stay alive.' Then 'the first disaster' occurred—an eclipse of the sun that lasted for several days—after which those who survived had to think about the world around them.

> The people began to think about how they might need better tools ... they talked it over ... they thought about how to use rocks and stones ... they thought how to use driftwood ... they thought about all the parts of the animals ... . They experimented and soon made useful things (Oquilluk 1981:8).

Many origin stories relate how people learned to follow the proper ways to do things by listening to spirits or animals who taught them practices that had to be replicated exactly. The story of the

---

[1] In Barrow, Alaska, Iñupiat speak of two kinds of stories. *Unipkaat*, often translated as 'legends,' tell stories of an ancient past, of the time when animals and humans could talk with each other and even, at times, might transform themselves into the other. *Quliaqtuat* are stories based on personal knowledge and experience (Leona Okakok, Raymond Neakok, Sr., pers. comm.). William Oquilluk was born in Point Hope but soon moved with his family to Kotzebue. His book, *People of Kauwerak*, pulls together his view of the history of the Iñupiaq people in his region, as it can be learned through the stories of the region (Paul Ongtoogook, pers. comm.).

Eagle–Wolf Messenger Feast, also related by Oquilluk (1981:150-156), is one such account. The above narrative, in contrast, is one that recounts ways in which people observed the world around them, thought about it, experimented with different techniques, modified their tools, and told others about what they were doing so they could engage in the process as well. It is a story that enshrines the value of survival through change.

Although I have never been told this story in Barrow, it has been striking to me how often, when I ask people why they have done something 'against the grain' of custom, that the answer has been consistent: 'for survival.' This chapter is, in part, an examination of some ways in which Iñupiaq tradition should be thought about as the customary practice of change—of the constant modification of the things people do when whaling, and of the technology they incorporate to do it. It is important to highlight this notion given that so many challenges to Iñupiat claims to the right to hunt are based on assumptions that 'legitimacy' and 'authenticity' are somehow grounded in unchanging conditions. The fact that Iñupiat may assert that they have whaled 'since time immemorial' does not necessarily imply that real traditions are static ones.

It is easy to read this answer as evidence of *flexibility*—long one of the terms used to characterize Inuit social action (see, for example, Wilmott 1960). Whether one is talking about the division of labor, marriage strategies, or the many ways one can become kin, pan-Arctic social organization can seem infinitely adaptable and is often spoken of by the people concerned as a survival strategy in an environment that is complex, unpredictable, and potentially unforgiving.[2] Flexibility in this case, however, is not another way of saying 'anything goes,' but is rather a way of achieving an agreed end that is mutually recognized as a culturally valued constant. Iñupiat comments about change and continuity in Barrow and elsewhere suggest that one constant centers on a moral imperative rather than on a particular set of fixed cultural practices—namely, to act on a respect for the whale by whaling and by sharing. How one does this is, of course, subject to influence from changing conditions.

The goals of this chapter are twofold: to show some of the interrelationships between changing conditions and whaling decision–making in Barrow, and to discuss the roles of technological shifts in these processes. Through a series of case examples, I examine the very complex processes through which decisions about

---

[2] By 'pan-Arctic,' I include the circumpolar peoples extending from Greenland to Siberia: Kalaalit in Greenland, Canadian Inuit, Iñupiat, Yupiit, and Siberian Yupiit of Alaska and Siberia. I am by no means suggesting that Arctic peoples conduct their lives in exactly the same manner—important differences exist and it is crucial that they are recognized. Nonetheless, where commonalities exist, they have often been glossed as evidence of flexibility.

whaling continue to be made in response to a variety of changing conditions. I concentrate on fall whaling in Barrow because it is different from spring whaling (differences that are often under-reported) and because in recent years it has provided significant amounts of whale meat and *maktak* (black skin and blubber) for the entire North Slope region. One overarching goal, then, is to bring the notions of tradition and technological change into a focus that does not set them in opposition.

This chapter examines three instances of shifts in whaling practices that took place in Barrow during the fall whaling seasons in 1996 and 1997. In part, I suggest that these shifts show how strategies may be generated institutionally at different levels: individual, crew, Barrow Whaling Captains' Association, and Alaska Eskimo Whaling Commission. It is important to note that this multi-layered set of decision–making pathways is not a new invention in itself, although it may well have become more elaborate with the need for Iñupiat to deal with national and international institutions. As with so many other aspects of social life in Barrow, Alaska today, change and continuity are closely interrelated. In a related way, these cases illustrate clearly how people experience developments in material technology as both constraining and facilitating. Taken together, I suggest, this material may be used to consider more general discussions concerning the way we model the process of change itself.

OVERARCHING FRAMEWORK

The general framework for this discussion is set out in two ways. First, with the help of information provided by North Slope Elders, some of the ways in which rules about the proper conduct around whaling influence Iñupiat social life in general are considered.[3] Keeping this generally social context in mind, we then turn to the formal organization of whaling, examining the different institutional levels at which strategic decision–making occurs. With this

---

[3] Research on which this paper is based combines archival work at the Iñupiat History, Language and Culture Commission (IHLC), interviews and group discussions with local experts, informal conversations with individuals, and participant observation. Fieldwork was conducted during summer visits in 1995, 1996, an extended stay between January and September 1997, and short follow-up visits in April and August/September 1998. The present paper relies particularly on the translated transcripts of the 1991 IHLC Elder's Conference held in Barrow (1991 a,b,c,d). The overall theme of the conference was whaling; the topics discussed in the men's and in the women's sessions placed heavy emphasis on the obligations and responsibilities of crew members in general, and of the whaling captain couple in particular. These transcripts are currently in draft form and were translated by Mabel Paniegeo. An earlier version of this paper was published in Cambridge Anthropology (Bodenhorn 2000/1).

framework in place, we have a context within which to examine the case examples in detail.

'Iñupiaq Food is Social Food:' Whaling as a way of Organizing the Social

The social organization of whaling is often talked about in terms of its division of labor: how people organize themselves to hunt, butcher, distribute, and share the whale. This is an exceedingly important part, but nonetheless only a part, of the story.

'I am Iñupiaq; I eat Iñupiaq food' is an equation that I have heard many times over the years—in many contexts and by many different people. But it is not just the nature of the food itself that is so important. 'Iñupiaq food is social food,' Fannie Akpik recently told a conversational Iñupiaq class and indeed, that is very literally true in many cases. People who may be forced to eat Iñupiaq food alone (in hospital rooms, for instance) talk of how incomplete that experience is. What makes it social, however, is not just that it tastes better when you can eat it together, but that it is a consequence of many social relationships—between humans, between humans and animals, and among animals themselves.[4]

It is common to hear that whales may give or withhold of themselves. In the women's session of the IHLC 1991 Elders' Conference, for instance, Ida Koonuk spoke of this explicitly: 'The bowhead is a very distinguished mammal. It can give itself up, which can make it very easy for the captain and crew, or it can withhold itself from another captain and crew and can be struck and lost ... (IHLC 1991a:12).' The gift relationship between whales and humans is a social one that depends on two other kinds of social behavior: generosity among humans and communication between the whales themselves. Kirk Oviok, from Point Hope, remembered his own upbringing:

> Like my aunt said, the whales have ears and are more like people. The first batch of whales seen would show up to check which ones in the whaling crews would be more hospitable to be caught. Then the whales would come back to their pack and tell them about the situation stating, 'we have someone available for us.' ... . This is what my wife and I have heard from my aunt

---

[4] The intensely felt connection between eating hunted food and identity is not a new observation. Brody (1975) says of Inuit in eastern Canada, that to be a real person, one must eat real food; Stairs and Wenzel (1988) elaborate on the centrality of shared food for the concept of *inummariit*, or genuine people. For Alaska, see, amongst others, Bodenhorn (1993, 2000) Fienup-Riordan (1983, 1990) and. Hess (1999). The connection is always worth repeating however, since its profundity often continues to be underestimated.

Negovanna. I firmly believe this is true, that whales have ears (Oviok, IHLC 1991b:4-5).

This is echoed by Mary Aveoganna, from Barrow. 'Always be ready with hospitality,' she instructed, 'so the whale will see an inviting place' (IHLC 1991a:tape 2). In his 1985 address to the Alaska Eskimo Whaling Commission, Patrick Attungana, another Point Hoper and ordained minister of the Episcopal Church, related an *unipkaaq* that expresses this even more explicitly:

> When the whales come, ... one of them stops, like it was camping, being caught by the people ... . That one that is like camping, ... it knows when its relatives are coming back ... . Those that are returning want to listen to what that one that is like camping has to say. That one tells them the stories, that he had a good host with those two, the married ones [the whaling captain couple] ... . That one that talks about having good hosts, starts looking forward to going back to those hosts when they return the following year. And the other one, that said it did not have good hosts, said that it will not camp again, but will go to another host (Attungana 1986:5-6).

Thus, assumptions about whale/whale sociality also have consequences for the moral weight Iñupiat place on human social behavior. This does not simply concern the way the whale hunt is conducted on the ice, but extends to the generosity with which humans treat one another throughout the year.[5]

Let us return to the 1991 Elders' Conference to examine some of these behaviors in more detail. The comments included here refer to ways that these Elders learned to behave in preparation for whaling, during the whale hunt itself, in the immediate aftermath of a successful hunt, and during the year overall.[6] The fact that the

---

[5] Although the moral nature of sharing seems most deeply felt in Barrow when people talk about the whale/human relationship, this feeling is neither restricted to whaling itself nor to whaling societies. In their exploration of what it means to be *inummarik* ('a most genuine person') in eastern Canada, for instance, Stairs and Wenzel (1988:6) claim that to be *ilgasuuq* (or generous) is not simply to give large amounts of material goods away. Rather, it is, 'the manner and attitude by which [people] expressed their generosity. *Ilgasuuq* is not simply an action; it is also an aspect of a genuine person's entire demeanor.' Fienup-Riordan (1991) discusses Nelson Islanders' emphasis on mind when considering the moral nature of individuals' relations with others. See also Wenzel *et al.* (2000) for an edited volume focused entirely on the moral importance of sharing.

[6] Sponsored by the North Slope Borough Iñupiaq History, Language and Culture Commission, Elders' Conferences are annual events that provide opportunities for Elders from all North Slope member villages to participate in several days' intense discussion of specified issues. The women's session took place on July 11, 1991. The session transcript includes comments from

context was an Elders' Conference should not lead readers to assume that this information was simply being recorded for posterity. One purpose of these conferences is to document information that is thought to be as useful in the present as it was in the past. It is with this in mind that they have been positioned here to form a central part of the present discussion.

We begin with references to cleaning ice cellars because this was mentioned so many times as crucial preparatory activity; we should bear in mind, however, that the cycle has no clear beginning and no absolute end. The whaling celebration is at once an opportunity to give thanks for a successful season and a chance to provide the hospitality that, as we have heard, will encourage whales to give of themselves again in the future. Indeed, in a discussion about how best to begin a film presenting the process of preparing an *umiapiaq*, or skin-covered whaling boat, Dorothy Edwardsen, a Barrow whaling captain's wife with a sure sense of the visual and the dramatic, suggested a good opening shot would be one from *Nalukataq* (or blanket toss). This would be followed by men setting out to hunt *ugruk* (bearded seal), the animal whose hide is used to create the skin-boat covering.[7] Both events occur in June when the sun is at its height and the ice continues to provide attractive resting spots for the bearded seals. As Maggie Ahmaogak has discussed, the seasons as well as the hunt of the whole range of Iñupiaq subsistence resources are interconnected.[8] Whaling is never just about whales. 'Always getting ready' is as apt a description of whaling activities in Barrow as it is elsewhere among many hunting peoples of the Arctic and sub-Arctic.[9]

'During the month of March, the first thing we get into is the ice cellar,' explained Ida Koonuk of Point Hope. 'We are told the one we are so expectant about does not like to be laid to rest in a messy cellar.'

(in order of their appearance in the transcript): Rosemary Oviok (Point Hope), Lora Oyagak (Barrow), Berna Brower (Barrow), Mary Lou Leavitt (Barrow), Alice Solomon (Barrow), Terza Hopson (Barrow), Dorcas Tagarook (Wainwright), Jennie Ahkivgak (Barrow), Ida Koonuk (Point Hope), Carol Omnik (Point Hope), Louise Ahkiviana (Barrow), and Mary Aveoganna (Barrow). Jana Hacharak and Emma Bodfish presided.

[7] This discussion was part of a class in which students collectively assessed existing work about whales and whaling and then designed projects for further investigation of their own.

[8] See Maggie Ahmaogak's AEWC: Overview and current concerns, posted on our website, *Traditional Whaling in the Western Arctic.* The full website address is listed in the bibliography. Alternatively, readers can click onto the Google search engine and simply type in maggie ahmaogak.

[9] Echoing Barker (1993).

That was one of the foremost teachings we have always heard. My mother-in-law would tell me, 'When you are done with cleaning up of the cellar, before you climb up, say verbally, 'you can now expect to be filled.' It has gotten to be a habit with me now, saying it inside the cellar before I climb out of it (I. Koonuk, IHLC 1991a:12).

Carol Omnik, also of Point Hope, concurred:

It is exactly like one of the former speakers, Ida, said: That the first thing that has to be done is the ice cellar. I, too, grew up when I would see people work to clean out cellars. It has always been a practice from time immemorial, a piece of whale meat from last year cannot be saved until a later time. It has to be taken out. Because the anticipated whale always sees and hears all that goes on (Omnik in IHLC 1991a:13).

Mary Aveoganna, of Barrow, expanded this somewhat:

We, the Aveoganna crew, my children and their spouses, started on an ice cellar, for the proper storage of the mighty bowhead. For it is common knowledge, as we are told from time immemorial, that the bowhead would discern what/how they are to be handled, distributed and stored as they give themselves up to this particular captain and crew. Every one of the crew members gave of their time and labor to get the cellar done in time for whaling. I kept encouraging them, telling them that it will anticipate to be filled with what we all are hoping, praying for. It is so clean, and prepared for what we all have awaited for (Aveoganna, IHLC 1991a:15).

The ice cellar, then, must be prepared as an inviting resting place for the whale. The meat that has been removed in the process cannot be hoarded, as Berna Brower of Barrow related:

When my daughter-in-law asked: What are we going to do with all the meat and fish that were put out from the ice cellar? Shall we put them back into the cellar? [I told her] No, ... Just leave enough fish and meat for the whaling season, for your whalers, then take the extras to the widows and the Elders. So all day she went out and gave away all the meat and fish to different homes that she knew of.
    Then one of the Elders spoke and said may she receive something tender and delicious to give away. So that is why we should always give priority to widows and Elders (B. Brower, IHLC 1991a:2).

The hope is not that one gives food away in order to get something tasty for oneself, but that by being generous, one will receive something that is worthy of being an appreciated gift again. The exhortation to share and be generous, especially to those less fortunate, appeared in almost every participant's testimony. Not only

Berna Brower, also but Ida Koonuk ('take widows and orphans under our wing'); Alice Solomon ('feed the hungry, the orphans, the poor'); Terza Hopson ('what the Elders have said about 'taking under your wing, so to speak, the poor, the orphans' fits right in with the commandments of our Lord'); Dorcas Tagarook ('we were taught not to omit anyone'); Jennie Ahkivgak ('feed the poor'); Carl Omnik ('take responsibility for the orphans and the Elders'); and Mary Aveoganna ('don't hoard') reflected the consistency of this message from across the North Slope region (IHLC 1991a:2-15).

The need for cleanliness was echoed in the men's session.[10] Eli Solomon drew an explicit parallel with the women's responsibilities on shore; just as the women keep the *qanitchat* (entry ways) clean, so everything around the tent should be clean—especially the left side of the boat (IHLC 1991b:9). This way, the whale will 'see' its way to its resting place and be more likely to give itself up.[11] Wyborn Nungasak, as well, was clear that one of the whaling captain's responsibilities was to ensure that 'the environment needs to be clean and acceptable by all' (IHLC 1991d, tape 2:3).

Words as well as deeds must be treated carefully. 'Watch your words!' Arthur Neakok, originally from Nuvuk (Point Barrow), said to the men; 'Refrain your tongue from backbiting,' exhorted Carol Omnik, while fellow Point Hoper Ida Koonuk emphasized, 'Harsh words do no good; the whale listens in [and reports back to other whales]. Because the whale we are all so eager for listens in, whoever we are dealing with' (IHLC 1991a:12).

Once the whale has given itself up, it is important to handle the meat with care. One should treat it 'tenderly' in the process of butchering, storing, and preparing the meat for the various feasts during which it will be consumed throughout the year, as Carol Omnik relates:

> It has always been said that as the whales gather together, they would communicate, one to the other, that this particular person does not work on me with tender, loving hands. Some would

---

[10] IHLC 01499tr1. The men's session also took place on July 11, 1991. Those present included, in order of contributing to this session: Ross Ahngasak (Barrow), Kenneth Toovak (Barrow), Noah Phillips (Barrow, Wainwright), Herman Rexford (Kaktovik), Alfred Leavitt (Barrow), Greg Tagarook (Wainwright, Point Hope), Sam Taalak (Nuiqsut, Barrow), Eli Solomon (Barrow), Roxy Oyagak (Barrow), Perry Akootchook (Kaktovik), Arthur Neakok (Point Barrow), Edward Hopson (Barrow), Levi Greist (Kuukpik River, Barrow), and Walter Akpik (Atqasuk). Wyborn Nungasak (Barrow) presided; Mabel Panigeo transcribed and translated the tape.

[11] This resonates strongly with Ann Fienup-Riordan's (1994) discussion of how important it is for Yup'ik hunters that the passage ways between animal and human worlds are kept clear.

> report that this one is the best person to be with. There are some
> others who would like to go to someone else, all because they
> want to be worked on tenderly. So the conversation goes on
> (Omnik, IHLC 1991a:14).

Above all, the women and the men reiterated, it is important to 'be in
harmony.' Levi Greist (IHLC 1991c:14) remembered a story in which
this was the central moral: 'Long ago, just before going out on a
whaling venture, man and wife had a dispute. It was not long after
they got to the open lead where they set up camp, that a whale came
up and started chopping the ice off with his flukes. Therefore, unity
is continually stressed.'

Patrick Attungana (1986:16ff) underscored the same message
on a wider level in his 1985 address to the Alaska Eskimo Whaling
Commission: 'When you hunt in harmony, you don't have problems
catching the animals. This is what needs to be thought about. If the
hunters from Barter to St. Lawrence Island hunt in harmony, the
animals will keep going. They will acquire, they will catch the animal.'
The whaling captain couple, the whaling crew, and the whaling crews
together are enjoined over and over to 'work together' and to 'be
grateful' to the whales for their gift of themselves.

In many important ways, the annual cycle is a marked
celebration of whaling. Nalukataq, Thanksgiving, and Christmas all
incorporate the distribution of shares, the communal consumption of
a feast centered on whale meat and *maktak,* and the clear expression
of thanks that whaling has made this possible. These feasts are
intensely social, celebrated in commensality, singing, dancing, and
prayer. The exhortations presented above, however, are about the
rules of sociality to be taken into account on a daily basis throughout
the year. And as I have already mentioned, the exhortation for
humans to be social in particular ways is quite explicitly to underpin
the sociality of the whale/human gift relationship. Very specific
things need to be done, but generally the decisions about how to do
them are left open.

I do not want to give the impression that either Iñupiaq beliefs
or practices are uniform, coherent, and somehow hermetically sealed.
They are not. How people talk about the social relationships between
whales, humans, and God varies from person to person, between
denominations, and between villages. As Isaac Akootchook (IHLC
1991c:1) pointed out, customs at times 'have to be altered to fit our
way of life ... . What will be effective in your village will not be good
for our part of the country.' Nor am I trying to reproduce a picture of
'pure' Iñupiaq ideas that can be viewed through intervening layers of
Christian doctrines. The participants in both the men's and the
women's sessions quoted above are, for the most part, practicing
Christians of several denominations. For many, an active reliance on
Christian prayer formed the backbone of proper behavior. Indeed,
Mary Lou Leavitt (IHLC 1991b:3) was firm that 'above all, the woman
of the house should be a praying woman.' Just as Terza Hopson drew

a parallel between pre-Christian and Christian Iñupiaq practices when she noted the injunction to help the needy, so Jennie Ahkivgak (IHLC 1991a:12) also pointed out that many of the ways in which Iñupiat talk about whaling—whether or not from an explicitly Christian viewpoint—are remarkably consistent. 'Especially when I read the Iñupiat Bible,' she mused, 'all that is written thereon is no different from the Iñupiaq customs and way of life. What we knew to be a fact, that our forefathers did not read, but by their verbal expressions, it would sound like they were reading the Iñupiat Bible as they gave instructions to live good moral lives.'

Whether from the spirit of God, or of the whale, or a combination of both, these moral codes thus provide a backdrop with reference to which a broad range of decisions is generated.

The Institutional Organization of Whaling

It is not enough to talk of whaling beliefs and the ways in which they might—and indeed do—structure social behavior. We need also to examine the institutional organization of decision–making insofar as whaling strategies are concerned. This chapter begins with cultural statements of beliefs because some of these formal decision–making pathways may look so familiar in non-Iñupiaq settings that it is easy to lose sight of the very specifically Iñupiaq ideas and beliefs that inform particular strategizing processes undertaken in Barrow and region–wide. Before turning to the case examples, then, let us look at several layers of institutions through which whaling strategies are created.

The *Alaska Eskimo Whaling Commission* (AEWC) is the largest and newest of these formal institutions. Created in 1978 in response to the International Whaling Commission's proposed moratorium on indigenous bowhead whaling, it comprises locally elected commissioners from each of the ten Alaskan whaling villages.[12] In important ways, the AEWC is a bridging organization. It represents regional interests to national and international bodies, and it ensures that local whaling captains' associations are kept 'in the loop' of information and decision–making. One major task of the AEWC is to negotiate with non-Iñupiaq organizations, primarily the U.S. Federal Government and, indirectly, the International Whaling Commission (IWC), about management plans in general and the quota system in particular.[13] It is also the AEWC's responsibility to monitor local implementation of the plans once they have been set. As part of this,

---

[12] See Maggie Ahmaogak's (2002) detailed description of the organization and the responsibilities of the AEWC. Ahmaogak is both the AEWC's current Executive Director and a whaling captain wife.

[13] The AEWC attends IWC meetings, for instance, but does not have voting rights.

it mediates the division of the quota among member villages and keeps track of inter-village quota transfers.

The *Barrow Whaling Captain's Association* (BWCA) is, as it sounds, a community–based organization that has counterparts in all the member villages of the AEWC. Although not a formally incorporated organization like the AEWC, its status is one of a locally–recognized institution and, in fact, comes out of a long–standing association of whaling captains.[14] According to a number of Barrow people who talked to me during the 1980s, the *umialingat* (literally, the whaling captain collectivity) used to meet regularly to discuss many issues affecting the community as a whole. Since, as we have already heard, the gift of whales to the community was dependent on their perceptions of proper human behavior, it was important for the captains to keep track of what was going on in the village. They were a collective decision–making body that played a major role in social life.[15] Today, the BWCA decides on local best practice for matters affecting Barrow crews collectively, a process examined as a case example below.

The Barrow Whaling Wives Auxiliary (*Utqiaġvigmiut Aġviqsiuqtit Aġnaŋich*, or UAA) also meets regularly to consider matters that are pertinent to the efficient meeting of responsibilities on the wife's side of the whaling captain couple. In the past few years, the UAA has created a Documenting Committee, for instance, to ensure that needed skills in the preparation of the skin boat were not being lost; they have considered how best to prepare feasts for hundreds of people in the safest possible conditions; and they raise funds throughout the year to support the very considerable expenses whaling captains must incur as part of the whaling process.

Each crew is also an organized group unto itself. The whaling captain husband/wife couple, often aided by a co-captain, must make decisions on the intra-crew division of labor, whether to go out on a particular day, and the like. Many of these decisions were discussed as overall responsibilities in the 1991 Elders' Conference.[16] In another context, Martha Aiken (2000) asserted that

---

[14] Maggie Ahmaogak alluded to this long-standing association in an interview: 'The whalers at the time—way back—had meetings of their own, trying to decide what kind of policies and rules they had ... the way they would share with the community—the way they were to give to the poor ...' (July 7, 1997).

[15] To quote Raymond Neakok, Sr. (in Bodenhorn 1988, 1:26): 'Those people will determine exactly how we should treat a person that is breaking the structure of society's running ... how you behave—these were the Elders, the *umialigiich*' (see also Marie Adams and Raymond Neakok in Bodenhorn 1988, 2:254).

[16] See Ahmaogak's (2002) AEWC overview for a more detailed description of these responsibilities.

the major responsibility of the whaling captain's wife's was the safety of the crew. In many ways, as we have already seen, this theme of safety was reiterated by the Elders with respect to both the whaling captain husband and wife.

Whaling captains are responsible for deciding when and where to cut the ice trails out to the spring camp (most important in order to be able to return to shore quickly if the ice begins to break); they need to monitor ice conditions constantly, checking for cracks and for flooding. They also must recruit the crew, organize the equipment, and prepare the boat. Above all, they need to make sure that the crew knows how to do things properly—as emphasized both by Wyborn Nungasak ('Another thing I want to add to the captain is to discipline, not only to discipline but also to show them how to do it right'—IHLC 1991d:3) and by Isaac Akootchook:

> When we start getting ready for fall whaling we [captains] teach/talk to everyone involved in whaling, especially to crew members and what is expected of them, also what gear to work on and to bring along during whaling. We also let our crew members know what to do within the boat ... (I. Akootchook IHLC 1991c:2).

'The wife's job is more broad and varied,' Wyborn Nungasak added (IHLC 1991c:3) and indeed, it often seems that whaling captains' wives need to be at all places at all times. As we have already heard, they are responsible for the cleanliness of the house, the ice cellars, and the entryways. They make sure the crew has adequate clothing, arrange for the sewers to prepare the skin–boat cover, prepare the food for the crew as well as for the community feasts, and help with the butchering, storing, and preservation of the meat. Like the whaling captains, Jennie Ahkivgak suggested, their job is also to instruct the younger wives and, as Mary Aveoganna pointed out, both husband and wife are responsible for being hospitable—'good hosts' in an echo of Patrick Attungana. It is their job to make sure help is provided if they discover that people are in need. This is carried out by individual wives as well as through the regular meetings of the UAA, which take place throughout the year.

Individuals control their own labor and may offer it in a multiplicity of contexts—moving from crew to crew; working with more than one crew; or deciding how much to work in any one season.

Barrow whaling, then, is affected by *local* decision–making on at least four related but recognizably separate levels: individual, crew, community, and regional. At all levels, decisions are constantly being made in response to many different kinds of changing conditions. To take a closer look at this process, I present a series of case examples from recent fall whaling seasons. I do this for several reasons: Barrow is the only North Slope village with a virtually unbroken practice of both fall and spring whaling. Although the social, cultural and political importance of Iñupiaq whaling is by now well documented, most of the ethnographic material has, in fact, dealt with spring whaling.[17] In part, this is because spring whaling remains the more highly marked season in terms of emotional and ritual importance; fall whaling nonetheless introduces a significant amount of meat, not only into Barrow, but into the region overall. In fact, 1990 was the first year since the imposition of IWC quotas that Barrow landed more whales in the fall than in the spring, a pattern that has largely held since then.[18] In addition, fall whaling is also conducted under very different conditions. As Roxy Oyagak (IHLC 1991c:6) pointed out, 'there's a vast difference during fall whaling and spring whaling' and it seems important that some of these differences should be noted. Finally, there are the variations in practice within the fall season itself. During the nine fall seasons that I have been in Barrow, each year has been very different from the others. The inter-seasonal variation seems to me to be much more marked in the fall than in the spring. Some of these differences warrant further consideration.

CASE EXAMPLES

(1) Where to butcher fall whales and why: a collective response informally adopted.
The nature of fall whaling requires that whales are landed and butchered on a gravel beach (Fig. 1) rather than on the ice that is present in the spring.[19] One consequence of this is that it is almost impossible to prevent gravel from being ground into the whale skin,

---

[17] See, among others, McCartney 1995; Marquette and Bockstoce 1980; Nelson 1969; Worl 1980; Worl and Smythe 1986; and Worl *et al.* 1981. This does not mean that fall whaling has gone entirely unremarked. Maguire's (1988) journal, for instance, provides a mid-nineteenth century account of the activity in Barrow. Chance's (1990) work in Kaktovik treats only autumn whaling.

[18] J.C. George *et al.* (1995). It should not be assumed that this increased harvest simply reflects local reactions to the quota system; the fall migration is longer than the spring one, for instance, and presents more opportunities for successful harvests. In *Hunters of the Northern Ice*, Richard Nelson (1969: 213) noted that Barrow took ten whales in the fall of 1964 and three the following spring. Nevertheless, the quota certainly is a factor under present conditions.

[19] This was already the case when I first arrived in Barrow in 1980. However, pictures from the IHLC indicate that whales were clearly butchered on the tundra during the 1970s. At that time, the tundra was much closer to the edge of the water. Over the past few decades, human activities and coastal erosion have resulted in the expansion of beach, increased amounts of gravel and the separation of beach and tundra by a road that follows the shoreline.

or *maktak*, one of the most prized parts of the whale for Elders as well as for other members of the community. Many Elders say they do not like fall whale meat for this reason; it hurts their teeth and takes the pleasure out of the meal. This general feeling was stated clearly by Wyborn Nungasak (IHLC 1991d:6), 'to those who participate in the fall whale hunt, there should be a better way to handle the meat. Since one bites into the meat and some rocks and gravel go with it, many are not pleased about that.' It is a sentiment echoed by many others, young and old.

Figure 1. *A fall whale being dragged up on the beach. One can see how gravel becomes a problem if whales are butchered on this surface.*

For some time, the solution was to put down plastic sheeting on the beach before beginning the butchering process. The result was not satisfactory, and in recent years increasing numbers of Barrow whalers decided to shift what had become the traditional butchering site. By 1997, virtually all of the landed whales were transported by front–end loader from the beach to plastic sheeting laid down anew for each whale on gravel–free tundra a short distance away. This solution, it was generally agreed by the Elders I have spoken to, produced much more pain–free results and people were pleased with the effort. By the end of the 1997 season, a platform had been erected on the beach, which resolved the problem even more effectively.

This collective response to a collectively agreed–upon problem deserves some discussion, primarily for the implications it has about the ways tradition, technology, and innovation are interrelated. The customary moral injunctions that were upheld in this example were 'listen to the Elders,' and 'treat the whales tenderly and with respect.' Neither the mental (i.e., knowledge) nor the material technology to maintain these values remained unchanged. The fact that there are recognized traditional sites on the beach for butchering did not prevent people from actively considering—and accepting—alternative sites. The fact that this change in practice is not necessarily a change in cultural values is not always clear to non-Iñupiat. One person's reaction to seeing a whale transported down the street was, 'Well, that's the end of traditional whaling!' Another's was, 'That doesn't look very respectful to me.' It seems to me that both of those reactions are based on serious misperceptions of the decision-making processes that went into the adoption of these strategies. Using a front-end loader to transport a whale is of course not in the least 'traditional'; but it certainly allows the whale to be handled more gently that if it had been dragged down dirt streets to the new butchering site. The impression of this observer was that the front-end loaders shown in the photograph (Fig. 1), proceeded in a kind of slow dance, in complete synchrony, so that the whale was not jostled unduly. In a similar way, plastic sheeting is obviously a newly imported substance. It has been adopted because it is effective in allowing younger people to respect the needs and wishes of Elders. I am not arguing that ways of thinking about whaling remain fixed while whaling practices shift. I am suggesting that the relationships between the doing and the thinking are complicated. As in so many situations, the use of the same technology in Iñupiaq and non-Iñupiaq cultural contexts—whether we are discussing front-end loaders, snow-machines, high-powered rifles, CB radios, or aluminum boats—does not necessarily indicate that the actions being undertaken mean the same things to the actors involved.[20]

---

[20] On reviewing this chapter, George Wenzel urged me to point out that this is by no means restricted to whaling—or to Iñupiaq practices. His own work in eastern Canada provides another example of North American hunters who have long been incorporating, modifying and using the tools they discover in their contacts with others (see, for instance, Wenzel 2000).

Figure 2. *Fall whale being landed on the beach in Barrow, September 1997. Echoing Oguiiluk, some of the useful things, shown here, with which Iñupiat have been experimenting include heavy equipment, rubber boots, pick-up trucks, and rain gear made of gortex.*

(2) Responses to perceived problems raised in the 1996 hunt: complex problems generating individual and formal collective actions.
Nineteen whales were taken between September 10–26 during the relatively short 1996 fall whaling season in Barrow. On five days during the week of September 12–19, between two and five whales were landed on the same day; a large bowhead produces about a ton of useable food per foot in length.[21] Even 'small' whales (5–10 m, according to J.C. George) thus demand considerable labor in the processes of butchering the animal, transporting and storing the meat, preparing the feast that is offered to the entire community, and cleaning up the butchering site. In 1996, the whales landed in Barrow ranged in length from 9–15.8 m; with a mean size of 13.85 m Significantly, only three of the 19 fell into the 9–10 m range that reflects Barrow preferences; 11 of the 19 were 15 m long or more (see Fig. 3).[22]

---

[21] This rule of thumb is so general that I cannot give an exact citation for it. However, I have discussed it with Craig George (pers. comm.) who confirmed its general accuracy.

[22] Historically, smaller whales have been generally preferred in Barrow. They are easier to land, to butcher, and their meat is considered to be tastier. All of these advantages have assumed increasing importance in the discussions under consideration. These preferences are village specific. Wainwrighters, for instance, prefer larger whales as a rule.

In fact, on September 12, the four landed whales were 15, 15, 15.8 and 14.4 m long, respectively. Just two days later, before people had been able to recuperate, three more whales—15.8, 13.3, and 16.9 m long—were landed within hours of each other. The serious strain on manpower (tempting—even forcing—people to cut corners) and the sudden glut of meat combined to generate the serious threat of waste. This was a clearly perceived risk for several reasons that were explicitly recognized and discussed on multiple fronts:

* it is disrespectful to whales;
* it attracts polar bears;
* it discourages potential helpers from offering assistance, thus creating further pressure for the people who are willing and able to help;
* it attracts the negative attention of outsiders, potentially weakening the negotiating position of the AEWC *vis-à-vis* the IWC.

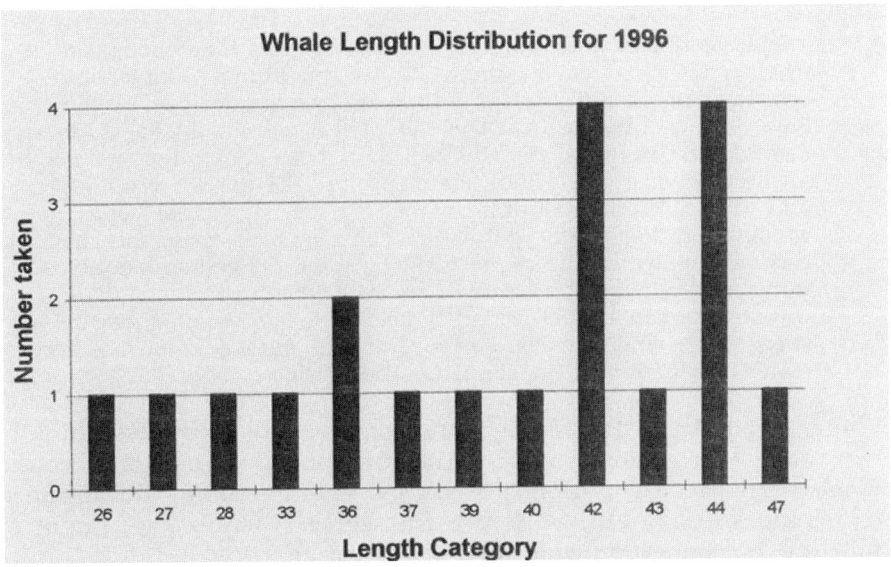

Figure 3. *1996 fall harvest, sorted by length of bowhead (provided to the nearest foot, but converted in the text to meters).*[23]

---

[23] Harvest data on which these figures are based as well as all graphs were provided by Craig George, N.S.B. Department of Wildlife Management. As with everything else, these practices may shift with environmental conditions. During the 2003 spring whaling season, for instance, Wainwright whalers opted to go after smaller whales; these were safer to butcher on the ice that is losing density due to global warming (H. Bodfish, pers. comm..)

We have already heard about the general importance of cleanliness. The specific issue of waste is often discussed in formal institutions. Sam Taalak, Edward Hopson, and Walter Akpik all spoke at the 1991 men's session about what they perceived as the larger implications of incomplete storage of whale meat and *maktak*—implications for both customary practice and for political action on national and international levels. According to Taalak,

> Since I came back to Barrow [from *Nuiqsut*] this last spring, I have gone out to the dump and have found some *maktak* in plastic bags, real thick slices of flukes, and meat ... . I say the captain is to blame for such wastes—I know the *uati* meat has a lot of tendon on it ... . We should be more careful how we do things ... because one picture like that, when it is shown, they will see to it that our quota is taken all away. Like one speaker said, during Christmas all the *uati* should be cut up and taken to the feast (Taalak IIHLC 1991c:8).[24]

Edward Hopson, from Barrow, made a similar point,

> These meetings I understand are to improve anything about our whaling system ... . About the whale, we the Iñupiat ... make our own rules about the whale ... . Therefore, we are our own Public Safety about whaling ... . Therefore I feel we should be more careful what we do to the whale. This is not only for Barrow; it involves *Nuiqsut* too, where one does not take the top layer off only and leaves the rest of it because it froze. Those that oppose whaling will do something drastic when they hear about this. In this meeting we should request AEWC to tell the captains not to throw away any meat or *maktak*, because when one puts *maktak* with blubber in the ice cellar, it keeps, no matter how long it stays there. When the top portion from the *maktak* is cut off, it tastes even better (E. Hopson IHLC 1991c:13).

Walter Akpik (IHLC 1991c:15), of *Atqusauk*, said succinctly, 'Think of what the oil companies can do if they find out ... we do not store the whale meat like we're supposed to. And another thing I do not need to remind you, we all love to eat the whale. Therefore we ought to take into consideration what we do with our share ... .'

As mentioned above, the Barrow Whaling Captains' Association meets to generate and implement local policy; the Elders' Conferences held by the Iñupiat History, Language and Culture Commission, on the other hand, cross many institutional boundaries. Individuals from around the North Slope talk—*quliaqtut*—as Elders, and/or as whaling captains, for instance, about their knowledge of the past—what they've learned through teaching and observation—and about what they perceive now, about waste or

---

[24] *Uati* is that portion of the whale generally defined as part of 'the community share' and reserved for feasts.

technological issues, for example. Some of their comments are meant to make individuals think about their own actions; others are concrete proposals for collective decisions they think should be undertaken formally at the village or the regional level.

(2a) Immediate solutions: individual level, collective practices

As mentioned above, decisions about the division of labor are the responsibility of the whaling captain, as opposed to an official institution, and can vary quite a bit from crew to crew.[25] During the 1996 fall season itself, a broader-than-usual variation in the way people carried out the tasks at hand seemed to occur in response to the much greater-than-usual demand for labor and the intense time pressure. Many men, for instance, did not simply help their own crew, but moved from crew to crew, offering assistance with their specialized skills. James Nageak, a member of the crew headed by his brother Roy Nageak, spent the entire evening of September 19, for instance, moving from whale to whale and sharpening the butchering implements of the hard–working members of several crews. Most striking to this observer was the active participation in the entire butchering process by adolescents, young women as well as young men. Many of them stayed on the beach virtually all night, lifting whale meat from the plastic sheeting to a truck waiting to carry the meat to captains' houses, where further division of shares would occur, carrying coffee, and running other errands. Each of these young people earned a *ningik* (a share) from the crew's portion by dint of his or her efforts. Their individual labor was accepted and appreciated by individual captains who treated them as they would treat any adult member of the crew. Again, this is an example of change and continuity working hand in hand. The customary division of labor was modified on the spot; the custom of recognizing the work that individuals do by allocating them shares of the meat and *maktak*, however, continued.

But that is not the end of the story. As mentioned above, the end of a successful take is marked by the provision of hospitality on the part of the whaling captain couple and the crew for the entire community. When the captain's flag is flown over the house, cooked portions of all parts of the whale will be offered, along with stewed fruit, bread, cake, and tea to anyone who shows up. Today, many people do not eat their portions on the spot, that space often being reserved for Elders. Instead, they collect 'to-go' portions for their children as well as for themselves. The customary practice is simply

---

[25] Marie Neakok, for instance, related to me that her experience with three different crews reflected three different strategies for the division of labor after the whale meat had been divided into shares for the crews. These differences included variations in the tasks the whaling captain couple took it upon themselves to do, as well as variations in responsibilities assigned to the men and women crew members.

to state, for example, 'two adults, two kids,' and to receive four portions of corresponding sizes. During the 1996 feasts, it was not unusual to hear servers ask if any of the 'kids' were teenagers. If the answer was yes, they received an adult share whether or not they had participated directly in the season's whaling. Individual labor was recognized individually when participating adolescents earned their shares. The importance of that contribution was also recognized when all teenagers were reclassified as 'adults' when receiving their portion of the captains' hospitality. This practice was a response to modifications of the moment and was not, as far as I am aware, institutionalized in 1997. None of these were decisions that were formalized at the collective level, but they were put into general practice by individual whaling captains and adopted by others who made decisions for their own crews.

(2b) Formal, collective decisions: 1997 responses to 1996 events
The risks outlined above were the subject of serious discussion as Barrow whaling captains prepared for the 1997 fall hunt; several strategies were formally adopted by the Barrow Whaling Captains' Association as a whole.

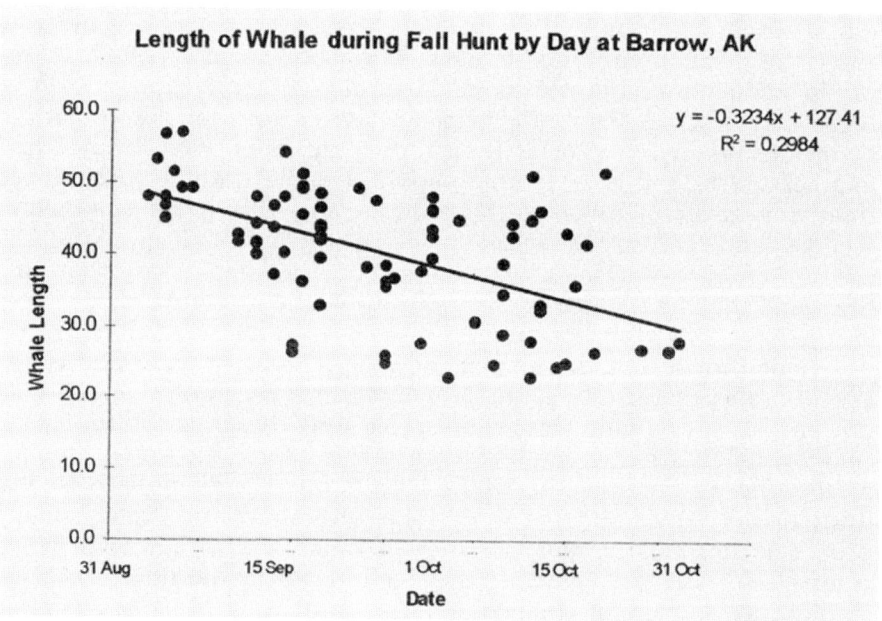

Figure 4: *This diagram (recording the 1996 fall harvest) shows clearly how it is possible to predict that smaller whales are more likely to appear later in the season. Craig George (1997) states in his memo, 'The relationship between body length and date is highly significant from a statistical standpoint (p<0.001). The probability of this happening by chance is less than one in a thousand.'*

1. Whales migrate in waves, grouped roughly by size. The population of the first wave is, for the most part, significantly larger than that of the second wave, a fact long known by experienced whalers and corroborated by harvest data generated by the NSB Wildlife Management Department (Fig. 4).[26] The BWCA decided that Barrow whalers should wait for the second migration wave before beginning the fall hunt. The whales would be the smaller, the preferred size, and the meat would be more tender. Butchering would thus demand less labor power, and Elders as well as youngsters would enjoy the meat more.[27]

2. In another effort to limit the demands on available labor, the BWCA instituted a daily–take limit of two whales.

3. Furthermore, the Association instituted a moratorium on crews resuming whaling until the previous catch had been completely butchered and the site had been cleaned up.

With all of these formally instituted strategies, the need for cohesion among the crews was plainly expressed to me by a number of captains. If one crew went out before time, the pressure on others to begin whaling as well would be intolerable. As with so many aspects of whaling, the view that 'there has to be agreement' was keenly felt.

The photograph below (Fig. 5) in illustrates all of the issues I have just outlined. The whale has been transported from the beach and deposited on plastic sheeting laid out on the tundra. The number of people on hand to help is not large, yet this whale was butchered and the site left virtually spotless within a few hours. The entire task was completed well before the loss of daylight made working difficult and before the crew members were exhausted by their efforts.

With all of these formally instituted strategies, the need for cohesion among the crews was plainly expressed to me by a number of captains. If one crew went out before time, the pressure on others to begin whaling as well would be intolerable. As with so many aspects of whaling, the view that 'there has to be agreement' was keenly felt.

---

[26] Craig George was invited to present his data to the Barrow Whaling Captains' Association at their pre-fall season meeting. The figure reproduced here is from the memo prepared for them (George 1997). In the memo, George notes that this pattern is reversed in the spring when the smaller whales are the first to appear.

[27] This would solve more than one problem simultaneously. 'I'm tired of getting big whales no one can eat!' one whaling captain said to Craig George (pers. comm.)

Figure 5. *Plastic sheeting has been spread on the tundra and the butchering begins with the careful removal of the outer layer of* maktak. *In the foreground, it is possible to see* maktak *laid out in shares (with meat to follow) for later distribution among the crews . September 1997.*

Table 1. Comparative results of the 1996 and 1997 Barrow whale hunt.

Season	Harvest level	Hunt length	Size range	Mean size	Mode
Spring 1996	05	May 25-29	9-18 m	12.6 m	9.4 m
Fall 1996	19	Sept. 10-26	9-15.8 m	13.85 m	15, 15.8 m 50%>15 m
Spring 1997	09	May 4- June 4	9.7-19.8 m	15.1 m	66%>14.4 m 10%=9.4 m
Fall 1997	21	Sept. 10 - Oct. 21	9.7-18 m	11.3 m	10 m 76%< 13 m
Spring 1998	09	May 5-27	6.3-17.9m	12.4 m	56%<13.3 m

Quantitative data gathered by the AEWC and by the North Slope Borough Wildlife Management Department provide comparative information, which illustrates the degree to which decisions taken in the fall of 1997 resulted in a strikingly different harvest that year (Table 1).

Barrow crews harvested 21 bowhead in the 1997 fall hunt, slightly more than the 19 taken in 1996. The size range was also slightly larger (9.8–18.1 m compared to 9–15.8 m for 1996), but the mean size dropped significantly, from 13.85 m to 11.3 m. The mode is even more revealing.[28] In 1996, just three of the 19 harvested whales were less than 13 m and more than half were longer than 15 m. The mode was in fact split between 15 and 15.8 m—four whales in each of those categories. The contrast in 1997 is striking; the modal size was 10 m with 16 of the 21 (76%) harvested whales measureing less than 13 m; only three whales were longer than 14.4 m.

Table 2. Comparative 1996 and 1997 whale harvest, by number per day and by length of season

Sept	10	11	12	14	16	17	19	20	22	24	26	27	28	30
96	42		42 42 44 39	44 37 47	27 26	36	42 33 44 43 44			39	25 36			
97		42	50	39				28 29 27	34 35 34		28	34 31 30	28 38	44

Oct	2	17	18	21
96				
97	31 27	28	26	29

Table 2 displays the comparable daily take in both fall seasons. Despite the BWCA decision to wait, the two seasons began on virtually the same day with the catch of a 15 m whale each year. However, from then on, the tempo of the hunts was strikingly different. In 1996, whalers landed 16 whales (the *vast* majority of the season's take) before September 20. Ten of these were 15 m or longer and 15 of them were landed in the space of a week. By contrast, Barrow whalers only took three whales in the comparable 1997 period. Instead, the period of most intense whaling in 1997 did not

---

[28] The mean indicates the average size of the overall harvest. Statistically this can be misleading, since it tells us very little about the actual whales caught. The mode tells us more about the spread of the harvest as well as the most common size of whales caught in a single season.

begin until September 20; only two whales exceeded 12.2 m in the period from September 20 to 28, when 12 of the season's 21 whales were taken. Although the two-per-day limit was not strictly followed, no more than three whales were ever landed on one day, a radical reduction from the previous year.[29] Following the guidelines suggested by the Barrow Whaling Captains' Association, the 1997 season extended almost to the end of October, increasing the likelihood of landing smaller whales; the number of landed whales was also much more evenly and reasonably distributed throughout the season.

The patterns shown in Table 2 cannot be understood simply as a function of arbitrary decisions to take or not to take whales on a particular day. Many factors go into the decisions whaling captains must make during the course of any single whaling season. Weather conditions in the fall, as in the spring, are critical factors. In the fall, days may go by during which the wind is too high for whalers to go out at all. George's article in this volume examines a number of environmental factors—wind speed and wind direction, for instance— that are taken into consideration by captains since they influence the likelihood of whales being taken (see also George *et al.* 1998). Differences in weather conditions from year to year will be reflected in different annual harvest patterns. By the same token, however, whale harvest patterns cannot be interpreted purely as a response to weather conditions. The harvest patterns of 1997 suggest strongly that the strategies put forward by the BWCA were consistently followed by Barrow whalers throughout the season. In 1998, the BWCA decided to wait until September 20 to begin their fall season in order to maximize their chances of landing primarily small whales, and that decision was honored in full.

DISCUSSION

The examples presented above clearly bring out some of the complex pathways along which adaptive strategies may be constructed. They also highlight the continuing importance of thinking about these decisions in a culturally informed way that does not freeze 'culture' into static 'tradition.' All three examples show evidence of explicit attention to values defined by Iñupiat *as* Iñupiat—attention that is

---

[29] Perhaps the clearest way to illustrate the enormous reduction of labor this implies is to translate the number of whales into tons of meat that must be butchered. On September 14, 1996, three whales were landed; their combined weight totaled approximately 130 tons. The following year, the three whales caught on September 20 totaled 'only' 86 tons (a reduction of 44 tons of meat requiring butchering). Two days later, Barrow whalers had their heaviest day of the season—three whales totaling 37.5 m. Even this day was a reduction of 24 tons in comparison to the relatively light catch of three whales on September 14 of the previous year.

made evident through innovation. The relationship between the two was generally one people found easy to produce and discuss.

The overarching goal of the project out of which this volume developed has been to think about the relationships between shifts in environmental conditions and the conduct of whaling in the Western Arctic through time and across space. Adaptability, not surprisingly, lies at the core of our understanding. Adaptability, however, cannot be modelled simply as a bio-response mechanism. Instead, as we have seen, it needs to be recognized as a product of social action that is generated out of thought processes which are culturally informed. This, in turn, often produces choices that may seem counter intuitive to some non-Iñupiat.

At a recent meeting of the Wainwright Whaling Captains' Association, the discussion turned to the ever-increasing costs of whale bombs and whalers' frustration that they are not allowed to use cheaper alternatives. This constraint is, in part, because the price of the International Whaling Commission's support for traditional Iñupiaq whaling has been that Iñupiat use what the IWC recognizes as 'traditional technology'—technology that has been rendered more efficient, but that has not changed radically in form. In this case, the IWC has assumed that culture, authenticity, and indigeneity must converge on their idea of 'tradition'; as it is manifested in the weapons whalers use. That is, 'culture' is about things and the things should be 'old.' Rossman Peetook, a whaling captain who has been involved in Iñupiaq struggles to maintain the right to whale since the 1970s, said in some exasperation. 'They think it's what we use that makes whaling traditional; they don't see that what it is about is *sharing*!'

Decisions to use one form of equipment over another have never been justified to me either in terms of cultural value or in terms of innovation. Neither antiquity nor 'new-ness' is automatically thought to be valuable in itself. An *umiapiaq*, or skin boat, may be used in the spring because one can navigate it through the spring ice more easily than an aluminum boat. The heavy autumn seas, on the other hand, are much more safely negotiated in a sturdier boat with an outboard motor. Gortex may produce excellent rain gear; however, skin parkas continue to provide the warmest possible protection for winter hunting. The best equipment is neither valued because it represents powerful 'Western' knowledge, nor preferred because it expresses 'traditional Iñupiaq values.' Instead, in an echo of Oquilluk's story, people 'work in their minds to stay alive,' making, trading, buying, or adapting useful things. Iñupiat were doing this long before Europeans showed up in the 19th century, and they have continued to experiment with others' knowledge into the present.

People are explicit that it is important not to lose the technology—the knowledge of how to turn the products of hunting into food, clothing, tools, or boats. The technology is certainly cultural, but it never stands in one place. The strategies we have been reading about, from the opening *unipkaaq* to our final example, have all incorporated the flexible experimentation with things into the moral values that underpin Iñupiaq social life. What this material should invite non-Iñupiat to consider is that the overwhelming conflation of 'culture' with 'things' so common in mainstream Euro-american culture does not necessarily carry over to others' conceptual categories. If we put more emphasis on the literal mean-ing of technology as the *study* of how to do things and less stress on the idea of technology as the continuing development of *objects* that do things, then the assumption that 'new' or 'old' technology is necessarily better or worse is more difficult to take for granted.

The strategies presented above have included other counter-intuitive elements. Decisions about selection criteria have opted for quality over quantity, for instance. Barrow whaling captains made decisions to go after smaller whales—even though the total number taken remained fixed, thus lessening the total amount of food that might be harvested. Given that smaller whales provide more tender meat and are more easily butchered, it was felt in Barrow (although not necessarily in all whaling villages) that smaller whales provide more useable food. In addition, however, these decisions meant whaling during a time in which the risk of bad weather was higher than it would be earlier in the season. Given that, to my knowledge, whaling captains do not decide to risk their crews' safety by choosing to whale in bad weather, the agreement to wait implies an acknowledgement that whales may be missed. The increased risk of encountering weather that might prevent whaling was outweighed by social gains whalers assumed would be made. We thus have to pay close attention, not only to cognitive processes such as the perception and assessment of risk, but also to how these are weighed with reference to social values.

This presents some monumental challenges, not least because flexibility is not a category that lends itself easily to systematic analysis—more often being used as an endpoint rather than starting point for explanation. Nor is it made easier by acknowledging that environmental conditions include social and political as well as physical factors, for this simply compounds the numbers of variables to be included when considering variation.

In this chapter, we have held one factor constant—the season—and have examined some fall whaling practices comparatively over a relatively short time span. We have put forward a small number of case examples to show how, in specific contexts, individual and collective decisions have been generated and enacted on both the formal and informal levels. Although I have by no means presented the full complexity of the politics of whaling in the late 1990s, it should be clear that the decisions people reach include a lively awareness of environmental conditions and an equally lively sensitivity to the political implications of their actions. To return to our opening quote from Fannie Akpik, 'Iñupiaq food is social food,' a clear statement that whaling is about carrying on social relations that extend far beyond the actors in the whaling boats.

Acknowledgments. People on the North Slope have been talking to me about whaling for a very long time and I owe a general debt of gratitude for all the hospitality, the opportunities to help, and the information I have been offered for just over two decades, most particularly to Raymond and Marie Neakok and to Mattie Bodfish. For this specific paper I would like to express special thanks to Maggie Ahmaogak, Executive Director of the AEWC, for detailed discussions about the workings of the AEWC, as well as about decisions local whalers have to make for themselves; Martha Aiken, head of the Documenting Committee of the Barrow Whaling Captains' Wives Auxiliary, for general discussions about the social responsibilities of whaling captain couples; Craig George, North Slope Borough Department of Wildlife Management, who was always ready to talk about harvest data, who provided me with raw data, and who took the time to create the sorts of tabular versions I kept requesting. Arlene Glenn, Oral Historian at the Iñupiaq History, Language and Culture Commission, provided me with access to photographs and to Elders' Conference transcripts; Frances and Emma Mongoyak took the time to help me learn through the Iñupiaq transcriptions of the Elders' Conference material as well as the English translations. Thanks, too, to the students who participated in the Research Methods Class I had the good fortune to teach at Iḷisaġvik College, especially Fannie Akpik, Dorothy Edwardsen, and Arlene Glenn. Their discussions were always acute, far–reaching, and informative— their sense of humor was never very far behind. Judy Sceeles, of Iḷisaġvik College, very generously gave me access to the word processing equipment I needed to write an early version of this paper before leaving Barrow. Maggie Ahmaogak, Craig George, and Roger Harritt all read the penultimate draft version; George Wenzel provided thoughtful reviewers' comments. I am most grateful to all of them.

REFERENCES

Aiken, Martha
    2000   'Presentation to Research Methods Course,' pp. 7-16 in F. Akpik and B. Bodenhorn, eds., *Piḷgallasiñiq Ivalupianik: Learning to Braid 'Real' Thread*. Joint publication. Copyright by the Barrow Whaling Captains' Wives Auxiliary. Barrow, AK: Iḷisaġvik College.
Ahmaogak, Maggie
    1997   Interview, AEWC office, Barrow, Alaska, July 7.
    2002   The Alaska Eskimo Whaling Commission: Overview and Current Concerns, *Traditional Whaling in the Western Arctic* website, www.uark.edu/misc/jcdixon/Historic_Whaling/ AEWC/aewc_maggie%20presentation.htm.

Attungana, Patrick
    1986   Address to the Alaska Eskimo Whaling Commission. Reprinted
                in *Uiñiq: The Open Lead* 1(2): 16ff, translated by James
                Nageak; ammended (unpublished) version, 1998.
Barker, James H.
    1993   *Always Getting Ready 'Upterrlainarluta:' Yup'ik Eskimo
                Subsistence in Southwest Alaska.* Seattle: University of
                Washington Press.
Bodenhorn, Barbara
    1988   *Documenting Iñupiat Family Relations in Changing Times.*
                Unpublished report in two volumes. Prepared for the North
                Slope Borough Commission on Iñupiat History, Language and
                Culture and the Alaska Humanities Forum, Barrow, AK.
    1993   'Gendered Spaces, Public Places: Public and Private Revisited
                on the North Slope of Alaska,' pp. 169-204 in Barbara Bender,
                ed., *Landscape: Politics and Perspectives.* Oxford: Berg Press.
    2000   "It's good to know who your relatives are, but we were taught
                to share with everyone': shares and sharing among Iñupiaq
                households,' pp. 13-41 in G.W. Wenzel, G. Hovelsrud-Broda
                and N. Kishigami, eds., *The Social Economy of Sharing:
                Resource Allocation and Modern Hunter-Gatherers.* Osaka:
                National Museum of Ethnology, Senri Ethnological Studies 53.
    2000/1 It's Traditional to Change: A Case Study of Strategic Decision-
                Making. *Cambridge Anthropology* 22(1): 24-51.
Brody, Hugh
    1975   *The Peoples' Land: Eskimos and Whites in the Eastern Arctic.*
                London: Penguin Books.
Chance, Norman A.
    1990   *The Iñupiat and Arctic Alaska: An Ethnography of Development.*
                London: Holt, Rinehart and Winston.
Fienup-Riordan, Ann
    1983   *The Nelson Island Eskimo: Social Structure and Ritual
                Distribution.* Anchorage: Alaska Pacific University Press.
    1990   *Eskimo Essays: Yup'ik Lives and How We See Them.* New
                Brunswick, NJ: Rutgers University Press.
    1991   *The Real People and the Children of Thunder: The Yup'ik Eskimo
                Encounter with Moravian Missionaries John and Edith Kilbuck.*
                London: University of Oklahoma Press.
    1994   *Boundaries and Passages: Rule and Ritual in Yup'ik Eskimo
                Oral Tradition.* London: University of Oklahoma Press.
George, J. Craig
    1997   Body Length in Relation to Date of Harvest during the Fall
                Bowhead Whale Hunt. Memo to Van Edwardsen, Vice
                President, Barrow Whaling Captains' Association, Barrow,
                Alaska.
George, J.C., R.S. Suydam, L.M. Philo, T.F. Albert, J.E. Zeh, and G.M.Carroll
    1995   Report of the Spring 1993 Census of Bowhead Whales,
                Balaena mysticetus, off Point Barrow, Alaska, with Observations on
                the 1993 Subsistence Hunt of the Bowhead Whales by Alaska
                Eskimos. *Report of the International Whaling Commission* 45: 371-
                384.

George, J.C., M.O. Todd, H. Brower, Jr., and R. Suydam
    1998   Results of the 1997 Subsistence Harvest of Bowhead Whales
            by Alaskan Eskimos with Observations of the Influence of
            Environmental Conditions on the Success of Hunting Bowhead
            Whales off Barrow, Alaska. *Report to the International Whaling
            Commission*, July 1998, prepared by the Department of
            Wildlife Management, North Slope Borough, Barrow, Alaska.

Hess, Bill
    1999   *Gift of the Whale: The Iñupiat Bowhead Hunt, a Sacred
            Tradition.* Seattle: Sasquatch Books.

Iñupiat History, Language and Culture Commission
    1991a Women's Session, Elders' Conference, Barrow, Alaska, July 11.
            Day 2, tape 1; Iñupiaq transcription and English translation by
            Mabel Hopson.
    1991b Men's Session, Elders' Conference, Barrow, Alaska, July 11.
            Day 2, tape 1; Iñupiaq transcription and English translation by
            Mabel Panigeo, IHLC ref: 01498tr1.
    1991c Men's Session, Elders' Conference, Barrow, Alaska, July 11.
            Day 2, tape 2; Iñupiaq transcription and English translation by
            Mabel Panigeo, IHLC ref: 01499tr1.
    1991d General Session, Elders' Conference, Barrow Alaska, July 10.
            Iñupiaq transcription and English translation by Mabel
            Panigeo.

Maguire, Rochefort
    1988   *The Journal of Rochefort Maguire, 1852-1854: Two Years at
            Point Barrow, Alaska Aboard HMS Plover in the Search for Sir
            John Franklin.* Vols. 1 and 2, edited by J.R. Bockstoce. London:
            The Hakluyt Society.

Marquette, Wilman M. and John Bockstoce
    1980   Historical Shore-based Catch of Bowhead Whales in the
            Bering, Chukchi and Beaufort Seas. *Marine Fisheries Review*
            42(9-10): 5-19.

McCartney, Allen P., ed.
    1995   *Hunting the Largest Animals: Native Whaling in the Western
            Arctic and Subarctic.* Edmonton: Canadian Circumpolar
            Institute (CCI) Press, University of Alberta, Occasional
            Publication No. 36, Studies in Whaling No. 3.

Nelson, Richard
    1969   *Hunters of the Northern Ice.* Chicago: University of Chicago
            Press.

Oquilluk, William A.
    1981   [With the assistance of L. L. Bland.]*People of Kauwerak:
            Legends of the Northern Eskimo.* Anchorage: Alaska Pacific
            University Press.

Stairs, Arlene and George Wenzel
    1988   'I am I and the Environment': Inuit Hunting, Community, and
            Identity. *Journal of Indigenous Studies* 3(1): 2-13.

Wenzel, George W.
    2000   'Sharing, Money, and Modern Inuit Subsistence: Obligation
            and Reciprocity at Clyde River, Nunavut,' pp. 61-85 in G.W.
            Wenzel, G. Hovelsrud-Broda, and N. Kishigami, eds., *The
            Social Economy of Sharing: Resource Allocation and Modern
            Hunter-Gatherers.* Osaka: National Museum of Ethnology,
            Senri Ethnological Studies 53.

Wenzel, G.W., G. Hovelsrud-Broda, and N. Kishigami, eds.
    2000   *The Social Economy of Sharing: Resource Allocation and Modern Hunter-Gatherers.* Osaka: National Museum of Ethnology, Senri Ethnological Studies 53.

Wilmott, W.E.
    1960   The Flexibility of Eskimo Social Organization. *Anthropologica* (NS) 2(1): 43-59.

Worl, Rosita
    1980   'The North Slope Inupiat Whaling Complex,' in Y. Kotani and W. Workman, eds., *Alaska Native Culture and History.* Osaka: National Museum of Ethnology, Senri Ethnological Studies 4.

Worl, R. and C. Smythe
    1986   *Barrow: a Decade of Modernization.* Anchorage: Minerals Management Service.

Worl, R. and T. Lonner
    1981   *Beaufort Sea Sociocultural Update Analysis.* Anchorage: Technical Report 64, Bureau of Land Management.

When Whaling Folks Celebrate: A Comparison of Tradition and Experience in Two Bering Sea Whaling Communities

Carol Zane Jolles

Abstract. In the north Alaskan Bering Sea communities of Inalik (Little Diomede Island) and Gambell (St. Lawrence Island) bowhead whaling is anticipated, much discussed and, when occasion allows, enthusiastically celebrated. Celebrations themselves mark and distinguish in powerful ways the structures and systems of meaning of these quite different marine mammal hunting communities. Both communities claim a whaling tradition as critical to their local identity, to their ability to survive, and to their traditional definitions of self that continue into the present. Yet, in one of the two communities no bowhead whales were taken successfully between 1937 and the spring of 1999, while in the other it has been common for two or three whales to be taken in a season. The discourse around whaling in Inalik and Gambell reveals theoretical dimensions of whaling associated with these two Eskimo communities, as well as the pragmatic activities of whale hunting and immediate experience that evoke the recent histories of whaling in each community. In this essay, I compare narrative accounts of whaling that characterize the two communities and discuss how celebrations illuminate and contrast the nature of the whaling experience.

INTRODUCTION

Marine mammal hunters who live along the shores of the northwestern coastal waters of North America seek out a variety of animals. These include seals, walruses, sea lions, sea otters, and when they make themselves available, whales and polar bears. Alaskan hunters focus their attention especially on animals whose meat will feed their families and, in supplementary fashion, will provide the bone and fur resources used in local craft production to generate the cash necessary for survival in any contemporary American community.

Whaling is undoubtedly the most extraordinary aspect of such circumpolar hunting. While technically it is one component of the mixed-market, subsistence-based economies of northern communities, it is premised on the notion that 'whaling, as a life system, embraces both men and women in its web of responsibility... providing a pragmatic association of deep meaning with taken for granted acts of living' (Jolles 1995b:336). In the pages that follow, I discuss two modern Alaskan Eskimo hunting communities and the men within them who hunt for a living. I focus particularly on the whaling boat captains and their crew members who ply the cold waters of the north Alaskan coast, often in locally crafted vessels of walrus hide and wood.

Of the animals sought in the north Alaskan hunt, whales are certainly the most spectacular, and several whale species are pursued regularly, including gray whales, minke whales, beluga or white whales, orcas or killer whales, and the great bowhead whale, *Balaena mysticetus*. My concern is the hunt for bowhead whales and the distinctive characteristics of that hunt. My approach is both pragmatic and theoretical and is based on observations and intermittent conversations with whale hunters between 1987 and 2002. I focus on two traditional Alaskan whaling communities: the Yupik Eskimo community of Gambell, on St. Lawrence Island, and the Inupiaq Eskimo community of Inalik, on Little Diomede Island.

Figure 1. *Gambell, Alaska, in mid-summer, 1998 (photo by the author).*

In my own experience in the Alaskan Arctic since 1987, Gambell and Inalik hunters most often refer to bowheads when they speak of whaling, regardless of other types of whales that they may have taken during the year. For instance, if one asks a Gambell hunter if the community took any whales the previous year, he might automatically reply 'No' and then add parenthetically and almost apologetically that someone in the community took a beluga whale[1] or a minke whale in late spring or summer.

In Gambell, in Inalik, and in other Alaskan hunting villages as well, whales embody much more than mere prey animals pursued as food. To grasp the essential meanings tied to whales and their pursuit, one must examine local beliefs concerning human-marine mammal interactions and the impact of such beliefs on concepts and practices associated with whale hunting. Beliefs regarding human-marine mammal interaction are dramatically illustrated in hunting behaviors directed toward bowhead whales, polar bears, killer whales, and wolves. Whales once figured importantly in the traditional religious systems. The religious rituals associated with those systems revealed both the symbolic meaning assigned to bowheads and the resulting configuration and direction of marine mammal hunting. Because of their inherent spiritual value, certain whales were not hunted while others were hunted in accordance with principles that expressed their special value. Among the Yupik hunters of Gambell, for example, there is some evidence that bowheads were considered metaphoric reciprocals of polar bears. Under the pre-Christian religious system, only men who had killed bowheads or polar bears could be placed at the top of the bluff behind the village when they died, a positioning that signified their status. Men who had killed bowheads were encouraged to take polar bears and *vice versa*, thus cementing their own status as fine hunters while also completing the metaphoric and reciprocal relationship of black and white through hunting success.

All along the northwestern coast, killer whales, known further south as black fish or devil fish, are not hunted at all, although these largest members of the dolphin family regularly attack the whales that northern hunters seek for food. According to a respected Savoonga (St. Lawrence Island) hunter, for example, killer whales have a special bond with humans because they too were once human. A very long time ago, these killer whale men learned to prefer tobacco and other human delights.[2] The same hunter related that a man of his acquaintance was guided home safely by killer whales after having been lost at sea. An elder from Gambell told how she and her family were fed by killer whales that left neatly cut piles of meat for them along the shore when they were desperately hungry.

---

[1] Gambell hunters often by-pass large numbers of beluga when they search for bowheads in early spring.

[2] John Kulowiyi, pers. comm., 1988.

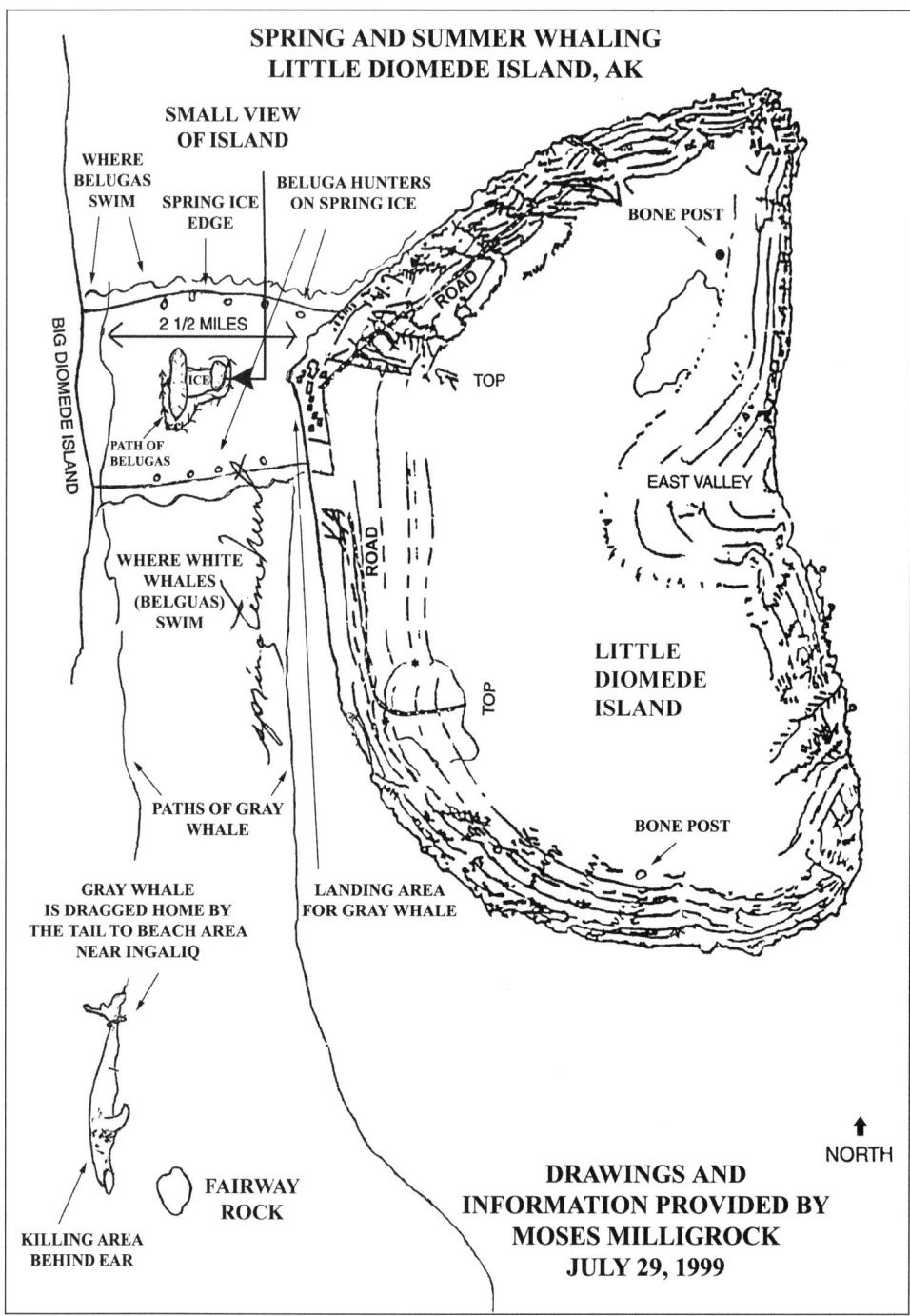

**SPRING AND SUMMER WHALING
LITTLE DIOMEDE ISLAND, AK**

SMALL VIEW
OF ISLAND

WHERE
BELUGAS
SWIM

SPRING ICE
EDGE

BELUGA HUNTERS
ON SPRING ICE

BONE POST

ROAD

BIG DIOMEDE ISLAND

2 1/2 MILES

ICE

TOP

PATH OF
BELUGAS

EAST VALLEY

ROAD

WHERE WHITE
WHALES
(BELGUAS)
SWIM

TOP

LITTLE
DIOMEDE
ISLAND

BONE POST

PATHS OF GRAY
WHALE

GRAY WHALE
IS DRAGGED HOME BY
THE TAIL TO BEACH AREA
NEAR INGALIQ

LANDING AREA
FOR GRAY WHALE

NORTH

FAIRWAY
ROCK

**DRAWINGS AND
INFORMATION PROVIDED BY
MOSES MILLIGROCK
JULY 29, 1999**

KILLING AREA
BEHIND EAR

Figure 2. *Diomede spring and summer whaling activities.*

Further north, among the Inupiat of Inalik and other northern Inupiaq communities, killer whales are considered to hold a symbolic and interactive relationship with wolves. Thus, when killer whales come ashore, they are transformed into wolves. Wolves entering the sea, on the other hand, are transformed into killer whales. For this reason, whale hunters take care not to wear a parka trimmed with wolf fur because bowheads are thought to sense the killer whale (wolf) presence and may refuse to give themselves to a boat that carries a hunter wearing the offensive fur. For these reasons and perhaps others, the presence of killer whales in the near vicinity of hunters does not mean that they will be hunted.

Figure 3. *A bowhead mandible rests near a Gambell boat captain's house, (photo by the author)*

In April, 1987, on my first trip to St. Lawrence Island, the whale hunting season was well underway. Gambell, where I was a guest, had spent the previous nine years trying to adjust to the International Whaling Commission (IWC) rulings and the strike quotas that had been set for the community. In 1978, after several difficult hunting seasons, the community had believed that both its livelihood and its identity were severely at risk.[3] That was no longer the case in 1987. Community members knew then that they would survive and continue to whale successfully, although the rules of the hunt and some hunting strategies and practices had irrevocably changed.[4] As Burton Rexford and others pointed out in 1993 (Jolles 1995b:318-319), in each affected community the strike rules imposed by the IWC generated a sense of urgency about the hunt and changed the nature and intensity of the pursuit.[5] One characteristic change to which a number of Gambell whaling captains referred was that prior to the 1978 ruling, boat captains felt more comfortable taking other animals when the opportunity presented itself. A boat might bring in seals or walruses, taken in the course of pursuing whales, in order to keep family freezers or, in some cases, community freezers full. By 1987, it had become habit to pursue bowheads almost exclusively until the Boat Captains' Organization voted collectively to end the whaling season and to turn to walrus hunting, especially the hunt for *qaseghqaq* (Yupik word meaning 'young walrus'). A major reason for this was and is the sense of urgency to which Rexford and others have referred. As long as quotas regulate the hunt, whales have to be pursued while they are available. Not to take a whale and not to take every opportunity for a whale means that the few strikes open to a community might be wasted. Whales are extraordinarily sensitive to sound, as almost every hunter with whom I have ever spoken has been quick to impress upon me. Thus, the whale hunt is conducted in relative silence, without the noise of motors or the periodic firing of rifles, unlike routine hunting for walruses and seals and, in fact, encounters with minke, beluga and other whales.

In 1987, Gambell had been assigned three strikes. Just two weeks after my arrival, a very large and magnificent whale was brought in by whaling captain Ralph Apatiki and his crew. This was the first time I witnessed the tremendously positive impact that a successful whale hunt has on a modern hunting community. Gambell, like other hunting communities, depends substantially on

---

[3] Between 1973 and 1983, Gambell took eight whales (Braham 1995:288-295) or less than one whale per year.

[4] Between 1984 and 1994, Gambell took 18 whales (Braham 1995:288-295; Braund and Moorehead 1995:274), more than doubling its harvest of the previous decade.

[5] As the writing of this chapter was nearing completion at the end of summer, 2002, the IWC had just voted to eliminate the bowhead quota for Alaska's traditional whaling villages. The future of the bowhead hunt and its importance as a cultural, religious, and economic mainstay was once again uncertain.

this very ancient resource as well as on the expensive imported groceries that fill the shelves of the village's two grocery stores.

In northern communities, meat is the centerpiece of every meal. In Gambell and in Inalik as well, the most common meals are designed around walrus and seal meat. Bowheads, which swim close to St. Lawrence Island in spring, are welcomed especially for their fat or blubber, the *mangtak* (*muktuk* or *mattak* elsewhere) many inches thick that insulates their bodies. When Gambell people speak of whales, they speak especially of whale blubber. Bowhead whales[6] supply modest amounts of meat in comparison with large amounts of *mangtak*. The fat is eaten fresh, cooked, aged, and pickled. It accompanies most celebration meals (Thanksgiving, Christmas, Easter, birthday parties, and other, more personal family events), and it is sorely missed when meat storage areas have been emptied. Bowhead meat and fat together provide substantially larger amounts of protein and essential minerals and vitamins than does an equivalent amount of "store-bought" meat such as beef and its fat (Freeman *et al.* 1998:46). On those occasions when I lived for many months in the Gambell community, meals of *mangtak* and locally collected and frozen greens such as low bush willow (*ququngaq*), roseroot (*nunivak*), and dwarf fireweed (*anguqaq*), for example, were much appreciated not simply for their taste, but also for their ability to satisfy hunger far beyond a meal of purchased foods.[7]

But the real story of whales is not simply that they feed people or that in earlier times their fat heated homes. Whales represent much more than that. The ceremonial aspects of their treatment speak to the extraordinary respect with which they are regarded and to their structural position within a northern physical and socioreligious universe mentioned above. Men's conversations about whales reveal their preoccupation with the whale hunting endeavor and the enormous significance that they assign to the hunt in all its aspects.

Perhaps first and foremost, whale hunting in the traditional manner requires men to face an animal that is as large as a house and as long as a great sailing ship. The waters in which men hunt bowheads are unbelievably cold and ice-filled, unbelievably stormy, and definitely unpredictable. The hunt is conducted from small boats. Some boats are modeled after the wooden dories that were introduced into these communities during the great whaling-fleet days of the late 19th century. Some are more recent, modified

---

[6] Belugas, in contrast, have only a modest amount of fat in comparison to meat. Beluga fat tastes similar to beef fat and, in Gambell, is not nearly so sought out as bowhead fat.

[7] Plant names are given both in English and in St. Lawrence Island Yupik.

versions of traditional skin boats, and still others are purchased directly from mail-order catalogs--a variety of fiberglass and aluminum models. Alaska Native men hunt whales with harpoons and, since the latter days of "industrial" whaling, with harpoon guns and bombs. Unlike contemporary industrial whaling (carried out, for example, by the Norwegians, the Japanese, and until the 1990s by the former Soviet Union on behalf of its northernmost citizens), traditional whaling employs the smallest of weaponry and transport to encounter the largest of animals. It is truly a 'David and Goliath' enterprise. Perhaps a further litany of the obvious is in order here: whales are unusually intelligent mammals; they communicate with one another; and they are dangerous. Whales have been known to throw themselves across a boat, splitting it in half; to come up under a boat, casting the occupants into the fatally cold waters of the Bering, Beaufort and Chukchi seas; and to dive deep under the sea ice with harpoon line and hunters still attached. All this is to say that the hunt for bowhead whales is an extraordinary endeavor.[8]

Figure 4. *Cut* mangtak *(whale blubber), in Ziploc bags, ready for a home freezer (photo by the author).*

---

[8] These generalized vignettes are based on actual accounts recited to me by hunters in Gambell and Inalik.

For all of these reasons and others that the ancestors of these communities presumably took with them to their graves, the people of the North came to regard whales as larger than life. This should surprise no one since the idea of whales has captivated large numbers of people who have never even seen a living whale. Prior to the 20th century, entire religious systems were premised on their appearance, their behavior, and on their contributions to community larders, fuel supplies, and building materials. How they were regarded—that is, the nature and extent of the religious activities that communities once used to encourage the return of whales— depended on the circumstances of the individual community's location, the waters surrounding the community, and even the language and other aspects of local tradition and social organization through which the hunt was formulated and carried out. At one extreme, we find the great Inupiat whaling communities along the north coast such as Barrow and Point Hope. These communities based their social divisions upon a *kashgi* or ceremonial 'men's' house or 'community' house system.[9] Toward the south, we find the great Yupik whaling communities on St. Lawrence Island focused originally on small patrifocal communities scattered across the island and, by the early years of the 20th century, on the *ramka* or patriclan system. Somewhere in the center of the strait is the Inalik community. It was also based on a *qagzriq* system (*kashgi*) consisting of five 'community' houses surrounded by smaller, individual, nuclear family-type dwellings (*ennughaq*).[10]

WHALING ON ST. LAWRENCE ISLAND

The whaling season in the north Bering Sea really begins with Gambell (or Sivuqaq, which is its Yupik and, to my way of thinking, 'proper' name), and its offspring community, Savoonga.[11] Gambell has been a whaling site for centuries.[12] Savoonga, located on the

---

[9] For a recent discussion of the *kagri (kashgil)*, see Larson 1995.

[10] There are several Iñupiaq spellings for 'men's house.' The approximate Inalik spelling of *kagri* or *kashgi* is *qagzriq*. Inalik Iñupiaq spelling was provided by Inalik resident Arthur Ahkinga. This reflects differences in word pronunciation and possibly a difference in lexicon.

[11] Savoonga has only been an established community since 1916.

[12] For documentation of Gambell's long history as a whaling community, see Apassingok *et al.* 1985, 1987, 1989; Leighton 1940 (Dorothea C. Leighton Collection); Jolles 1995a, Jolles with Oozeva 2002; Jolles and Kaningok 1991; Geist 1927-1934 (Otto W. Geist Collection); R. Silook, 1976.

northern coast of St. Lawrence Island, is generally icebound during the whaling season. Since 1972, Savoonga residents have traveled south to the once thriving community site of Pughughileq or Southwest Cape to set up whaling camps. Prior to that time, Sivungamiit (people of Savoonga) who wished to hunt joined their relatives in Gambell. Extensive research has been carried out by members of the Gambell community and by a series of researchers starting in the early years of the 20th century to document whaling traditions.[13] Over that same century, of course, there have been dramatic changes in the nature and kinds of ceremonies performed, in the overall religious and social traditions of the community, and in the technologies employed in hunting itself. I continue this discussion with a further exploration of the present, and add descriptions of past tradition and past practice to illustrate ways in which these affect and shape present hunting thought and behavior.

In Gambell, sometime in early spring,[14] a few older hunters gather to share a meal of thanks[15] in accordance with their own immediate family (clan– or lineage–based) traditions. The meal anticipates the spring hunt, but technically it has been described to me as a meal meant to honor God for what has already passed, the

---

[13] For descriptions of whaling ceremonies, see, for example, the writings of Gambell recorder and historian, Paul Silook. Silook's work appears in the Geist Papers and in the Dorothea C. Leighton Collection, housed in the Alaska and Polar Regions Archives at the University of Alaska, Fairbanks, and in the Henry Bascom Collins Collection in the National Anthropological Archives, National Museum of Natural History, Smithsonian Institution.

[14] Meals for giving thanks correspond roughly to the time that a lineage group would have carried out its traditional seasonal thanksgiving activities in the past. These celebrations, incorrectly termed 'moon' worship by early researchers, took place during distinct moon phases, starting with the new moon. Each of the moons of the year is named according to the subsistence events it heralds or, alternatively, as with the month of November, the actual appearance of the moon in the Sivuqaq landscape. The lineages which performed these ceremonies first, toward the end of January and throughout February, were generally members of the large Aymaramket group, associated especially with more recent late nineteenth century immigration from Chukotka. The other very large group, the Pugughileghmiit or people of Southwest Cape, performed their ceremonies later in the spring, around the beginning of April.

[15] Qualifications in these descriptions are the result of the general reluctance among St. Lawrence Islanders to attribute traditions and practices within their own families and clans to anyone else. Circumlocution is a characteristic of much Yupik discourse (see for example, Morrow 1990). Not surprisingly it characterizes these descriptions and acknowledgments of traditional practices within the community.

previous year's hunt. The meal takes place regardless of whether the community was able to land a whale the previous year. Prior to the introduction of Christianity, most hunting celebrations, including those associated with whaling, took place at the lineage or clan level. Each clan or *ramka* had its own specialized method of giving thanks. Celebrations occurred periodically throughout the year, but especially in early fall and again in early spring. These thanksgiving celebrations were combined with other traditional ceremonies to honor ancestors. Some of these have taken on new forms and can no longer be considered aspects of the hunting tradition itself. For example, Memorial Day services and prayers are similar in some respects to older prayer and thanksgiving traditions, but these are now associated with the continued well-being of deceased relatives within a Christian framework. Almost all older celebrations included ways to offer food to entities or powers within the spirit world.

The name for such ceremonies performed in conjunction with whaling was *eghqwaaghusallu*. Other ceremonies, also carried out in conjunction with whaling, had names that spelled out their specific purposes, such as *uyvasghqellgha* or ceremonies performed only after a whale had been taken. (For detailed descriptions of these and other related ceremonies, see Apassingok *et al.* 1985:204-245). The ceremonies were performed by each individual *ramka* (clan group) and by lineages within a clan group. Each boat captain was responsible for the formal religious obligations required during the whaling season, but lineage heads were the ones who scheduled the performance of religious duties. In general, small fires were built into which men placed specially prepared foods in anticipation of the hunt. According to Gambell elder and retired boat captain, Pelassi (Lincoln Blassi), who died in 1980, 'the boat captain would sing ceremonial songs in the evening. The ceremony of singing was called *ivaghghulluk*' (Apassingok *et al.* 1985:215). Pelassi recorded this prayer used at the beginning of the whaling season:

> The time is almost here
> The season of the deep blue sea . . .
> Bringing good things from the deep blue sea.
> Whale of distant ocean . . .
> May there be a whale.
> May it indeed come . . .
> Within the waves
> *(Apassingok* et al. *1985:215).*

The verses were then repeated for other sea mammals.

The ceremonies themselves highlight the excitement and the sheer pleasure of spring hunting after a long winter and the severe weather that characterizes it. Ceremonies were scheduled from late January through early April. Each occasion brought people outside, to the beach, to join together in meals that in today's terms might be

Figure 5. *Gambell crew members contemplate* mangtak *shares, May 1988 (photo by the author).*

thought of as 'winter picnics.' This is not to downplay the very religious nature of these activities, but the atmosphere in some respects paralleled the kinds of anticipation associated with Christmas or Easter. These were times of thanksgiving as well as times to ask for spiritual assistance. Some groups invited 'anyone' on the beach to join in the festive meal after the initial ceremonies had taken place. Others reserved participation to their own *ramka*. Each group had its own preferred array of ceremonial foods, although all foods were known and consumed by all groups on a regular basis. In the past, whale hunting awaited the completion of these ceremonies (see Jolles 1995a:247-252 for a partial calendar of these events for 1928-1929).

Today, as in the past, hunting begins in Gambell as men hurry to the beach in the early hours of a spring morning. Unlike whaling further north, camps are not established on the ice. Waiting, which is one element, takes place in homes, in the long *sayuuraq* (outer storm entrance) of the Gambell Native Store, and until a few years ago, in the waiting room of the post office. These were and are places where men gather when the weather is too severe to venture out in boats. If the weather is clear enough and calm enough to sail, boat crews set out—hurrying, out of respect for the whales, to put their boats into the water. Every hunter in Gambell speaks of this hurry. This contrasts sharply with the camps set out on shore fast ice and on vantage points near the shore in hunting communities further north. There, the characteristic which parallels this hurry is the waiting, on the beach, until a whale has been sighted. In the past, those who waited did not consider returning home. Women brought food to the men in the camps along the shore during their vigil.

Gambell men are at sea long before whales have shown themselves; they sail their boats in open leads, often traveling miles from shore in search of whales. The whale mentioned above that was landed in April, 1987, took almost 15 hours to bring home. Under such conditions of harvest, the focus of the community is on the *mangtak* (whale fat), not on the meat that may spoil in the long and difficult haul home.

WHALING ON LITTLE DIOMEDE ISLAND

Whale hunting in Inalik has been affected adversely by the island's geographical location at the very edge of the Russian-American border. While both the St. Lawrence Island community and the Little Diomede community suffered when the border was closed in 1948, the Diomeders' traditional hunting grounds were severely curtailed by this move. In the past, according to elders Oscar Ahkinga, Charlie Iyapana, and Moses Milligrock, whales were hunted on the Chukotka side, on the Wales side, and through Bering Strait when it was open. It was common practice to take bowhead whales along the Chukotka side. Today, in summer, one can watch the tantalizing gatherings of gray whales in Russian waters from viewing places in Inalik. (Figure 2 illustrates the movement of beluga that precede or accompany bowheads on their journey to their breeding grounds.) Bowhead whales, which appear in late spring, cannot in general pass through the center of the strait—the waters that rush between Little Diomede and Big Diomede are shallow and likely to be frozen too far down for a bowhead to pass under the ice between the two islands. Hunting is always in the southern waters just beyond the ice edge, and it must stop one and a half miles from shore at the watery border where Russia begins and the United States ends.

Moreover, it seems very likely that bowhead whale hunting in Inalik has always been more problematic than whaling in Gambell. While whales could be pursued along the Russian and American shores, the waters surrounding the island are said to be so clear and the sight and hearing of the whales so sensitive that approaching a whale demanded extraordinary discipline and care. Retired whaler Oscar Ahkinga (now deceased) and his boyhood friend Moses Milligrock (also deceased), described the hunt for me as they remembered it from their youth, along with the demands that local hunting conditions placed on the hunters. Hunting commenced from the ice edge, a moving piece of landscape along the southern boundary of the two Diomedes, joined by the common ice shelf that

319

Figure 6. *Aerial view of Inalik (Ingaliq) in summer
(picture courtesy AeroMap, Inc.).*

creates the illusion of geographic unity in winter. Whaling camps were generally set up along this southern boundary in anticipation of the arrival of the spring whales.

As Oscar's son, Arthur, pointed out to me, the ice edge itself changes from year to year, depending on the weather, thus the camps have no permanent anticipated locations. According to Moses, whales approach the island under water until they encounter the ice edge. As they come close, the whales must choose to turn either east around the southern boundary of Little Diomede or west around the southern boundary of Big Diomede (Figs. 7 and 8). Hunters used to wait silently in the windbreaks constructed along the ice edge where the whales approached. It is not clear how much of this practice is still followed today, since whaling is carried out on a much more modest scale now than it was in the past.

Because the whales are so perceptive and the conditions themselves so precarious for the hunters, every precaution is (was) taken to hide both the hunters and the boats from the approaching whales. According to Arthur Ahkinga, 'When they build that windbreak, when it's real calm out, when there's no wind at all, they put ice around the boat, too ... because the whale sees the boat, when it's close by ...' (Jolles 2001:142). Silence is also demanded and camouflage is an essential component of the Inalik hunt. Both men stressed these points again and again.

> **Carol**—So, the men are dressed in white ... And the boat is surrounded by white.
> **Arthur**—Yes, by white ... When there's no wind at all, everyone stays quiet. They don't talk much ... Because the whale has a very keen hearing ... They don't walk around too much, because the whale could hear.
> **Carol**—And that sound transfers through the ice (Jolles 2001:142).

This contrasts sharply with the Sivuqaghhmiit hunters who must leave the shore from an ice dock (*kenileq*) shoveled out from the shoreline in order to locate whales from their boats (*angyapiget*). The equivalent act of camouflage in Gambell is the use of white or light gray marine paint over the finished walrus hide boat. The paint increases the life expectancy of the boat skin, but it also acts as a kind of camouflage in ice-strewn waters. The technique is also used in Inalik and other whaling communities.

The whale, like the men who engage in the hunt, is extraordinarily careful and cautious. Arthur Ahkinga explained that the whale 'keeps its body under water. Just its snout comes up, out of the water.' His father added that the 'whale know[s]—already. That's why he come[s] 'that' way. They hide, with their body down' (Jolles 2001:143). Only when the whale decides to present itself to the hunter is it possible to take a whale. Again, here are Oscar and Arthur Ahkinga's words:

> There is an Eskimo saying that if a whale wants to be caught by *this* boat, then it comes up in front of them. In other words, it gives itself to the crew, to that certain crew (Jolles 2001:145).

The recollection of past whaling traditions in Inalik is fast disappearing. Only a few men and women are old enough to remember a time when the older ways were practiced. On research trips to Diomede between 1997 and 2002, one of the underlying themes and worries of the community has been the loss of most of its elders. There is already a split between those who do and do not speak Inupiaq; for the most part, only community members older than forty are likely to speak Inupiaq. While losing one's language is not equivalent to losing one's traditions and history, the combination of language loss, a shrinking community of elders, a small population overall,[16] and the fact that the community took its very first bowhead whale in 62 years in the spring of 1999, means that whaling is regarded quite differently in the two communities.

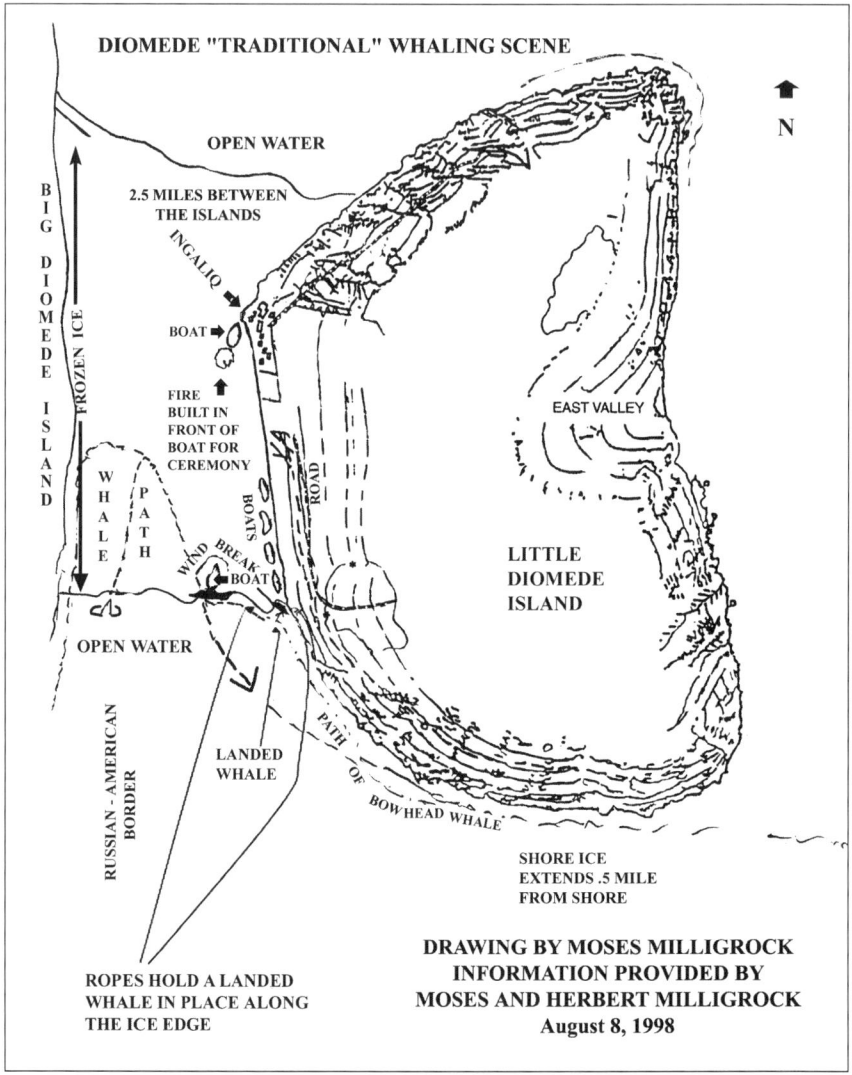

Figure 7. *Traditional whaling activities in Inalik (Ingaliq).*

---

[16] In 2002, the Inalik community had approximately 150 local residents, excluding school teachers, while Gambell had approximately 700 residents plus another 700 relatives living in the neighboring community of Savoonga.

A relatively large group of elders from St. Lawrence Island has actively participated in community history projects that document the island's whaling traditions. In contrast, only a few elders in Little Diomede remember the details of past whaling that can then be compared with and complemented through present practices. Whaling was never easy from Little Diomede since extreme weather conditions, the unusually clear waters in the strait, and the shallow reach between the two Diomedes combined to make the taking of a whale quite a feat. The political realm has simply added to difficulties. Because whaling, in effect, has not been experienced as a continuity except in theory, Inalik hunters today search for the reasons behind their difficulties in ways that the St. Lawrence Islanders do not. Diomede hunters, for example, tend to attribute a loss of older spiritual or traditional practices to their lack of current success. This perhaps would not have occurred to them had their experience of the hunt been continuous. St. Lawrence Islanders describe what they do in terms of past practice, but in fact, much of what they do now in terms of ceremony is Christian, either directly or indirectly. This relates not only to what people do but to the reasons they give for what they do.

In Inalik, only a few elders remembered the procedures that were once followed to bring in a whale. In this excerpt, Oscar and Arthur Ahkinga describe traditional ways to prepare the boat for hunting as well as special preparations which the boat captain must make. Arthur translates and interprets for his father.

> **Arthur**—Before they go hunting, they get a sod [*a clump of dirt mixed with urine*] and put it under the seats of the boat, because in the old times they put sod under each seat of the boat because they say, if it's not done, the boat is out of the water.
>
> **Oscar**—That's what they do. They [*the elders*] always tell us, they don't stay in the water, that new boat. A long time ago, the old people know that.
>
> **Arthur**—The shaman told that. Because if you don't put sod in the boat, under each seat, then the whale sees the boat because the boat is out of the water.
>
> **Carol**—Yes. So, it's almost like a spiritual camouflage as well.
>
> **Arthur**—Yes.
>
> **Carol**—So, you dress yourself to hide yourself from the whale?
>
> **Arthur**—And the boat itself, too.
>
> **Carol**—And the boat is camouflaged both by the color of the boat and by the things that you do to it. ... What did the boat captain have to do himself to be spiritually ready to hunt? How did he prepare himself in here? (*I place my two hands over my heart.*)
>
> **Arthur**—Yes. The captain cleans the house before he goes.
>
> **Carol**—He himself?
>
> **Oscar**—Because he has to get the place in order, put it in the clean. That's the way they are. That's the meaning about it.

Carol—So he—how should I ask this? So he cleans a pathway almost? To the whale?
Arthur—He cleans the house.
Carol—Outside the house too?
Oscar—Yes, even the boat ... they always clean them like that. The first boat, or even the old boat. When they are going to go whaling, they always--it's like a blessing, you know (Jolles 2001:145-146).

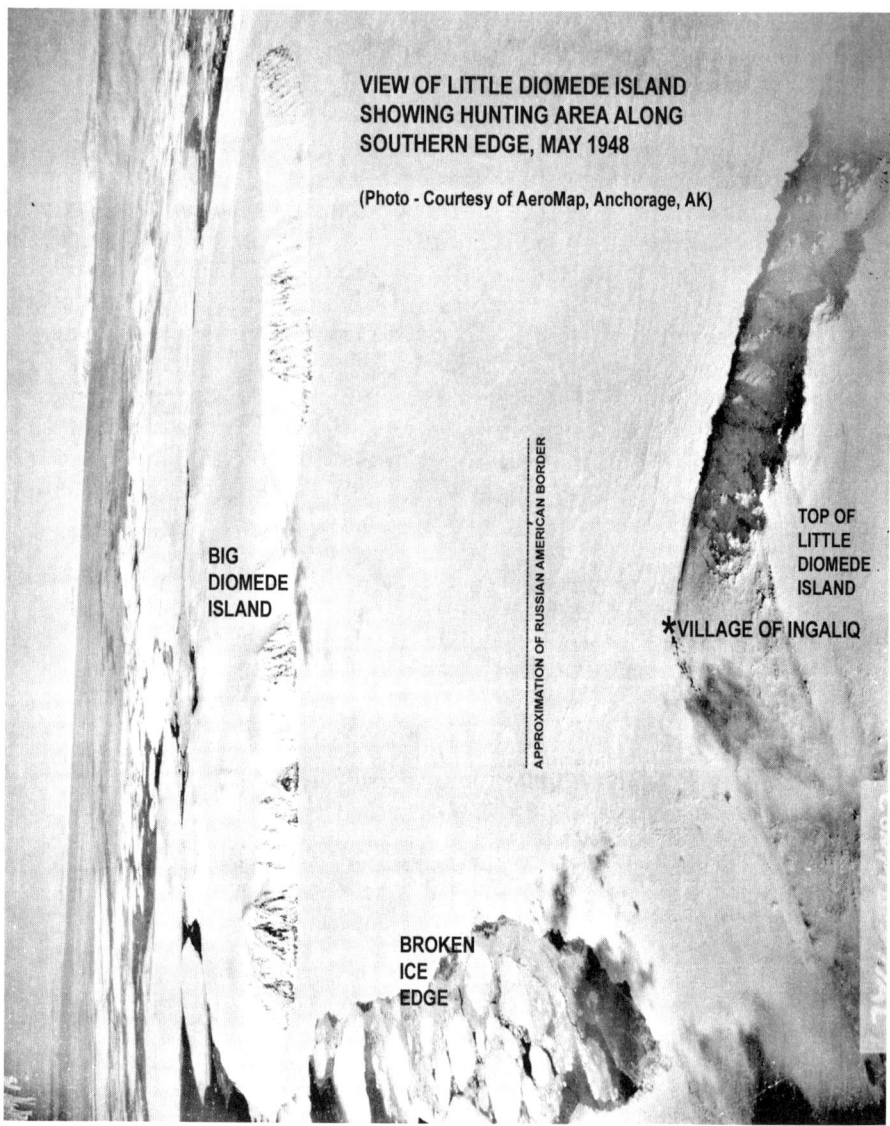

Figure 8. *Little and Big Diomede Islands, May 1948 (photo courtesy of AeroMap, Inc.).*

In Diomede, it becomes obvious that a lack of continuity of practice has meant that a few elders such as Oscar and Moses have articulated and recorded a tradition not only grounded in the past, but substantively located in the past. Because bringing in a bowhead whale has not been a lived experience for anyone born after 1937 until the spring of 1999, the successful harvesting of a bowhead whale in Inalik was situated somewhere between the realms which anthropologist Clifford Geertz (1973) calls 'models of' and 'models for' human action. Or, what Victor Turner (1967) might have called the movement between sensory and ideational poles of performance. Hunters in Gambell, on the other hand, have a continuous practical and theoretical knowledge and experience of whaling.

WHALING AS METHOD, THEORY, AND PHILOSOPHY

Whaling itself highlights very important values in Gambell and Inalik expressed through hunting discourse and practice. It reveals patterns of respect for elders, of community-based cooperation and food distribution, and the kinship systems that underlie community social organization. In both communities, the practical methodology and the spiritual or philosophical bases of this particular hunt, the bowhead whale hunt, are therefore significant topics of conversation. However, the content of those conversations and the contextual representation of the hunt itself identify whalers quite distinctively with one village or the other. Men from Gambell, with whom I and my colleague, Herbert Anungazuk, spoke in 1997, for example, related initial stories similar to this one offered by Leonard Apangalook, a past vice-chairman of the Alaska Eskimo Whaling Commission.

> Of course, I come from a traditional whaling family, although my grandfather never did catch a whale. It all began with my father. He was a very successful whaler. I began participating on whale hunts when I was about twelve years old. I guess you might say as a cabin boy when I first started, kind of a handy boy around the boat. That's how I began, but like most other jobs, I progressed on to operating the motor, crew man, striker, and then, finally, captain. I went through all the progressive steps a whaler goes through as he grew up all the way to captain. Like I said, my father was a successful whaling captain. He has harvested I don't recall how many whales, but quite a few. I, myself, since I took over, have taken only two–two whales. So that's the extent of my tenure as a whaler as I progressed from my beginning to present (Anungazuk 1997, Gambell Interviews: Leonard Apangalook).

Leonard Apangalook's[17] story is characteristic of those told by other Gambell men. First, he honors his elders, men of his father's side—his grandfather whom he never actually met, and his father, a man with many whales to his credit. He outlines his own instruction, that is, his credentials and experience. Implicit in this recitation is that he came to his present position as captain because of his extensive training and his willingness to wait his turn. Finally, while he mentions his own success, he does so humbly. Mr. Apangalook also touches on an element of hunting discourse that becomes the focus with other Gambell men—kinship. Kinship, in fact, is a major structural component of the spiritual or philosophical basis of whaling here. Tied to it is the responsibility, expressed by all of the men interviewed, to share—across the community according to specific sharing blueprints. The whale's body itself represents both hunting theory and hunting practice, becoming a map or embodiment of a community's whale hunting experience and a carved or written discourse and expression of food distribution practice. Both Gambell and Inalik have such maps—most often drawn by elders—a product perhaps of the ethnographic and oral history methodologies in which elders are sought out because of their knowledge of community tradition. Different community experiences produce different results, however. In Gambell, any man who has participated extensively in the hunt could probably draw such a map. Young men learn these maps through instruction by their elders while actually engaging in the butchering and distribution process on the beach.

In most years, someone on St. Lawrence Island does manage to take a whale. At the end of spring in Gambell, the Boat Captains' Association formally declares that the whale hunting season is over. The hunters are free to pursue walruses and other marine mammals that feed the community. The decision reminds one that modern technology as well as access to a huge array of imported goods and services have changed not only the methods but also the structure of whaling in modern marine mammal hunting communities.

---

[17] In Gambell and other Alaska Native villages, the Yupik or Inupiaq surnames assigned originally by government officials were taken directly from a single individual's given or 'essential' name. In these villages, indigenous or first language names are still in use. Thus, a person's 'last name' or surname may also be in use as a first language 'first name.' In Leonard Apangalook's case, for example, Apangalook is in use in his community as an indigenous name as well as a last name. To refer to Leonard Apangalook as 'Apangalook' is to give him a name that is not his own. To solve this problem, I refer to Leonard Apangalook either by his full name or by Mr. Apangalook to distinguish him from the person in his community known as Apangalook.

Figure 9. *Gambell men work on a whale, May 1987 (photo by the author).*

In Inalik, when a bowhead whale was landed on May 4, 1999, after an hiatus of 62 years, the crew sought out elder, Peter Oscar Ahkinga, because he had the most extensive experience of bowhead whaling, having whaled not only in Inalik in his youth, but also having spent nearly a quarter of a century whaling in Barrow, Alaska, as well. Such an interruption of knowledge and experience affects not only practice, but discourse as well. For the delighted Inalik crew members, the recollection of the hunt, while extremely emotional, did not seem to bring the experience of ancestors—of fathers and uncles, themselves successful whalers—into immediate conversation. Instead, it was the excitement of having participated in that single event, the historical moment, that ended the community's long wait for a bowhead whale. In 1999, the boat captain, Thomas Menadelook, and his crew were still young men; the oldest were perhaps in their forties. Their personal histories as hunters included successful gray whale and beluga (white whale) hunting, but never a bowhead. Until that year, the hunt for a bowhead hovered between dream and reality.

In the words of Patrick Soolook, who served as 'balloon man,' the person in charge of the floats that must be thrown in to slow a bowhead and keep it from going under the ice when it is struck:

> It was like every second counts, every move, every move you make—every second, everything you do. It's critical. We were really happy, really happy, after Raleigh shot it [*the whale*] with the shoulder gun. We were hoping we would get it, you know. We all were real happy. We're full of enjoyment, you know. Finally we started making noises ... and Ron shot a flare [*Ron belonged to the second boat that later assisted Menadelook's boat to bring the whale to the ice. The flare notified Inalik folks of the amazing catch*] ... . When we got that whale, we cut a piece right from the whale ... right from the middle. ... Then we started cooking in the tents—coffee and food. ... I went back. ... I had to go back and when they get out of the ice, I had to go back [*to the village*] and get information on how to butcher this. I had to go look for Peter [*Peter Oscar Analuk Ahkinga*] (Jolles 1999, Little Diomede Interviews: Patrick Soolook).

When the butchering was finally finished, small piles of meat and blubber lined the shore ice in front of the village where women could lay claim to them for their families. The whole whale, except for parts specifically assigned to the captain and his crew, went to the village at large, divided equally. As Patrick Soolook exclaimed later with great enthusiasm, 'We iced it!' They brought it home to the ice. There had been only two boat crews out that day, in a community that currently can support up to six boats when all are in working order, that is, covered with sound walrus hides and supplied with fully functioning equipment. Captain Menadelook's boat accomplished all of the killing tasks that day and shared the hauling tasks with Ronald Ozenna, Sr.'s boat.

Dragging the whale onto the thick ice shelf that holds hunting boats and tent shelters during whaling season required the entire Inalik community. From this ice edge, the hunt traditionally begins, and here, in the past, ceremonies were held to make boats spiritually and physically less visible to the whale; later, pieces of meat sat out for four or five days, depending on the sex of a whale, to honor a harvested whale 'visitor' who would soon take leave and return to the sea. In 1997 and again in 1998, Inalik elders wondered if their failure to perform ancient ceremonies could be the reason they had been unable to take a whale for so long. At the same time, elders recognized that no one still knew the entire ritual sequence that had once been followed. Phillip Ahkinga's words are similar to other elders:

> Some of the things they did drop is a ceremony of preparing the boat itself, the bow, and taking it down and make sure it is the traditional and customary way where the bow is facing one way. Those things are not followed. Therefore, we thought, us older people, that is why it gets so hard to strike a whale here now, along with the ocean we have to share with the other side (Jolles 2001:102).

Only the broad stages of ritual remain in memory, and that spring, the captain and his crew, following elders' instructions, shared the *muktuk* or whale blubber with crew members before landing the whale, butchered and shared the whale according to elders' instructions, brought the meat to the village edge for distribution, and finally, held a dance in which the successful crew members danced again and again to honor the whale that had come to the village. From the moment the community brought the whale to shore to the dance celebration in the whale's honor, all was recorded on video-tape by the wife of one of the crew members for all to see. Taking a whale was a source of great pride, not so much discussed as demonstrated: in the certificate issued by the Alaska State Legislature and hung in the homes of the successful crew, in the emotional responses of individuals, and in the replaying again and again of the videotape. One man from the second boat spoke for the rest, I believe, when he told me: 'I think this was the happiest moment in my life. I even shed a tear, got all choked up.' Taking a bowhead whale, an experience so long denied, evoked strong feelings in these men, although none with whom I spoke mentioned his ancestors; rather, the men were struck by a sense of history—they had ended the long hiatus.

In 1997 and 1998, Herbert Anungazuk and I interviewed eight men in Gambell who were or had been either boat captains themselves or members of whaling crews. The first obvious difference in Gambell was that in 1997, Gambell had 26 registered boat captains. Each captain was in charge of a crew made up almost exclusively of his own close relatives. In Diomede, the process of collecting enough men to crew the boat occupied the first two hours of the day's hunt. Only two of the six men in the boat were very closely related, in fact brothers: the captain and the striker. The others were men, loosely related, who wanted to hunt. In Gambell, a boat captain generally knows who will hunt with him from the moment of the first open water to the beginning of June when, by common decision, men cease to hunt bowheads and begin the spring walrus hunt. Six to eight men, most of them close relatives (sons, younger brothers, first cousins from the father's side) will constitute a crew. While there are exceptions, even the occasional outsider will regularly hunt for an entire season with a particular crew. Men in Gambell who plan to whale await their captain's word. Once notified, the men hurry—out of respect for the whales—to prepare the *angyapik* (walrus hide boat) for the hunt. Gambell men hunt under sail. Some years ago, Inalik men did too, but according to Arthur Ahkinga, they gave it up. In Gambell, when a man talks about whaling, he talks about his family and about the tradition of whaling within his family. These are the words of Job Koonooka, a hunter from Gambell:

> We started out with my own dad's crew, in those days. And, we
> [*Mr. Koonooka's brothers and cousins*] were all young. And then,
> after he passed away, in my sophomore year in high school ... we
> joined up with his cousin. And, we whaled with him for a while, I
> forgot how long it took us. And then, one day, we finally decided
> we should revive our own skin boat. ... (Anungazuk 1997,
> Gambell Interviews: Job Koonooka).

At one point in the conversation, Herbert Anungazuk, who was
conducting this interview and whose own association is with a
slightly different whaling tradition, asked: 'I know with your
knowledge of whale hunting, you wouldn't hesitate if the position of
captain was given to you?'

Mr. Koonooka quickly answered: 'I think of that often. But I
have to wait my turn, eh?' In Gambell, unlike further north perhaps,
whaling itself, the crew members' positions within the boat, and even
shore responsibilities are ordered by extensive kinship or *ramka*
hierarchy. During the same interview, Mr. Koonooka responded to
the question, 'What's the most astounding story about whaling that
you've heard in Gambell?' with an anecdote illustrating the
underlying spiritual or philosophical basis of the whaling experience,
one that stresses kin or family above all else:

> One crewman had a kind of mishap ... [T]he waves hit
> unexpectedly and threw him overboard ... and he was under ...
> chunks of ice ... fortunately small ice. And ... I was working with
> [*that*] guy [*later*]... and he told me, 'My family was the ones that
> gave me strength to swim to the boat.' His family saved him ...
> gave him second strength. And it turned out okay (Anungazuk
> 1997, Gambell Interviews).

In Gambell, participation in a whale harvest, whether as a
member of the successful crew or as an adjunct, highlights family
and emphasizes kinship in the broadest sense. Surrounding the
whale's body, the boats come together to eat 'first fruits,' the cuts of
whale blubber given immediately to all of the boats present for a
particular kill. In Mr. Koonooka's words, 'I really enjoy it because we
get a chance to be with our brothers, our nephews, and the other
guys, close together.'This deep connection with family becomes even
more central to discourse when the experience of taking a whale is
fresh in a man's mind, revealing the close connection between the
theoretical 'models of' and the lived practice evident in 'models for'
the whaling experience. Such models illustrate the pragmatics of
kinship in which boat crews not only consist of closely related men,
but they are further distinguished by their exclusively patrilineal or
patrilateral affiliations and ultimately by their clan or *ramket*
memberships. They also illustrate the cyclical nature of the hunt

itself, because kin associations, extending backward in time at least three generations, pass through a man's mind when he takes a whale. Here these relationships are articulated and beautifully expressed by boat captain, Branson Tungiyan, who took his first whale in the spring of 1997. Mr. Tungiyan begins the story of his whale by acknowledging the man to whom he will turn all through the course of that eventful day.

Figure 10. *Whale flipper, a Gambell striker's share, sits on shore*
(photo by the author).

Ralph Apatiki is my uncle, but I look upon him as my oldest brother, [be]cause he is the oldest man ... in our family group ... we were together the day before and we missed by one breath a small whale ... the next day, my brother called and ... said that Ralph ... asked him if ... we would be able to go out that day and my brother . . . told him yes ... and that traditionally, culturally ... goes back . . . if we had gotten close to a whale then we have to be prepared ... (Anungazuk 1997, Gambell Interviews).

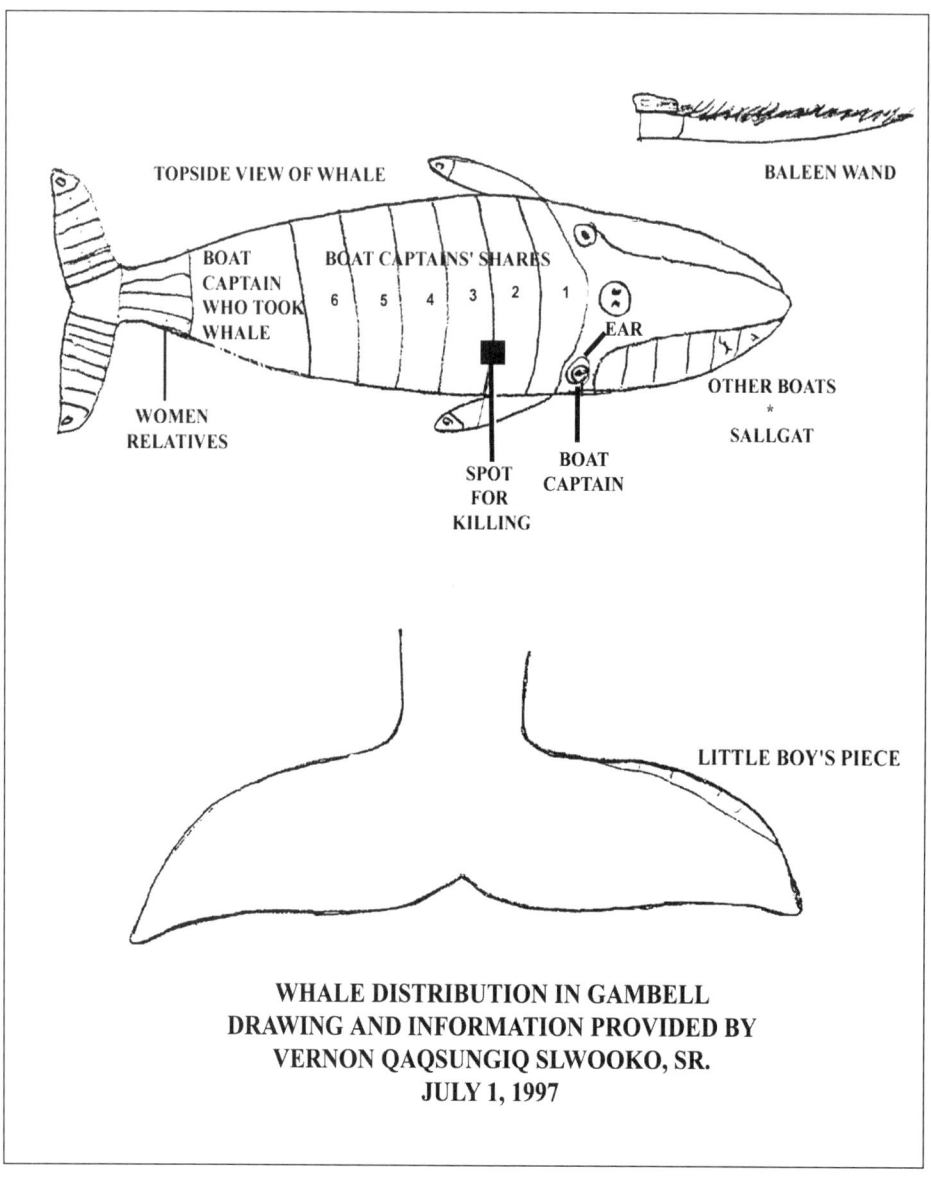

TOPSIDE VIEW OF WHALE

BALEEN WAND

BOAT CAPTAIN WHO TOOK WHALE

BOAT CAPTAINS' SHARES

6 5 4 3 2 1

EAR

WOMEN RELATIVES

SPOT FOR KILLING

BOAT CAPTAIN

OTHER BOATS

SALLGAT

LITTLE BOY'S PIECE

**WHALE DISTRIBUTION IN GAMBELL
DRAWING AND INFORMATION PROVIDED BY
VERNON QAQSUNGIQ SLWOOKO, SR.
JULY 1, 1997**

Figure 11a. *Distribution of bowhead whale shares (topside view), Gambell.*

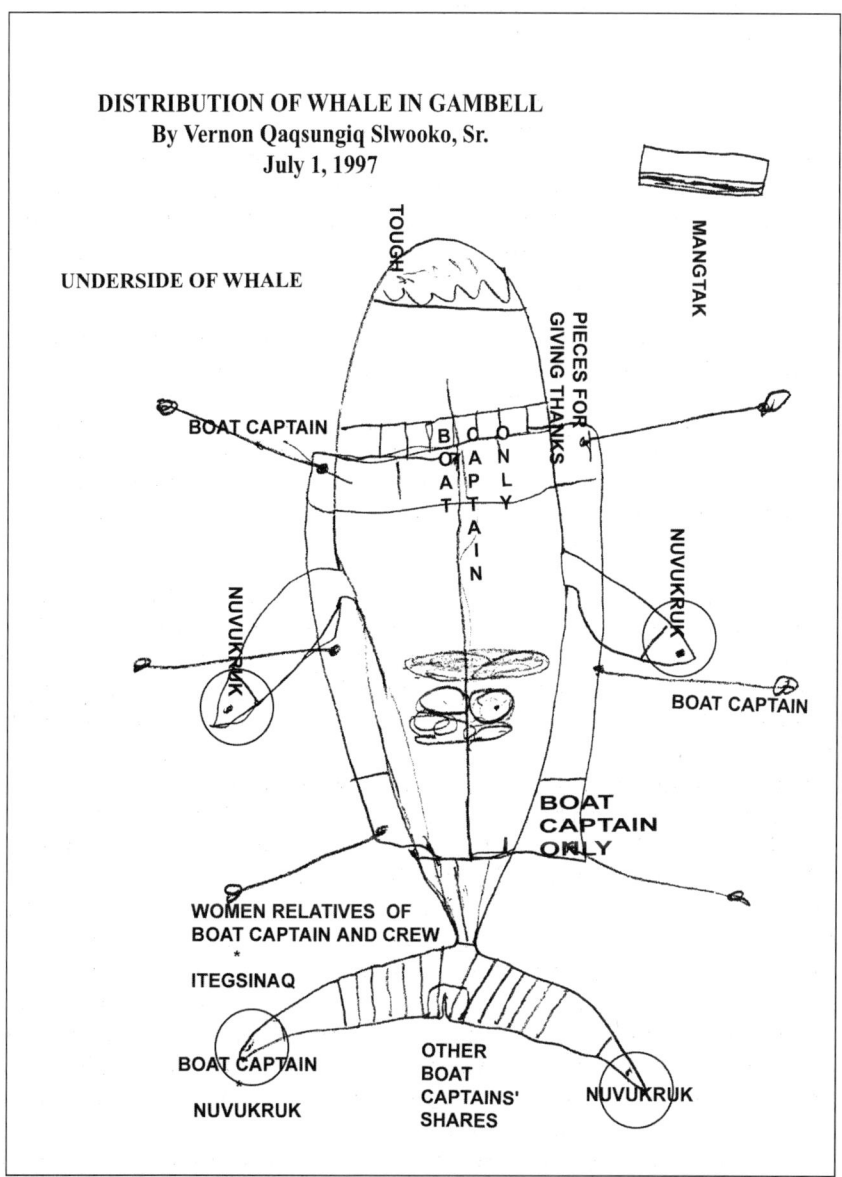

DISTRIBUTION OF WHALE IN GAMBELL
By Vernon Qaqsungiq Slwooko, Sr.
July 1, 1997

UNDERSIDE OF WHALE

TOUGH

MANGTAK

PIECES FOR GIVING THANKS

BOAT CAPTAIN

BOAT CAPTAIN ONLY

NUVUKRUK

NUVUKRUK

BOAT CAPTAIN

BOAT CAPTAIN ONLY

WOMEN RELATIVES OF
BOAT CAPTAIN AND CREW
*
ITEGSINAQ

BOAT CAPTAIN

NUVUKRUK

OTHER
BOAT
CAPTAINS'
SHARES

NUVUKRUK

Figure 11b. *Distribution of bowhead whale shares (underside view), Gambell.*

Implicit here is that inquiry is similar to an order. If one 'can' comply then one 'ought' to comply with one's oldest brother's request. Mr. Tungiyan's crew that day consisted of his closest male kin: his nephews and his son. Mr. Tungiyan later said:

> I never realized how emotional our first whale[s] would be, and, of course, all those people that have passed on from my family came

into my mind, wondering what they would have thought ...
(Anungazuk 1997, Gambell Interviews).

While it was happening, Mr. Tungiyan was caught by the power, the
deep emotion and meaning of the occasion: 'I haven't even talked to
my brothers on the CB yet ... I couldn't talk. I couldn't talk for a while
... I told the boys ... .' Let's wait for Ralph ... our closest relative's
boat.' Only minutes later, Mr. Tungiyan made a decision which
epitomizes the depth of the feeling and meaning made manifest
through the whaling experience:

> I called Ralph, since he was ... just like our oldest brother. I told
> him ... 'I'm giving you all—every decision regarding that whale—to
> you ...' And ... that's when ... I let it go from there. It was Ralph's
> decision from there on what had to be done or where it should be
> taken or whatever, because I told him that decision is his, since
> he was our oldest family member. And I could see him smile,
> making him feel good that I hadn't lost our culture (Anungazuk
> 1997, Gambell Interviews).

The final expression of tradition comes with the actual
distribution of the whale to members of the community. Each
whaling community does so following specific physical diagrams
imposed on the whales's body, mentioned above. These diagrams give
the actual history of the whaling event: they identify the successful
boat captain and crew, the arrival sequence of other participating
boats, and the kinship ties of the major participants.

In the whale distribution diagrams drawn (Figs. 11a & 11b) by
retired boat captain Vernon Slwooko, we see the blueprints or models
for distribution in Gambell. In the third diagram (Fig. 12), drawn by
Moses Milligrock, we see a similar design for Inalik. In the actual
description of the two whaling events, however, Mr. Tungiyan
stressed the need for the boat captain to exercise restraint and
humility while at the same time acknowledging the path of the whale
distribution to his various kin. His description emphasizes the
traditional role of the patriclan system, modified in the 1950s by an
elder from one of the two largest patriclans to share more
ecumenically across clan boundaries.

Mr. Soolook, on the other hand, as a crew member in Inalik,
stressed the community-wide nature of the distribution and worried
that the distribution was not as egalitarian as it could have been. The
distinction here is between the patrifocal pattern in Gambell and a
more bilateral emphasis in Inalik. The latter evolved in conjunction
with the loss of the *qagzriq* or 'men's house' social organization that
once would have relied on *qagzriq* or men's house membership with
its networks of in-marrying kin, and on the primacy of the *qagzriq*
owner's own family. I believe, however, regardless of the
distinctiveness, each man who has participated in a hunt eventually

wrestles with the models held up by his community that call attention to relationships of blood and marriage and to the generational elements that emerge and flourish over time.

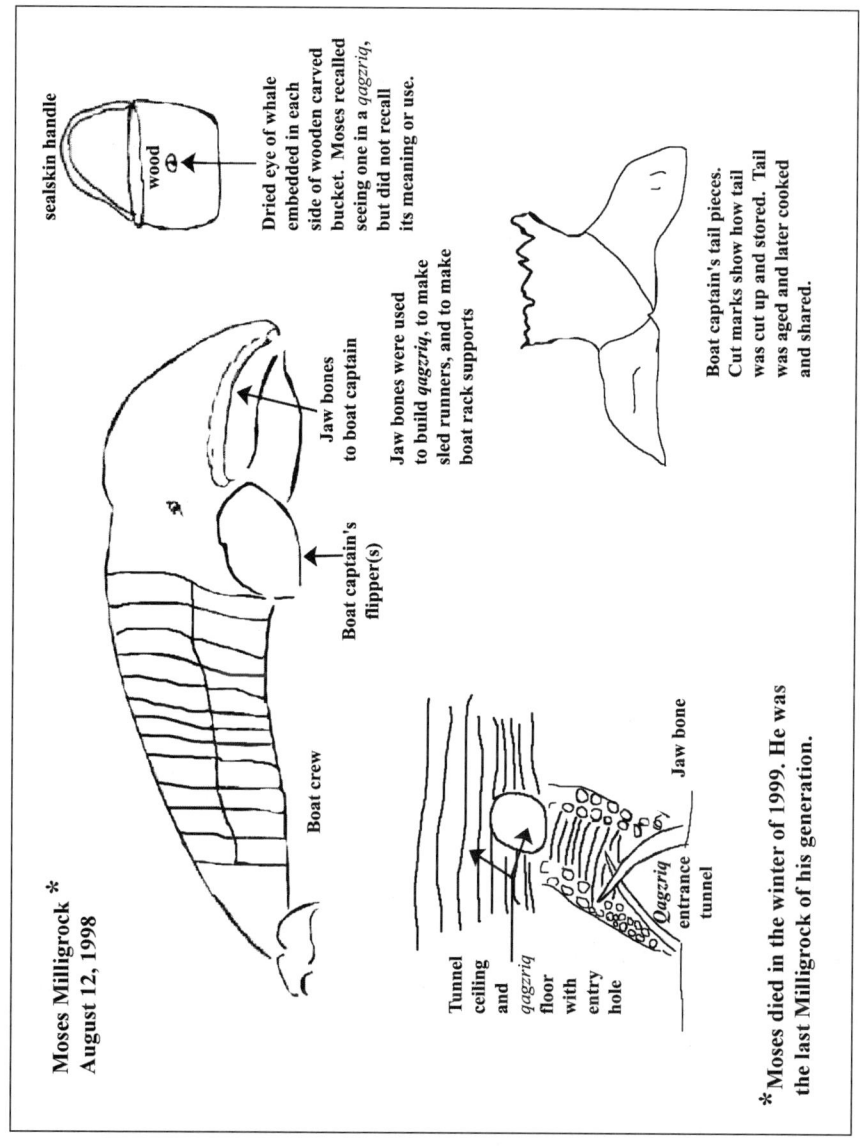

Figure 12. *Distribution of bowhead whale shares, Inalik.*

SOME CONSIDERATIONS

The traditional whaling villages recognized with strike permits until the summer of 2002 stretch along hundreds of miles of Alaska's northwestern and northern coastlines and embrace the ancient villages of Gambell and Inalik. The two island communities are comparable in a number of ways. Both are physically isolated from Alaska's mainland Eskimo communities. Both rely on, and are devoted to, a subsistence-based domestic economy. Both draw most of their sustenance from locally hunted walrus and seals, but build a subsistence life-way philosophy on socio-religious concepts of the universe that imbue bowhead whales and polar bears with special meaning. Since the 1940s, each community has been impacted by international politics governing movement across the Russian-American border. For the Yupik people of Gambell, the border has limited travel and interaction with Siberian Yupik relatives from Chukotka. For the Inupiaq people of Diomede the nearness of the border has essentially curtailed travel along the southern spring ice edge from which they regularly hunt whales. Yet, while the effects of imposed boundaries cannot be denied, the ice and weather conditions, the pathways generally taken by bowheads, and the separate cultural traditions and histories of Gambell and Inalik appear to have shaped the whaling experiences of the two communities in distinctive ways even more than the politics of the globalized world in which they are embedded. Within living memory, men from Gambell have always hunted from open water, sometimes traveling miles from shore to take a whale. In contrast, Diomede men have encamped on the southern ice edge and have ventured into nearby leads that appear around the southern and western edges of the Diomedes and the close Chukotkan shore. In essence, because of differences in geographic location and ice conditions, Gambell men are in a position to seek bowheads while Diomede men are in a position to await bowheads. In addition, ceremonies that were once performed in the two villages appear to emphasize differences in traditions and in actual harvest rates. Traditionally, Gambell men honored a harvested whale by abstaining from the hunt for five days. Diomede ceremonies following a successful harvest sometimes lasted for days or weeks as *qagzrit* members exchanged dances and gifts. These differences in ceremonial behavior suggest that perhaps the expectation of few whales or many whales shaped ritual behavior.

Current hunting conditions in Diomede reflect international border politics, loss of traditional knowledge and training systems, and gradual loss of spoken Inupiaq. The *qagzriq* system that once shaped hunting patterns and provided an important locus for hunting education no longer functions. (There were five *qagzrit* at the end of the 19th century and two still functioning in the 1940s and 1950s.) While the bilateral kinship system with a strong patrilineal bias does operate, it is not as effective as the older system. It also seems likely that the approach of bowheads in spring has been affected by community use of the ice bridge between the two Diomedes as a landing strip for commuter airlines. The whale harvest in May, 1999, took place during a two week period of extreme weather when planes could not land. In Gambell, whales appear regularly, international political boundaries do not often prevent a whale harvest, and the patriclan kinship system forms a strong foundation on which rest local community social structure and boat crew organization. Unlike the bilateral kinship system minus the *qagzriq* system in Diomede, the *ramka* system with its patriclan structure continues to provide a basis for hunting education and for boat crew organization.

Whaling narratives in the two communities inevitably reflect the different social, cultural, political, and physical conditions outlined above. At their heart in each community, however, is a continued devotion to a whaling tradition regardless of differences in harvest success and in the conditions that create those differences. Community narrative recollections slip easily from a single life history into oral histories that document the more general experiences of a whole community. At the same time, such discourses, as the words of the hunters, quoted hear illustrate, have unique abilities to reveal the power of individual renditions of events. These renditions, taken together, allow us to see into the social structure of each community, to note the values and deep meanings expressed in the thoughts and actions of men who hunt, and to glimpse and perhaps to grasp, however briefly, conceptions of the sacred implied by all of these elements taken together. Subsistence ideologies permeate these two communities, but each has developed in concert with the physical, social, and historical circumstances through which each has evolved. What is clear at every level is that whaling is an activity that moves across boundaries of survival and mere economics to affect the social and religious well-being of all who participate.

Acknowledgments. I would like to thank the communities of Gambell, St. Lawrence Island, Alaska, and Inalik, Little Diomede Island, Alaska, for their support and cooperation in the endeavor to document whaling traditions in northwest Alaska. I would especially like to thank the hunters in both communities who agreed to be interviewed and to contribute their time and the benefit of their knowledge and experience. The work has received extensive funding support from the Arctic Social Sciences Program, Office of Polar Programs, National Science Foundation. All photographs were taken by the author. Individual author-artists of figures are indicated on the drawings.

REFERENCES

Anungazuk, Herbert
   1997   Gambell Interviews with Whalers. Taped interviews in author's
           possession.
Apassingok, Anders, Willis Walunga, Raymond Oozevaseuk, and Edward
           Tennant, eds.
   1987   *Lore of St. Lawrence Island: Echoes of our Eskimo Elders, Vol. II:*
           Savoonga. Unalakleet: Bering Strait School District.
Apassingok, Anders, Willis Walunga, and Edward Tennant, eds.
   1985   *Lore of St. Lawrence Island: Echoes of our Eskimo Elders, Vol. I:*
           Gambell. Unalakleet: Bering Strait School District.
Apassingok, Anders, Willis Walunga, Raymond Oozevaseuk, Jessie Uglowook
and Edward Tennant, eds.
   1989   *Lore of St. Lawrence Island: Echoes of our Eskimo Elders, Vol.*
           III: Southwest Cape. Unalakleet: Bering Strait School District.
Braham, Howard
   1995   'Sex and Size Composition of Bowhead Whales Landed by
           Alaskan Eskimo Whalers,' pp. 281-314 in A.P. MacCartney,
           ed., *Hunting the Largest Animals: Native Whaling in the Western*
           Arctic and Subarctic. Edmonton: Canadian Circumpolar
           Institute (CCI) Press, University of Alberta, Occasional
           Publication No. 36, Studies in Whaling No. 3.
Braund, Stephen and Elizabeth Moorehead
   1995   'Contemporary Alaska Eskimo Bowhead Whaling Villages,' pp.
           253-280 in A.P. MacCartney, ed., *Hunting the Largest Animals:*
           Native Whaling in the Western Arctic and Subarctic. Edmonton:
           Canadian Circumpolar Institute (CCI) Press, University of
           Alberta, Occasional Publication No. 36, Studies in Whaling No.
           3.
Freeman, Milton M. R., Lyudmila Bogoslovskaya, Richard A. Caulfield,
Ingmar Egede, Igor I. Krupnik, and Marc G. Stevenson
   1998   *Inuit, Whaling and Sustainability.* Walnut Creek and London:
           AltaMira Press.
Geertz, Clifford
   1973   *The Interpretation of Cultures.* New York: Basic Books.
Geist, Otto William
   1927-1934 Otto W. Geist Collection (Dr. Otto Geist Papers). On file.
           Alaska and Polar Regions Archives. Fairbanks: University of
           Alaska, Fairbanks.
Jolles, Carol Zane
   2001   *Our Stories: Whaling and Subsistence Traditions in Ingaliq, Little*
           Diomede Island, Alaska, Volume I: 1997 and 1998.
           Unpublished report to The Native Village of Diomede and the
           National Science Foundation, Office of Polar Programs, Arctic
           Social Sciences.
   1999a   1997-1999 Gambell Interviews with Whalers. Taped interviews
           in author's possession.
   1999b   1997-1999 Little Diomede Interviews with Whalers. Taped
           interviews in author's possession.
   1995a   'Paul Silook and the Ethnohistory of Whaling on St. Lawrence
           Island, Alaska' pp.221-252 in A.P. MacCartney, ed., *Hunting*

the *Largest Animals: Native Whaling in the Western Arctic and Subarctic.* Edmonton: Canadian Circumpolar Institute (CCI) Press, University of Alberta, Occasional Publication No. 36, Studies in Whaling No. 3.

1995b   'Speaking of Whaling: A Transcript of the Alaska Eskimo Whaling Commission Panel Presentation on Native Whaling,' pp. 315-338 in A.P. MacCartney, ed., *Hunting the Largest Animals: Native Whaling in the Western Arctic and Subarctic.* Edmonton: Canadian Circumpolar Institute (CCI) Press, University of Alberta, Occasional Publication No. 36, Studies in Whaling No. 3.

Jolles, Carol Zane, and Herbert Anungazuk
1997   Little Diomede Interviews with Whalers. Taped interviews in author's possession.
1998   Jolles, Carol Zane, with Elinor Oozeva
2002   *Faith, Food, and Family in a Yupik Whaling Community.* Seattle: University of Washington Press.

Jolles, Carol Zane, with Kaningok
1991   *Qayuutat* and *Angyapiget*: gender relations and subsistence activities in Sivuqaq (Gambell, St. Lawrence Island, Alaska). *Etudes/Inuit/Studies* Vol. 15(2): 23-53.

Larson, Mary Ann
1995   'And Then There Were None: The 'Disappearance' of the *Qargi* in Northern Alaska,' pp. 207-220 in A.P. MacCartney, ed., *Hunting the Largest Animals: Native Whaling in the Western Arctic and Subarctic.* Edmonton: Canadian Circumpolar Institute (CCI) Press, University of Alberta, Occasional Publication No. 36, Studies in Whaling No. 3.

Leighton, Dorothea
1940   Dorothea C. Leighton Collection. Alaska and Polar Regions Archives. Fairbanks, AK: Elmer Rasmuson Library, University of Alaska.

Morrow, Phyllis
1990   Symbolic Interactions, Indirect Expressions: Limits to Interpretations of Yupik Society. *Etudes/Inuit/Studies* 14(1-2): 141-158.

Silook, Roger S
1976   *Seevookuk: Stories the Old People Told on St. Lawrence Island.* Anchorage: Alaska Publishing Company.

Turner, Victor
1967   *The Forest of Symbols.* Ithaca: Cornell University Press.

Festival and Tradition: The Whaling Festival at Point Hope

Mary A. Larson

Abstract. In Point Hope, Alaska, the Whaling Festival is one of the year's major events, tied inextricably to the annual whale hunt in the community. It is a three-day celebration, as whaling crews prepare and share foods and other gifts, and people gather to give thanks, eat, and participate in various competitions. The festival has been an important part of community life for as long as anyone can remember. The original intent of this research was to discuss the changes that have taken place in this event over the years, but in looking through archival materials and talking to the people of Point Hope, what is really striking is how little of significance has really changed. The major parts of the celebration together with their affiliation with the two qalgit (ceremonial or community house groups) are very similar to what they were 100 years ago. The event as it now stands reflects the importance that the community places on whaling and the ceremonies and festivals that surround it.

INTRODUCTION

The original intent of this research was to investigate the Point Hope Whaling Festival, how it has changed over time, and the possible reasons for those changes. A review of the available documentation, however, revealed that the similarities are much more striking than the differences. Looking at materials representing people's recollections over a 100-year period, it is apparent that more has stayed the same than has changed during that time.

Before beginning a full discussion of the Whaling Festival, some background information is necessary. First, the Whaling Festival is generally referred to as *Nalukataq* in other Iñupiat whaling communities, but in Point Hope, *Nalukataq* refers solely to the blanket toss, which usually happens on the last day of the three-day celebration.

Second, many Iñupiat communities once had a building type known as a *qalgi* (also spelled *qargi, kadzgi,* or *kashim,* with at least 72 orthographic variations).[1] These served various purposes in

---

[1] Plural, *qalgit.* Alternative spellings for this and other Iñupiaq terms will appear in this paper when sources are quoted that use those spellings. Otherwise, standard orthographies are based either on the advice of Larry Kaplan of the Alaska Native Language Center (ANLC), University of Alaska Fairbanks, or on spellings provided in the *Abridged Iñupiaq and English Dictionary* published by the ANLC and the Iñupiat Language Commission (1980).

different locations, but in Iñupiat villages they were generally used as ceremonial or community houses. In larger villages (generally whaling villages) there were two or more permanent *qalgi* buildings, and people had affiliations with a specific one. The *qalgit* served as a form of social organization in the whaling communities, with many of the ceremonial, economic, judicial, and recreational activities being structured around *qalgi* affiliations. Although *qalgi* structures are no longer maintained in any of the villages, Point Hope residents still have very strong affiliations with one of two *qalgi* institutions— *Qagmaqtoq* or *Ungasiksikaq*.[2] This is important to this discussion because all of the Whaling Festival activities are arranged by *qalgi* membership. Whaling crews are affiliated with either *Ungasiksikaq* or *Qagmaqtoq*, and the festivities take place at outdoor *qalgi* grounds belonging to one or the other of the two *qalgit*. Prior to 1900, there were up to seven active *qalgit* in Point Hope at one time, and this had some bearing on the timing of Whaling Festival activities, as will be noted later.

This research draws on a number of sources covering different time periods. The documents with the greatest time depth are transcripts of interviews that Froelich Rainey conducted in Point Hope with Kuwana, Arveksinya, Iveqsiq, and Dives. Although recorded from the late 1930s until around 1940, the people speaking were specifically referring to the Whaling Festival as it existed prior to 1910. This is evident because they all spoke of the *qalgi buildings*, the last of which went into disrepair before 1910. In fact, Kuwana spoke of a third *qalgi* (*Kalgireruk*) that did not exist after approximately 1890, so his account refers to an even earlier time. These transcripts, along with the rest of the Rainey material cited, are housed in the Alaska and Polar Regions Department of the University of Alaska Fairbanks.

The second source is from the same collection. While Froelich Rainey was conducting interviews in Point Hope in the late 1930s, he was also observing and taking notes on the activities going on around him. A portion of his field notes discuss the Whaling Festival as he witnessed it during his time in the community.

Also in the Rainey collection is a Native herder's journal from 1941. All that is known about the author is that he was from Point Hope, but the journal is dated and gives a particular individual's recollections of the Whaling Festival in 1941.

In 1962, James VanStone authored his often-referenced account, *Point Hope: An Eskimo Village in Transition*, which documents information he gathered during his stay in the community

---

[2] This is the English orthography for these *qalgi* names. In Iñupiaq, a dotted 'g' would replace the regular 'g' in '*Qagmaqtoq*,' and a tailed 'n' would replace the 'ng' of '*Ungasiksikaq*.'

in the mid-1950s. This work, although heavily referencing Rainey's account, also gives a sense of how the festival had changed between the 1930s and VanStone's time in the community.

The final source of information is the author's own observations from the 1997 Point Hope Whaling Festival, held June 12 through 14 that year as well as follow-up conversations during a 2001 trip to Point Hope.

THE 1997 WHALING FESTIVAL

In order to give a complete outline of events, the description of the festival will start with the 1997 event. Understandably, not all sources were equally interested in all aspects of the festival and so do not describe them in totality. Beginning with the 1997 Whaling Festival, this essay then touches on the earliest sources and moves forward through time.

In 1997, the festival took place, as was traditional, over a three-day period. The first day, the boats were brought up from the beach to the respective *Qagmaqtoq* and *Ungasiksikaq* areas. This was done simultaneously by both of the *qalgit*, and the windbreaks for the area were set up.

Figure 1. *Flipper distribution at the Ungasiksikaq feast, day two.*

Figure 2. *Mikigaq distribution at the Qaqmaqtoq feast, day two. A mikigaq cask is visible in the center of the image.*

The second day started early in the morning with breakfast, which consisted of various organ meats. Later in the morning, *Ungasiksikaq* hosted the first gathering for the flipper distribution (the order alternates each year). Everyone sat behind windbreaks formed by upturned *umiat*[3] and *Visqueen* (plastic sheeting), with the *Ungasiksikaq* elders getting the most comfortable area in the middle of the windbreak where the sledge was. Members of *Ungasiksikaq* and their affiliated guests sat mainly on the west side of the windbreak, while *Qagmaqtoq* members and guests were seated on the east side. This is in keeping with the traditional association of *Ungasiksikaq* as the *qalgi* closest to the geographical location of Point Hope (i.e., the actual point of land, not the village).[4] The celebration was opened with prayer, then large buckets of *mikigaq* (fermented whale meat) were brought around, and everyone received a share. *Akutuq* (a mixture of fat and berries often referred to as 'Eskimo ice cream') was also distributed but primarily only to the elders, as it was a scarcer commodity. Out-of-town visitors were, however, treated to small tastes as a courtesy.

The distribution of slices of flipper (or flukes) followed the *mikigaq* (Fig. 1). It began with the captains giving slices to specific people, called up by name, and then to certain groups. The order is based purely on the preference of the captain. Guests from outside Point Hope were called early in the process when the flipper slices were passed out. After the distribution of the flippers, the children were called up and given oranges, then the celebration dispersed for a break. While some food was eaten on site, much was taken home in plastic bags for later.

The same process was repeated in the afternoon at the *Qagmaqtoq qalgi* grounds east of town (Fig. 2). This area is located on the other end of *Qalgi* Avenue, which was so named because the festival grounds were on either end of the road. The order of the events was the same, the main difference being that at the *Qagmaqtoq* celebration, the flippers were cut while people were gathered there, as opposed to at *Ungasiksikaq*, where they had been precut (again, according to the preference of the captain). Another difference was that two whales had been taken by a *Qagmaqtoq* crew that year (Henry Nashookpuk's crew), so there was more *mikigaq* and more flipper to be passed out. When the *mikigaq* from the two whales was distributed, (pieces from each whale were handed out to the crowd separately).

---

3 *Umiat* is the plural of *umiaq*, the Inupiaq skin boat.

4 The village of Point Hope moved to its current location in 1977. Prior to that, at the old town site, the *qalgit* had been oriented on a north/south axis. *Qagmaqtoq* was the northern *qalgi*, and *Ungasiksikaq* was the southern *qalgi*. When the community moved, the orientation for the *qalgit* became east/west, with *Qagmaqtoq* being the eastern *qalgi*. The basis for the orientations, however, did not change. *Ungasiksikaq* is the *qalgi* closest to the point, and *Qagmaqtoq* is farther from the point, regardless of where the community is or along which directional axis the *qalgit* are organized.

Figure 3. *The men's area at the Qaqmaqtoq feast, day three.*

Everyone dispersed for the evening. 'Singspiration' was held at the local Episcopal church, and walrus and seal hunting were both going on at that time, so many people were occupied with that when not at the festivities.

Figure 4. *The women's cooking area and windbreak at the Qaqmaqtoq feast, day three.*

The third day is the day of the cooked foods, and both *qalgit* hold their feasts simultaneously in their respective areas.[5] Also, men and women stay separated behind different windbreaks for this day of the festival, with the women cooking and taking the food back to the men when it was done (Fig. 3). Individual cooking took place along with some collective cooking, with large pots of duck soup simmering (Fig. 4). There was a collective stove in use in the women's area. Foods included cooked meat, soups, doughnuts, tea, and coffee. This feast continued most of the day, with people taking turns refilling coffee and tea water for the group.

Figure 5. *Nalukataq at Qaqmaqtoq (with crew flag and American flags visible).*

When everyone had finished eating, there were competitive races, and then the blanket toss (*Nalukataq*) began (Fig. 5). In 1997, it started at *Ungasiksikaq*, and when people were finished there, they moved over to *Qagmaqtoq*. It was different from many of the earlier descriptions in that all the people tossing gifts from the blanket (at least at *Qagmaqtoq*) were female, although Rainey mentions that it was mainly men who did this near the turn of the 20th century.

It is also important to note that not everyone was eligible to chase what was tossed. Other descriptions mention that all the older

---

[5] Because of the separation of the *qalgit* on the third day, the author can speak about the festival only as it was held at *Qagmaqtoq qalgi*, at whose feast she was a guest.

people scrambled for gifts, but in 1997, just the older women were allowed to try. Gifts included bedding, candy, shower curtains, towels, and fur for parka ruffs. The competition was fierce.

The final part of the celebration was a traditional dance, held in the gymnasium that night and attended by members of both *qalgit* and their guests.

EARLIER WHALING FESTIVALS

The question remains: how does this description compare with material provided by the other sources? The Rainey interviews with Kuwana, Arveksinya, Iveqsiq, and Dives discussed the Whaling Festival prior to 1910. Various narrators noted that the celebration started on the day the crews stopped whaling and came in off the ice. The *umiat* were brought onto the beach and were then sledged to the respective *qalgi* areas.

The second day of the festival[6] centered around flipper and *mikigaq* distribution. As Arveksinya described it:

> The night before the feast, they take out the flippers and put them on racks to thaw. Next day when the sun is still way north, the houses near the point start first carrying flippers to *qaligi* to give them away. The women tote them. After them comes a woman with the food bowl for crew full of baleen and follows the meat. The old people grab that and fight over it. This all takes sometimes till midnight when all six *qaligi* were working. At the same time they make sour meat and give it away to the people. And to anyone who has helped sew skins for boat or bring in meat or kill whale they call also by name and give something ... I think they call them by name to get the flippers, too ... The flippers they take home and eat some later (Arveksinya n.d.).

Kuwana described the festival similarly from the earlier days when he could remember three *qalgit*:

> Next morning early those who caught whale took canoe to *qaligi*...That morning his [Kuwana's] mother put sour whale meat in water bag (*mikiaq*) he took it to his council house (*Kalge iruk*) ... They cut flippers in the council house... After they eat, they move over to *Ungasiksikax* ... Then they move on to *Qagmaktoq*. Where they cut up flippers (Kuwana n.d.).

There were also proscriptions for seating, giving the elders the most honored places, as is the case today. Dives noted:

---

[6] Some sources also refer to this as the first day of the feast, not counting the day on which the *umiat* were brought ashore.

> There is a rule very important to them that the oldest ones who have had boats longest take the center place in the *qaligi* and when they pull the canoes up, even tho [sic] the young men are catching the whales, the old ones have the place of honor, and they boss the others (Dives n.d.).

The second day of the festival was similar to the 1997 festival, as far as what was being distributed, but a primary difference was that it took place at that time inside the *qalgi* buildings. As now, people went from one *qalgi* celebration to another, with the festivities for each *qalgi* being held in succession rather than simultaneously. Also, as in later years, the *mikigaq* was passed around, but people were called by name or group to receive slices of flipper. Food was also taken home, although not in plastic bags.

The third day involved visiting graves and leaving pieces of *maktak* strung on baleen—what many have referred to as 'feeding the ancestors.' Arveksinya noted that this took place when burials were still on racks and that the tradition was stopped a few years after the mission was located at Point Hope (Arveksinya n.d.).

> After they come back from the graves they go to their places in *qaligi* place and cook the good part of the meat they have saved for this, and they eat all day and are glad. Those who have made songs teach them to the people, dance songs, and everybody dances and is glad, those who have been crying about the dead in the morning they forget this and they all dance (Arveksinya n.d.).

Kuwana echoed the same sentiment:

> Everyone who has lost a child or other relative, the next day early in the morning goes to visit graves (not in ground but on racks). Lot of crying that day. After grave visit they start the fires in the three *qaligis* and start big feast. After feast they start to dance. Those who weep in morning—have a new life. Happy again forget, their lost ones (Kuwana n.d.).

The major difference between the festival then and now is obviously the lack of the traditional grave visitations and the leaving of *maktak*, which constituted a major part of the second day in earlier times. The rest of the day is the same as it is now in the respect that people spent most of the day cooking and eating, followed by *Nalukataq* and the dance.

The primary change in *Nalukataq*, as mentioned earlier, is that most of the tossing of gifts from the blanket formerly was done by men—mainly crew captains. Arveksinya said that, particularly with a man's first whale, people could request gifts from him from the time the whale was caught until *Nalukataq*.

349

People can ask for presents from him, and someone stand on the skin holding all the presents and called for the people they belonged to. Or he just wanted to give away something, he didn't care to whom, and he threw a lot of things out among the people, and all the old people fought over them (Arveksinya n.d.).

The nature of things distributed has changed over time as well. According to Iveqsiq, 'When they start to *Naluktk* [sic] the man who got his first whale gives something away while he jumps, and the old people wrestle for it, to get as much as they can.' She goes on to say that the gifts tossed from the blanket included, 'Calico, after the white people came' (Iveqsiq n.d.).

Froelich Rainey spent time in Point Hope in the late 1930s, with a long stay in the village in 1939 and 1940. As is apparent from his interviews with Arveksinya, Kuwana, Iveqsiq, and Dives, he was particularly interested in the Whaling Festival. Material from his publications is very descriptive of the festival, but because of references to *qalgi* buildings and no longer extant ceremonial practices, it appears that the information was based on the interviews with Point Hope residents rather than on his own observations.

Figure 6. *Photograph of a whaling festival from Froelich Rainey Collection, ca. 1940 (compare to Figs. 3 and 5); courtesy of the Alaska and Polar Regions Department, University of Alaska Fairbanks.*

There is a firsthand observation of interest, however, among his field notes, where he wrote, 'After *Nalukatak* we heard that they had good dancing in the store late one night' (Rainey n.d.). This is similar to the current practice of holding the dance in the school gymnasium, and echoes a circumstance that was common on the North Slope. Point Hope was the last community to lose its *qalgi* buildings, just prior to 1910. Barrow's *qalgit* had fallen into disuse or disrepair, reportedly due to missionary pressures, around the turn of the last century, and other communities suffered the same plight. These buildings had served as the community gathering places, and with their demise there were no locations in which to hold large events. In some villages, schoolhouses served as replacement buildings for feasts and dances, while in other areas stores and churches and, later, school gymnasiums, filled the void (Larson n.d.).

Also among Rainey's notes was a photograph of either the second or third day of the Whaling Festival (Fig. 6). When placed alongside Figures 3 and 5 from the 1997 festival, the similarity is striking. The main difference is simply the water tank that is under construction in the background of the 1997 images.

The 1941 Native herder's journal from the Rainey collection does not give extensive information about the Whaling Festival, but it does touch on the main points and their order:

> June 9—All the *oomiak* whose [*sic*] has caught whales landing on the land.
> June 10—Today they *Keh-vah-ruk-geroat*—means dividing the flippers to the people. They had one *Nalookutuk* at *Kukmuktook* [*Qagmaqtoq*] *cozgy*.
> June 11—Today is cooking day for the ones that get whales. It is *Nalookutuk's* Day. They danced also, Their foods, whale meat, *muktuk*, fruit, coffee, tea, doughnuts (Anonymous 1941).

The same three-day progression is noted with a more detailed description of the food, which is very similar to what one would see at the festival today.

James VanStone spent time in Point Hope in the mid-1950s and is the first to note a number of changes in the festival, including a scheduling change. All previous sources that have mentioned scheduling (i.e., the earlier interviews and the herder's journal) note that the three-day celebration began on the day the crews stopped whaling and came in from the ice. When VanStone was in Point Hope, the festival was being scheduled at a later date by the successful whaling captains (VanStone 1962). This change appears to have occurred sometime between the herder's journal in 1941 and VanStone's account.

Another difference in VanStone's observations is that he noted that the blanket toss occurred on the second day of the festival as well as on the third, whereas most older sources and current observations place it on the third day only. This could be due to a number of factors, including weather and just plain preference, but it might also be that the crew celebrating had caught its first whale that year. Elders in Point Hope have mentioned that in the old days

Nalukataq occurred right after the flipper distribution (traditionally on the second day) when a captain and his wife had caught their first whale. It could be that VanStone was in Point Hope on such an occasion, particularly because he also mentions a blanket toss occurring on the third day of the celebration as well.

VanStone is also the first to mention the change in who distributed gifts from the blanket. As noted earlier, Rainey's sources indicated that men tended to toss gifts during *Nalukataq*, but by 1997, only women were doing this (at least when the author was present at the *Qagmaqtoq Nalukataq*). VanStone observed:

> Usually the first person tossed in the skin is the successful captain ... Following the captain and his wife, anyone who wishes may jump into the skin and be tossed. Women who have had a male child since the last whaling feast often throw small gifts from the skin ... Men seldom seem to toss gifts, although they did in aboriginal times (VanStone 1962).

In reference to the time of mourning, he notes, 'Formerly, the morning of the third day was reserved as a time to mourn for the dead' (VanStone 1962). His use of the term *formerly* would seem to indicate that either the tradition was no longer carried out or he did not observe it in the 1950s. This would square with what earlier chroniclers had told Rainey about the cessation of grave visitations on the third day following the arrival of missionaries, but it also implies that, with or without the actual visitations, the emphasis on mourning had also disappeared from the festival. In fact, when the author spoke with some Point Hope elders in June, 2001, none recalled hearing about this practice.

CONCLUSIONS

What have been the major changes in the Whaling Festival over the last hundred years, and how can those changes be explained? The most notable differences involve location, the loss of emphasis on mourning, and the scheduling of the festival.

Regarding location, there are actually two differences between the earlier days and the present. First, the site of many of the festivities moved from inside the *qalgi* to outside *qalgi* grounds, and second, the *qalgi* grounds themselves were moved. The cause of the change of location relative to the interior of the *qalgi* was obviously the lack of *qalgi* buildings, but what was the reason for the lack? It would seem to relate to the missionary presence at Point Hope. John Driggs, M.D., was Point Hope's first missionary and served the community from 1890 to 1908. According to the best available information (the recollections of elders and Driggs's own writings and correspondence), Driggs did not discourage use of the *qalgi* buildings and, in fact, visited them and was interested in what went on inside them. Perhaps it was because he was a medical doctor rather than ordained clergy that he took this stance. The Reverend Augustus Hoare, who came to the community in 1908, had different views about the *qalgit*, which led to their abandonment by 1910 (although the *qalgi* affiliations for the last two remaining *qalgi* buildings are still in use to this day).

The *qalgi* grounds had previously been located near the beach west of what was then the community of Point Hope. The move to new *qalgi* grounds was due to environmental conditions. The community had experienced severe flooding and erosion for years. In fact, Driggs's correspondence in the 1890s chronicles the town being flooded out during fall storms. In 1977, the town was moved two miles to the east, and at that time the *qalgi Nalukataq* grounds also needed to be moved. They were then oriented east and west of town as opposed to their old north/south orientation. As mentioned earlier, this had to do with relative proximity to the point rather than with an absolute directional orientation.

The loss of grave visitations, or even just the loss of an overt emphasis on mourning, on the third day seems to be the most substantive change, and like the disappearance of the *qalgi* buildings, the cessation of this practice seems to have been based on missionary disapproval (Kuwana n.d.; Arveksinya n.d.). Unfortunately, there is little specific information on this matter. It may have changed when the missionaries introduced interments rather than rack burials, but this is unclear. It is also uncertain if the emphasis on mourning expired with the grave visitations and the stringing of *maktak*, or if it lasted for a longer period of time.

Scheduling of the Whaling Festival is also different. Although at one time the festival began when crews came in from the ice, the celebrations have been scheduled later in recent years. The change is based on practical considerations and have occurred so that out-of-town guests can get to Point Hope for the festival. This is based on increasingly available and regular transportation and the mobility of the community. Many Point Hope community members now live out of town or have friends and relatives who want to come to town for the festival. In order for these people to be able to get to the village, dates must be set that allow people time to make arrangements. So many people now come for the festival that even with scheduled plane service and extra flights, people are hard pressed to get in and out of Point Hope at that time of year. The extra friends and relatives arriving for the celebration add to demands on the organizers, who need to arrange for more food. The extra time between the end of the whaling season and the Whaling Festival not only allows travelers to make arrangements but affords whaling crews more time for preparations.

Some of the changes in the Whaling Festival have been major and based on outside influences (the loss of the *qalgi* buildings and the cessation of the practice of visiting graves), while others have been more minor and have revolved around practical considerations (the moving of the *qalgi* areas and the difference in scheduling). More striking than the differences over time, however, are the similarities. In many respects, the Whaling Festival in Point Hope today would be completely recognizable to a resident from the early 1900s. This is particularly true for the period after 1910. Photographs of the festival from around 1940 are practically indistinguishable from those taken in 1997, and descriptions of the daily events are almost completely the same, down to most of the fine details. This reflects the importance that Point Hope places on whaling and the ceremonies and festivals that surround it. The people of Point Hope are proud of their history and proud of their traditions, and that is evident in the continuity of the Whaling Festival.

Acknowledgments. This work would not have been possible without the assistance of the residents of Point Hope, who throughout time have been willing to share their recollections and history with others. Heartfelt thanks go to Steve Oomittuk, Connie Fredenberg, Andrew Tooyak, Jr., and Racheluz 'Pinky' Tooyak, who have helped me so very much over the years and who have been extremely gracious and giving in their hospitality. Thanks are also due to Ray Koonuk, Sr., Earl Kingik, and Elijah Rock, Sr., who were of great help to me during my most recent visit to Point Hope. Thanks also to the staff of the Alaska and Polar Regions Department at the Elmer E. Rasmuson Library, University of Alaska Fairbanks, and particularly to Rose Speranza, who has helped supply me with photocopies from the archives and who has also given me insightful and continuing feedback. Travel for the 1997 Whaling Festival was sponsored by the Ipiutak Heritage Project, while the National Science Foundation supplied funding for the 2001 trip and for background research. Quyanaqpak!

REFERENCES

Anonymous
    1941   Native Herder's Diary. Froelich Rainey Collection, Box 1, Folder 27. Fairbanks: Alaska and Polar Regions Department, Elmer E. Rasmuson Library, University of Alaska Fairbanks.
Arveksinya (Frank)
    n.d.   Interview with Froelich Rainey. Froelich Rainey Collection, Box 2, Folder 'Point Hope—Stories—B.' Fairbanks: Alaska and Polar Regions Department, Elmer E. Rasmuson Library, University of Alaska Fairbanks.

Dives.

n.d.   Interview with Froelich Rainey. Froelich Rainey Collection, Box 2, Folder 'Original Notes—Stories—A.' Fairbanks: Alaska and Polar Regions Department, Elmer E. Rasmuson Library, University of Alaska Fairbanks.

Iveqsiq.

n.d.   Interview with Froelich Rainey. Froelich Rainey Collection, Box 2, Folder 'Point Hope—Stories—B.' Fairbanks: Alaska and Polar Regions Department, Elmer E. Rasmuson Library, University of Alaska Fairbanks.

Kuwana

n.d..   Interview with Froelich Rainey. Froelich Rainey Collection, Box 2, Folder 'Point Hope—Stories—B.' Fairbanks: Alaska and Polar Regions Department, Elmer E. Rasmuson Library, University of Alaska Fairbanks.

Larson, Mary A.

n.d.   'Utkeavie Had Three Dance Houses ... : 'The 'Disappearance' of the *Qargi* in Northern Alaska' in, *Cultures in Contact: Proceedings of the 24th Annual Chacmool Conference*. Chacmool, Calgary. In press.

Rainey, Froelich.

n.d.   Froelich Rainey Collection. Box 1, Folder 21. Fairbanks: Alaska and Polar Regions Department, Elmer E. Rasmuson Library, University of Alaska Fairbanks.

1947   The Whale Hunters of Tigara. *Anthropological Papers of the American Museum of Natural History* 41(2) 231-283.

VanStone, James

1962   *Point Hope: An Eskimo Village in Transition*. Seattle: University of Washington Press.

'Story of a Whale Hunt:' Suzanne Rognon Bernardi's Photographs and Observations of Iñupiaq Whaling, Wales, Alaska, 1901–1902

Susan W. Fair

Abstract: *Suzanne Rognon Bernardi was one of the earliest Euroamerican teachers to live and work full-time in Wales, Alaska, at the turn of the 20th century. Following Tom and Ellen Kittredge Lopp, Harrison and Neda Thornton, Frances Kittredge, and Charles Kittredge, Bernardi used her position as a territorial teacher to compile a series of photographic albums called* Story of a Whale Hunt, *each of which is annotated with ethnographic commentary, sometimes forming an ongoing narrative. She also collected objects associated with whaling, now housed in the University Museum, Philadelphia, and described here. Bernardi's published writings are limited, but her photographs and captions provide an important early record of traditional Iñupiaq whaling ritual and ceremony. Contemporary Wales' residents may find Bernardi's photographs and descriptions useful in interpreting the traditional whaling activities of their ancestors.*

> Iñupiaq culture is very productive. The culture fuses itself into our minds through the creation of folklore, our pride in accepting family into our total being, and our unbiased respect to the animal beings by accepting them unashamedly into our universal realm. The cultures of indigenous man continue to be broken down into unique sections and some… rattle anthropological science when new finds unsettle previously found knowledge of northern people.
> *Herbert O. Anungazuk, 1999*

INTRODUCTION

The Reverend Sheldon Jackson, who established schools in southeast Alaska between 1877 and 1881, endeavored to construct them throughout the Alaska Territory. In 1884, the Organic Act provided funds for education in Alaska, and by 1889, Jackson, then general agent of education, was planning schools in Iñupiaq territories—each one to be overseen by a particular church. By 1890, Jackson was advertising new teaching positions, and, out of 24 applicants, he hired William Thomas 'Tom' Lopp and Harrison Robert Thornton for duties at the Wales school. A young woman, Ellen Louise Kittredge,

---

* Photo credits for images in Figs. 2, 3, 5 and 6 in this chapter appear as they were submitted by the author (the late Susan W. Fair) until the rightful copyright holder could be identified. Numerous enquiries have been made in an attempt to identify the copyright holder; if you have any information on the images that could assist in the search, please contact the CCI Press at ccinst@gpu.srv.ualberta.ca.

arrived there in 1892 with Thornton, who had been in Washington, D. C., for a year and was bringing his bride, Neda Sargent Pratt, a medical social worker, to Alaska (Smith 1997:8). Ellen, a teaching assistant, swiftly married Tom.[1]

Figure 1. *Suzanne Rognon Bernardi was outgoing by the standards of her day, even in Gold Rush era Nome and Teller. Here, circa 1902-1910, she poses after skiing with male friends and possibly, two brothers. Photo courtesy of Anchorage Museum of History and Art, Anchorage AK (B91.25.60).*

The Lopps and Thorntons were the first non-Natives to live year round in Wales. Hiring the Lopps, especially, proved to be fortuitous, for the couple admired Iñupiaq worldview and beliefs. Tom and Ellen Lopp provided night classes when subsistence activities pressed, shared Native food, adopted local traditions, attended ceremonies in village *kasgiit*, and learned the Iñupiaq language, encouraging their children to use it almost constantly. While many early territorial teachers felt that formal schooling for Iñupiaq pupils should be kept very basic, the Lopps regarded local people as talented peers and lifelong friends, seeming to learn as much from their students as they taught them. The Lopps also encouraged the artistic abilities of their

---

[1] Kathleen Lopp Smith notes that Ellen Kittredge was very open 'to people of different races and languages' (Smith and Smith 2001:6). Having lived on a homestead in Minnesota, she was also 'accustomed to cold and privation, enjoyed learning other languages, and was a strong advocate of the temperance movement' (Smith and Smith 2001:7).

students, a policy that resulted in several collections of Iñupiaq drawings, paintings, and woodcuts depicting regional festivals, traditional hunting methods, and local dress. Ellen Lopp would soon express considerable turmoil at the thought of introducing change into Iñupiaq traditions and of converting Wales' residents to Christianity (Smith 1997:4; Smith and Smith 2001:64).

By 1897, Ellen Lopp was exhausted from her duties as teacher, mother, and community hostess. During this time, she and her husband had also been posted to Teller Reindeer Station at Port Clarence for a year to learn and promote reindeer herding in the region. She requested through Sheldon Jackson that either another teacher or a governess for her five children be posted to Wales, and her sister Frances, then 26 and living in Minnesota, agreed to come north. Frances traveled with a younger brother, Charles Trowbridge Kittredge (Charlie), in order to avoid the appearance of impropriety. He had hoped to do some prospecting, and eventually did so, but it was Charlie who took the nine-month teaching contract in 1900 while Frances assisted Ellen (Smith 1997:90-91,98; Smith and Smith 2001:270).

A young Indiana teacher was subsequently appointed to be the successor of Charlie Kittredge. Suzanne Rognon Bernardi, who fancied herself somewhat adventuresome, if not footloose, arrived at Wales aboard the *Arctic* on October 17, 1901 (Fig. 1.). Although Bernardi was to stay less than a year, she took a number of photographs of Wales life in 1901 and 1902, assembling them into a series of albums, each entitled *Story of a Whale Hunt*. While Bernardi was not a trained ethnographer, she was a curious and observant individual who had been educated as a teacher. She interviewed several Wales elders and whalers whom she mentions by name, and each album features handwritten ethnographic and personal commentary (usually in a narrative/caption form) about whaling ritual. Inconsistencies exist in Bernardi's single published article and notes, so her comments should be judged with some measure of caution. Ideally, Bernardi's observations should, at some future time, be compared to and enriched by Iñupiaq oral historians in Wales. Inuit George Quviq Qulaut (1998:20) for example, addressing the *Imaging the Arctic* conference, discusses the importance of photography to residents of his home community: 'I often think of the other photographers that were here before. These people were traders, missionaries, teachers and scientists... . If I find some of these people I wonder if they would be willing to share their photographs with the people of Igloolik.'

This essay describes the arrival of this group of territorial teachers in Wales, discusses some of their impact on the village, and offers a brief biography of what is known about Suzanne Bernardi. Most of the text, however, is devoted to paraphrasing the ethnographic observations that run through Bernardi's albums along with some commentary from her 1912 *Courier-Journal* article. The selected illustrations focus mainly on whaling activities. Suzanne Rognon Bernardi's work is significant because the photographs, especially, are taken from an intimate perspective. They speak for themselves, revealing considerable detail. Bernardi also collected objects associated with whaling for the University Museum, Philadelphia, items which Wales residents may wish to examine or reclaim at some future point. The multiple copies of *Story of a Whale Hunt* offer a provocative glimpse of whaling practices in Wales a century ago. Along with 1890s photographs by Tom Lopp and John Justice, they are among the earliest photographic records of traditional whaling, whaling ritual, and Iñupiaq ceremonies in this area at that time.

EARLY WESTERN EDUCATION IN WALES

The relationship of Western formal education with traditional Iñupiaq knowledge transmission, the styles and goals of specific early territorial teachers, Christian missionization, the introduction of reindeer husbandry, and the production of localized (Seward Peninsula) Iñupiaq graphic arts are politically intertwined in an important chapter in Alaskan history. Records left by many teachers include photographs, letters, journals, and published works. In Wales at the turn of the 20th century, the legacy left by such outsiders is particularly rich.

The Reverend Sheldon Jackson, a Presbyterian missionary, was appointed general agent of education for Alaska in 1885. It was Jackson who introduced reindeer into the Alaska Territory in 1891. He was enamored with the theories of social evolution predominant in those times, theories that heralded the pursuits of agriculture and animal husbandry as more 'civilized' than the ancient hunting and gathering lifestyle of Alaska Natives. Thus, Jackson, most of the teachers he directed, and missionaries of various denominations set out to convert Native peoples to Christianity and capitalism simultaneously, schooling them in non-Native and nontraditional ways (Krauss 1980:95).

Dorothy Jean Ray (1975:205) has commented that 'the establishment of schools and the reindeer industry paralleled the beginnings of mining' on the Seward Peninsula. These activities also (except for large-scale mining) almost always conflicted with private sector, non-Native interests, including those of commercial fur traders and shop keepers. For the Iñupiat, most of these endeavors represented unprecedented, uninvited, long-term changes to Iñupiaq life, although in some communities—Wales among them—Iñupiaq traditions and formal education merged with somewhat less trauma than in many other villages.[2]

Wales was selected for a school because of its strategic location in Bering Strait, its proximity to Port Clarence and Teller, where reindeer were first offloaded from Siberia, and its large population of roughly 500 Iñupiat. Other schools were established at Point Hope and Point Barrow, also whaling villages. The 1892 arrival of the first Wales teachers, Tom Lopp and Harrison Thornton with their brides Ellen Kittredge and Neda Sargent Pratt, was described previously. Once settled, the Lopps participated enthusiastically in Wales life— their home was completely open to local people and, often, to others passing through. They were particularly interested in encouraging the production of Iñupiaq graphic arts, giving talented students a chance to express themselves while permanently recording local traditions (Fair 1995). Drawing was a regular part of the Wales' school curriculum (Smith 1997:79; Fig. 2). The first reindeer apprentices appear to have been the most prolific graphic artists. These young men became instrumental in diffusing the graphics tradition from Wales to nearby *Saniq* coastal and inland areas, and perhaps further, as they moved with the reindeer. They included James Keok, Thomas Sokweena (Sokeinna), Stanley Kivyearzruk (also Kayaruk, Kawerak), and George Ootenna,[3] most of whose drawings depict the details of a

---

[2] Some readers may disagree that Harrison Thornton's presence in Wales brought 'less trauma.' Thornton was a paranoid, status- and class-oriented individual who behaved erratically. He carried a pistol in Wales, against village policy, and attempted to force the Lopps to stop associating socially with Native friends and students. In March, 1893, he threatened to shoot local Inupiaq if they came to his home at night, a message that was understandably not received well by villagers or the Lopps (Smith and Smith 2001:53,74). At one point, Thornton had also tried to banish an elder headman, Elignok, from the community (Smith and Smith 2001:64). Later that summer, Thornton was killed by three young Wales men at his doorstep, a tragedy for all. The men were subsequently executed by order of village elders. The murder stemmed not only from Thornton's personality, but because of grievances about an incident aboard the *William H. Allen* under the command of George Gilley in 1877, in which 13 Wales Inupiaq traders and one sailor were killed. One of Elignok's sons had been among the dead (Bockstoce 1986:189-191; Smith and Smith 2001:1,73-74). The year after Thornton's death, no school was held in Wales because the Lopps were at Port Clarence attending to the new reindeer station. When they returned, formal schooling resumed and was attended enthusiastically by many locals.

[3] Bernardi's spellings of Iñupiaq personal names are no doubt phonetic and certainly inconsistent. Continued research on Bernardi materials should be reviewed by Wales' residents to correct these deficiencies. Kathleen Lopp Smith and Verbeck Smith's recent *Ice Window*, an annotated publication of Ellen Kittredge Lopp's letters (2001), uses these spellings, so I do not change them here. Names in parentheses are present-

mixed Iñupiaq economy, local clothing, and ceremonies.[4] They provide an important record of both traditional activities and cultural change. Only Keok's drawings include whaling activities.

Figure 2. *George Ootenna was one of Tom and Ellen Kittredge Lopp's early students. A young adult when he drew ethnographic scenes for the Lopps, Ootenna became an accomplished reindeer apprentice and herder. Various Iñupiaq hunts take place here. Men hiding behind a sea ice formation take migrating birds with bola and rifle, while at bottom, five gut parka-clad whalers successfully strike a breaching bowhead. Photo: Ootenna pencil drawings, circa 1890s, Wales, AK, courtesy Kathleen Lopp Smith Collection.*

---

day related family surnames supplied by Herbert Anungazuk (pers. comm., 2002). Also see Simon and Gerlach (n.d.) for alternative spellings.

[4] Collections of drawings from this period, not all of which were executed in Wales, are located at the University of Alaska Museum (Clowes Collection), the Smithsonian Institution, NMAI (see Phebus 1995 [1972]); and in the private collection of Kathleen Lopp Smith. The NMAI collection once belonged to the Bureau of Education and was used for expositions at the turn of the century (Phebus 1995:14). Often, these particular drawings display the day's academic lessons on the reverse side of the page (11, 14). Alaska Historical Library holdings include 'Eskimo Pictures and Maps,' ethnographic drawings in the Wickersham Collection produced by *Etoachina* of now-abandoned *Ipnauraq* near Shishmaref in 1901 (Wickersham n.d.b).

SUZANNE ROGNON BERNARDI, A PROFILE

Although most of this essay focuses on the photographic essays on whaling compiled by Bernardi, it is useful to know what can be gleaned about her background. She was a relatively unusual, unbiased, and adventuresome Victorian woman of about 30 years old when she traveled north, probably widowed, or perhaps, divorced. Until the age of 17, she lived near Valley City, Harrison County, Indiana, after which she taught in the south, mainly in Alabama, for eight years. The Lopp family of nine children also hailed from Valley City, and Ellen Kittredge (from Minnesota) had taught at black schools in the South as well (Smith 1997:8). Nevertheless, although Bernardi's early life experiences are somewhat parallel to those of Tom and Ellen Kittredge Lopp, she apparently was not acquainted with either of them although the small group discovered quickly, in Wales, that they knew some Indiana folks in common.

Suzanne Bernardi, like Ellen Lopp's sister, Frances, before her, traveled north with a younger brother, O. Jack Rognon, although Jack was not with her when she disembarked at Wales (McKinney 1981:134; Smith and Smith 2001:247). Between 1900 and 1902, Jack Rognon and another unidentified brother lived in Teller, trying to work the Gold Rush. By the time the Bernardi brothers arrived in Teller, however, though claims in other districts were doing well, the town was said to be 'on the bum.' A *Nome Nugget* headline (1901c) claims 'Teller A Dead One' and the article goes on to say that 'there will not be a corporal's guard left next winter,' noting that cafes there had closed during the winter and that 'inhabitants, unemployed gamesters...' and others were heading for other parts. A brief introduction to her cover story 'Living With Eskimos of Cape Prince of Wales' in the Louisville (KY) *Courier-Journal*, Sunday, October 20, 1912, indicates that Bernardi returned to the Midwest after teaching and performing missionary duties among Alaskan Eskimos for ten years. The record of her activities in Alaska during this period, however, is limited, and her personal journals have never been located.

We do know that Bernardi traveled from Nome to Tacoma, Washington, in 1905 carrying with her some Iñupiaq objects she had collected for George Byron Gordon at the University Museum, University of Pennsylvania, Philadelphia. When Bernardi first met Gordon, probably in Alaska in 1905, he was an assistant curator at the museum. He traveled to Alaska on collecting trips in 1905 and 1907, and in 1910, became the first University Museum director, after which he did not return to the North (Kaplan and Barsness 1986:13; King and Little 1986:16-53). While in Tacoma and, later, Louisville, Bernardi continued her correspondence with Gordon, requesting monographs that the museum was printing at the time, chatting about her painting and her health, and imploring him to write: 'I am painting in oils a little wood scene of Sitka, Alaska. I wish you could see it, it is a gem' (Bernardi 1906). She was confined to a Louisville hospital for six weeks with inflammatory arthritis in a knee and ankle in 1906, writing to Gordon that she could not 'go to Nome on first boats and am just heartsick. I expect to sail about July 1 and have 6 weeks kindergarten [in Kentucky]. I expect to go over to Siberia with a trading schooner party and hope to go to Point Barrow. Are you going up this summer and if not will you give me a commission to buy for you there?' (Bernardi 1906). In June, she wrote that she was still in Kentucky, though her heart, she said, was in Alaska.

Bernardi planned to sail in July, and by August 7, 1906, she had indeed returned to Nome and was preparing to depart on the *Corwin* for Point Hope the following day (Bernardi 1906). At some point, she made it as far north as the McKenzie River and in Southwest Alaska, to the lower Yukon and Kuskokwim Rivers, as her photographs in the University of Pennsylvania Museum Archive show portraits and architecture at those locations (Bernardi 1907). By October, 1906, she was back in Nome, having had 'a glorious summer.' During this period, she employed a former student, Thomas Illayook from Wales. The young man traveled aboard the *Thetis* with her, negotiating for artifacts: 'He was able to trade for me and secured almost a complete whale hunting outfit, that is, the ceremonial gear from King's [sic] Island, Cape Prince of Wales, Point Hope and Point Barrow. I send you everything I secured except two lamps' (Bernardi 1906). Gordon was interested in obtaining such ritual whaling items to fill in gaps in collections begun by E.A. McIlhenny, and Bernardi helped him do so (Kaplan and Barsness 1986:19, 28). She was still in Nome in the spring of 1907, as Gordon wrote that he expected to find her there (Gordon 1907).

Upon her arrival at Wales in the fall of 1901, Mrs. Bernardi was received warmly by the Lopps: 'She had more of the qualifications for this place than I had hoped to find in one person. She certainly seemed like a good woman' (Smith 1997:115; Smith and Smith 2001:318). The family opened up a lean-to with a separate entry at the back of their house to her. The Lopp children were immediately taken with her: 'She has invited the children to dinner tonight... We all like Mrs. Bernardi very much' (Smith 1997:118; Smith and Smith 2001:317, 325). She taught that winter, attended local festivals, and took photographs. She did not leave Wales again until February, 1902, when she drove off, pulled by sled deer, with Judge James Wickersham of Nome and his companion, Louis Lane, heir to and manager of the Wild Goose Mining Company fortune (Smith and Smith 2001:338).

The young teacher was gone for a month, and by mid-March, she had disappointed the Lopps by shirking her teaching duties for social pleasures: 'Mrs. Bernardi is tired of us and her school. If Rev.

Scruggs of Teller school will exchange, she will do that; if not, we will hire Miss Armond to finish' (Smith 1997:125; Smith and Smith 2001:343). The young woman was apparently enamoured of the faster life in Teller, Port Clarence, and Nome, and of the companionship of her brothers and their attractive friends. Frances Kittredge wrote home:

> You asked in two letters how we liked Mrs. Bernardi. I cannot say as much now... I knew that she was different from people we were used to being intimate with, but since we liked her so much that would only make her more interesting. Unfortunately she was more different than Ellen and I thought. I do not understand her; I give it up. Evidently she is used to being made very much of, especially by men, and the lonelier life at the Cape has worn on her. She likes a gayer style of society than we do. In her several weeks' visit at Teller, she was paid great attention. Three balls... were given especially in her honor—that is, besides the usual affairs of Teller society. She had numerous invitations to dinner, etc., etc. I fear that after all her good time there, the Cape will be duller than ever, but there are less than two months left. Then she plans to go to Nome" (Smith 1995:126; Smith and Smith 2001:344).

Bernardi finished the last two months of the 1902 school year and stayed on to photograph spring whaling, but did not return to Wales that fall. The Lopps reluctantly left Wales for Seattle in 1902 as well. Bernardi's whereabouts between 1902 and 1912 are unclear; she seems to have been, at various times, in Wales, Teller, Nome, Seattle, Tacoma, the Midwest, and Washington, D.C. (Bernardi 1905). Loose Bernardi photographs at the Anchorage Museum of History and Art show her in a Gold Rush wallpapered bedroom, then out skiing, accompanied by a number of young men. By April 5, 1915, Bernardi had returned to Washington D.C., where she married George Jeffrey, formerly of Nome: 'Jeffrey is well known here [in Fairbanks] where he was secretary to Judge Wickersham on the bench, all will join in wishing him good luck' (*Fairbanks Daily News-Miner* 1915).

BERNARDI'S FIELD COLLECTING

One copy of *Story of a Whale Hunt* is located at the University Museum, Philadelphia. A number of objects that once accompanied whaling rituals in several Alaskan villages, not all of which were obtained by Bernardi, are also in University Museum collections. These items include male and female mittens with chunks of graphite (described below), a mummified bird fetish bound in sinew and noted in Bernardi's photographic captions, a whaling belt from Cape Prince of Wales, and four similar belts—one with charms—from Point

Barrow. Other whaling artifacts in University Museum collections were obtained on Little Diomede, Sledge, and King islands, as well as from Barrow. These objects were collected by William B. Van Valin, also an Alaska schoolteacher who became an outstanding field collector (Kaplan and Barsness 1986:41,45).

Figure 3. *In Wales, open racks were surmounted with men's property:* umiaq, *kayaks, walrus hide rope, and hunting charms. Another Bernardi caption states: 'His [the* umialiq's*] Luck Pieces and Charms are kept in a box on the cache. Perhaps a skull of a wolf or caribou or dried bird.' Polar bear skulls crown these poles. Suzanne R. Bernardi photo, circa 1902, courtesy University of Washington Libraries, Special Collections, NA 3221.*

In her lead to the 1912 *Courier-Journal* cover essay, Bernardi (1912:1) casts herself as somewhat of an adventurer. It appears that she intentionally used her position as an Alaska Territorial teacher to write, accomplish amateur but useful ethnographic research, take photographs, meet like-minded people, and as a springboard for collecting artifacts for Gordon, with whom she corresponded frequently: '[We] traveled by reindeer in winter. In summer made journeys afoot totaling hundreds of miles, and used native skin boats, United States revenue cutters and trading sloops. In order to obtain accurate knowledge of the people, study their home life and record their history and legends, [we] lived directly among them' (Bernardi 1912:1; Fig. 3.)

Bernardi's field collecting was not uniformly successful, however, although she tried to negotiate for many objects, both artifacts and items in use at the time. 'Perhaps some of them seemed high priced, but you must remember how far I had to go to unfortunately get so few things' (Bernardi 1905). Gordon responded that the prices she was paying were excessive compared to what he had spent in Alaska previously (Gordon 1905). Unfortunately, some artifacts she recovered were taken from burial pits on the mountainside at Wales. She visited the burials with local people and at one point, with Judge Wickersham. Her account of one such journey is melodramatic:

> The getting [of the skull] is primarily responsible for my rheumatism and came near causing my death. A ship landed nine miles below those granite pits on Cape Mt. A young man aboard volunteered to cross the mountain with me and help me get a skull from one of the pits. The only pit I had found containing bones. We found it and starting back we were caught in a blinding snow storm, lost, and wandering and floundering over the mountain we heard the sea. It was then after dark. It was nearly midnight when we reached a mining camp. More than a dozen times I gave up exhausted. I was wet to the skin and cold; I couldn't walk nor stand on my feet for three days (Bernardi 1906).

Bernardi's letters to Gordon, which she sometimes signed 'Susie,' indicate that she wanted the museum to authorize an account for her to purchase ethnographic objects rather than simply reimbursing her. She also pleaded from Nome to be allowed to set up shop at the coming Alaska Yukon Pacific Exposition:

> On account of the [traveling] expenses and prices of the shipment I have to put them in as a whole. If you will send me a note authorizing me to collect for you next year I'll try to get the parka and better jade—I couldn't spare the money this summer to invest and wait so long. I could have had a very beautiful squirrel skin one [parka] at $60. At the Seattle Fair, that is the Alaska Yukon-Pacific Exposition, I am very anxious to have the concession to exhibit an Eskimo village. I wish you were here to go north with me. I am going principally to get subscriptions to the Sunday News, a new paper here. With warmest regards, I am yours sincerely, Mrs. S.R. Bernardi. (Bernardi 1906).

Bernardi received little money for the artifacts she did purchase, and although Gordon recommended her, she was not given a position at the exposition. She had already impetuously arranged for one East Cape family and another from Diomede to travel: 'Many of the good looking well-dressed wealthy deer owners at the Cape wish to go out to be seen as well as to see' (Bernardi 1906; Gordon 1908). In a letter sent from the Golden Gate Hotel in Nome, she referred to a fine drum she had purchased but was uncertain whether Gordon would want it: 'This one has birds [sic] heads with beads for eyes and mans [sic] legs (moveable) on the birds body. This old ivory is the handle. I've been promised this winter the small ceremonial bucket with little ivory walrus around it' (Bernardi 1906). In another letter from Tacoma, she allowed that she had not been paid: 'The small bill of goods you were to remit to me from Seattle, I have never received... . I have six of those quite large [trade] beads, if you care for them you may have them for what I paid for them, $4.00' (Bernardi 1905).

The *Courier-Journal* article states that Bernardi's brother served as photographer during Bernardi's travels. She comments, however, that she carried and used the camera in Wales and aboard the *Corwin* at other locations: '... . I sat, sphinx-like, on my pinnacle of ice, waiting patiently to get a good picture of the whale as they brought him to shore [in Wales].' And, 'my camera and I were still very fit, and I had the pleasure, by the aid of my field glasses, of giving the first news of a steam whaler... . Soon more than a hundred [people] were waiting at the edge of the shore ice' (Bernardi 1912:12).

The ship she refers to was the *Alexander*, which had sailed from San Francisco on April 2, 1902, returning there on November 1 that year with a cargo of 20,000 tons of whalebone, presumably baleen (Hegarty 1959). According to Bernardi, she and her brother Jack were invited to dine by the ship's captain, James A. Tilton, off Wales, possibly on July 15. An entry in a log kept by First Mate W. S. Varnum notes: 'Stopped at Prince of Whales [sic] to get the news of the ships' (Varnum 1902). Captain Tilton does not mention her on that date in his own journal, though, now in the collections of the Martha's Vineyard Historical Society. Bernardi records, however, that while they were aboard the *Alexander*, two *umiat* came out to trade (Bockstoce 1977:90). Her *Courier-Journal* article says the Iñupiaq *umiat*, which must have been tied up to a whale boat, were hauled along as the *Alexander*'s crew pursued a whale. Then, the whalers gave the carcass to the Iñupiat. The young woman's socializing on this and perhaps other ships may not have increased her popularity with the Lopps. There are also discrepancies in her various accounts, so she may have exaggerated some events.

'STORY OF A WHALE HUNT:' SUZANNE BERNARDI'S PHOTOGRAPHIC ALBUMS

Suzanne Rognon Bernardi left a provocative photographic record of Inupiaq traditional life in Wales at the turn of the century. While nationally known photographers like Edward S. Curtis and locals like the Lomen Brothers of Nome frequently arranged posed images, Bernardi captured people at home, at work, at school, and at ease. Although she was an amateur, she was perceptive, as were Tom and Ellen Lopp. Bernardi became particularly impressed with ritual and festival aspects of Wales life, particularly those associated with traditional whaling (Fig. 4). Taken together, letters penned by Ellen Lopp and her family (Smith and Smith 2001) and Bernardi's albums, ethnographic notes, and the *Courier-Journal* article compose a rich and unusual record of life in Wales during this time.

Figure 4. *From* Story of a Whale Hunt: *'A black mark of graphite is put around the boat at the water's edge to keep out evil spirits of the sea. The boat is launched and takes its turn patrolling Bering Straits with 6 or 8 other crews.' Here, a dancer performs near an outgoing boat. Suzanne R. Bernardi photo, circa 1902, courtesy University of Washington Libraries, Special Collections, NA 3229.*

Story of a Whale Hunt is a series of albums made, apparently, as gifts by Bernardi for colleagues, friends, and relatives. Each album is slightly different, brimming with handwritten ethnographic observations and personal notes. Photo captions in some of the albums create a linear narrative, demonstrated here by sentences taken from the University Museum album:

> American Continent, in the Eskimo village of Kingegan, Every Boy...; at a certain age lays aside his play things, begins to think; of a snug home of driftwood, half under ground with a cache near by in which may reset the fruits of his labor, ivory, skins, seals, etc.; before making a choice of the village girls for a wife, he must prove his courage and ability to fill a man's place in the world. 'Here is seal meat, eat it;' Having proved his fearlessness... while the Bowhead whale goes through... (Bernardi 1901-1902, *Story of a Whale Hunt*, University Museum Archives, Philadelphia).

Six known complete copies of *Story of a Whale Hunt* exist, all compiled *ca.* 1901-1902, at the following locations. One is in the Special Collections Division, FM-25, University of Washington Libraries, Seattle (Location K0069, No. 049, see also Native American microfiche NA3208-3266). Two albums are in the Archives, Anchorage Museum of History and Art, Anchorage, Alaska, as part of a collection that also contains some loose Bernardi photographs. One album, donated by Bertha W. and M. McKay (McKay Collection), is housed at the Carrie McLean Museum in Nome, which also has some Bernardi photographs in their general files, many of which are not attributed to her. Another album is in the collections of the Bancroft Library, University of California Berkeley (Alaska Pictorial Miscellany group), also with miscellaneous Bernardi photographs. Two Bernardi albums are said to be in the Alaska and Polar Regions section at the Rasmuson Library, University of Alaska Fairbanks, but could not be located for this essay. Other copies of the albums may exist.

A final copy of *Story of a Whale Hunt* is located at the University of Pennsylvania Museum Archives, Philadelphia, along with additional photographs and correspondence. The American Collection there also has at least 15 artifacts purchased by George Byron Gordon from Suzanne Bernardi between 1902 and 1907, although the items were not formally accessioned until 1915.

WHALING RITUALS PHOTOGRAPHED BY BERNARDI

Bernardi understood simplistically the complex connections between subsistence hunting and animistic beliefs that were still strong in Wales during her time there. What follows is her basic written narrative, paraphrased from captions and comments in multiple copies of *Story of a Whale Hunt*. At the outset, she comments that missionary intervention made it difficult to write a complete account of whaling rituals because local people had been taught 'that their observance of ceremonial rites to propitiate the evil spirits was wrong' (Bernardi 1912). This undoubtedly referred to Jackson, Thornton, and other missionaries or teachers, as Tom and Ellen Lopp had not perpetuated such views. Tom Lopp excitedly attended *kasgiit* celebrations himself, taking notes and photographs. Bernardi refers to Wales whalers as 'stay-at-home' hunters who had a monopoly on whale hunting because of their strategic location at Bering Strait. She calls *Ochavook* [sic] or Whale Mountain, behind *Kingegan*, a landmark or 'decoy' for migrating whales.

Bernardi says she was told the story of the mountain by a Wales elder, grandfather of Kituk. The clusters of hand-built stone pits on Whale Mountain, from which she removed bones and artifacts, were said by the elder to be inhabited by individual spirits. In one of these pits, Bernardi and her colleagues found 'stones, jade hammers, *adz* [sic], and much prehistoric pottery,' most of which are now at the University Museum, Philadelphia (Bernardi 1912). The mountain was incessantly hungry, the elder said, and could be placated only with whale meat. After being fed, the pits transformed themselves into Arctic foxes, then '...he be again spirit, no white fox. He sleep plenty when belly full whale meat. No bother Eskimo' (Bernardi 1912). 'Bothering' was less likely a reference to the evil spirits imagined by the missionary than it was an acknowledgment of the real possibility of a lean hunting season, which could result in starvation for the village. Bernardi also mentions a prohibition on the use of white fox fur in Wales for reasons associated with these beliefs.

At the time of Bernardi's observations, aspiring young whalers did just what the elder described. They transported whale meat to the mountain, feeding the spirits before they could participate in *kasgi* life with older men, dancing, singing, and making preparations for the hunt. When whale meat was scarce, such young men substituted stones (Bernardi 1912). It is unclear from Bernardi's accounts whether the stones may have been used to construct the pits themselves, thus "feeding" the mountain by literal increasing of its mass.

When the mountain had been propitiated, villagers prepared for the hunt as bowhead whales drew near. *Umiat* had been stored upside down on elevated racks over stretched polar bear hides that had cured all winter. Thongs strung intermittently with cubes of year-old whale meat amulets were threaded under each boat. The boats themselves were removed and re-skinned "in holiday dress" before the hunt (Bernardi 1912). One University Museum album caption states: 'On the last day of April, the nearly covered canoe with all whaling implements are put on exhibition' (Bernardi 1901–1902). Bernardi saw an *umiaq* belonging to the *umialiq* Kitsinna that had been outfitted with new paddles: '...each one decorated with birch [stain] painted figures and tufts of deer hair' (Bernardi 1901-1902). A small broom made of fine beach grass, about 10 in. in length, was placed in the boat so that after launching, says Bernardi, it could be used to christen the craft with ocean water.

Figure 5. *From* Story of a Whale Hunt*: 'The captain of a boat opens the ceremonies by taking from his meat cellar pieces of the whale he last killed and while sitting in charmed state under his whole charmed whaling gear he serves his pieces to each caller.' Suzanne R. Bernardi photo, circa 1902, courtesy University of Washington Libraries, Special Collections, NA 3234.*

Objects associated with the whaling rituals were never touched by human hands, according to Bernardi's comments in albums and essays. The bentwood feast bucket associated with Kitsinna's boat was handled with a long mitten, while a smaller mitt was worn by a shaman who would be inscribing with graphite during the ceremonies. Graphite was used at the waterline to outline a boundary around each *umiaq* before it was launched for the hunt 'to keep the devils of the sea out of the boat,' and each whaler was also marked with a bold graphite line running from his brow down the bridge of the nose, although the meaning of this tradition, Bernardi says, had been forgotten by that time (Bernardi 1912:1).

During these ceremonies, successful whaling captains seated themselves prominently at their meat cellars, probably large underground *situaliit* (Fig. 5). Each captain who had been successful the previous year went to his cache on May 1 to 'take a box [with] all kind of ceremonial whale charms, luck pieces, etc' (Bernardi 1901–1902). At the cache, each captain then distributed whale meat stored from the preceding season to visitors, portions of which were given out in small wooden dishes. Over the entryways to the caches, tall poles laden with symbolic objects were suspended, including the skull of a wolf as well as a bent and pegged driftwood container. Carved ivory whales were inlaid around the rim of the bentwood buckets. Lincoln Milligrock of Nome, who hunted whales as a young man at Little Diomede, recalls fancy buckets like those described by Bernardi being passed around Diomede *kasgiit* to thirsty dancers sometime in the 1930s. He remembers that the buckets, filled with ice, made a characteristic clinking sound (Milligrock, pers. comm. 1999).

A carved ivory chain, according to Bernardi's captions, was attached to the post above Kitsinna's cache, each link a record of the number of whales he and his crew had killed '...and so on down the pole were the charms used in capturing the whale' (Bernardi 1912:1). Nearby, elevated open racks were surmounted with men's property: *umiat* and kayaks in storage, walrus hide rope, and hunting charms. A Bernardi caption states: 'His [the *umialiq's*] Luck Pieces and Charms are kept in a box on the cache. Perhaps a skull of a wolf or caribou or dried bird.' Polar bear skulls crowned the posts of some prominent men's racks. Inupiaq creation tales sometimes tell of local heroes who build their homes near village *qasgiit*, mounting their caches with magical amulets that symbolize personal attributes (Kakaruk and Oquilluk 1964:4,14).

The whaling captain and all visitors wore gut garments, as though prepared for whaling. The *umialiq*, possibly Kitsinna,[5] is dressed in high, waterproof *ugzruk* boots and a showy labret. Items attached to the post were then transported to sea on whaling *umiat*. Bernardi also saw dancers wearing wolf head skins in the *kasgi* during what she refers to as the 'Whale Feast,' while other dancers wore regalia of ermine, beads, and feathers. What she saw was probably actually the Wolf Dance.

---

[5] A man named Kitsinna (also Kitsenna) was photographed with Captain Cochran and Tom Lopp in 1915, on Cochran's ship off Wales (Kathleen Lopp Smith Collection). The two are not the same man. Bernardi spoke at length about Kitsinna, so I suggest that it is Kitsinna featured in the portrait.

Figure 6. *From* Story of a Whale Hunt: *Inside the qasgi 'men dance and feast and sweat and fast for days before their first hunt. The woman may only go to the entrance of the dance house to carry food or material for work.' Suzanne R. Bernardi photo, circa 1902, courtesy University of Washington Libraries, Special Collections, NA 3228.*

As feasting concluded, all members of whaling crews retired to the appropriate village *kasgi*, where they prepared physically and psychologically for the coming hunt. For the preceding week, young boys going on their first hunt had taken their meals and slept in the dance house, where no women had been allowed beyond the entrance. Seal oil lamps were lit, fasting began, and a men's 'sweat dance' was prepared while elders told their adventures, successes, and survival narratives through song and dance (Fig. 6). The men, says Bernardi, danced for 24 consecutive hours before the first whale boats could be launched. Only those who participated in these 'fitness rituals' could go to sea:

> Perhaps a tale from Netaxite narrating his thrilling experience of thirty-two days out on an ice floe, or Tokwontnuk's grandfather, whose canoe was swamped by a monster whale after the tow line had tangle in the fins [provide examples]. On his grave is shown the mast of the boat that was broken by the tail of the whale (Bernardi 1912:1; Lopp 1902).

Figure 7. *Wales residents dressed in seal gut and canvas parkas haul a whale up onto the shore after a successful hunt. Suzanne R. Bernardi photo, circa 1902, courtesy University of Washington Libraries, Special Collections, NA 3232.*

When festivities in the *kasgi* concluded, Wales children gathered at the shore to dance and sing songs 'to the spirit of wind and sea' around each *umiaq* and her crew. This ritual was followed by the procession of an elderly woman, no doubt past menses, who distributed ashes gleaned from the shavings of newly made paddles in a path to the open water, usually about a mile away across shore ice. One *Story of a Whale Hunt* caption states: 'Before embarking—a piece of cooked whale skin is partaken of by each from a charm spoon and wooden pail after a fast of 24 hours—then with a piece of native graphite the tribal mark is drawn on the forehead of each [crew member].' In a caption for the same photo in another album, Bernardi notes that at the launch site, a shaman suspended 'the skull of some animal' over the boat, then danced around it. All physical reference to the hunt was either destroyed or transported from land to the sea, cleansing materials associated with land and human activities. Bernardi saw six to eight skin boats rolled on inflated seal pokes down this ashen pathway. The boats were blessed, she says, then sent out for a test run in the ocean. Spirits did not follow because they were said to fear the ashes.

As the hunt commenced, Bernardi watched closely from a rise in the shore ice. She was rewarded with the sight of shooter Sam Okbaok wounding a whale, and the ensuing towing of the whale to shore by several *umiat* under sail. The hunter in the stern of the victorious boat had cut a narrow strip of skin from tip to tip of the animal's tail, which he wrapped around himself and hung over his shoulder. This man then proceeded to the *umialiq's* [crew captain] home, where he spread the skin out. There, other villagers estimated how many sleds would be required to haul the meat to storage cellars. This underscored the hunter's success. After this, all able-bodied villagers proceeded to the shore ice to haul the whale for butchering while its head remained in the water, buoyed by seal pokes (Fig. 7).

Although Bernardi ably recorded the butchering process, she did not inquire about the meaning of many of the associated rituals. From her accounts, two 'whale butchers' clad in one-piece, waterproof, sealskin suits and wearing whetstones around their necks initially cut out the navel of the whale. The charm would be used as a record of the hunt, integrated into the captain's ceremonial gear by his wife. Then, the men stepped up on the backbone and into the body of the whale for butchering, which Bernardi assumed was accomplished 'without much system.' Actually, there was probably very distinct protocol to this process (Figs. 8, 9). As they worked, an elderly woman held two tightly bound birds above the men, the species of which were not noted by Bernardi. This woman had prepared the birds, alive, compressing their tails and feet and wrapping them, mummy-like, in strips of seal fat, leaving only the beak exposed. Bernardi explained that 'by confining the lives of these two birds in their bodies, the devils [would] take them instead of the butchers' (Bernardi 1912:12). One such artifact was collected by Bernardi for the University Museum in Philadelphia (NA3295). Ellen Kittredge Lopp commented on the butchering as well:

> A native man was getting into a waterproof suit—one of those peculiar garments, which is shirt, trousers, boots, mittens, and hood all in one. The front of the hood is so large the man gets into the suit through that, then with the draw string, which is in the edge, draws the hood snugly about his face. Five of the older men... stepped over onto the whale just back of the head where the blubber had been cut off. Sitting there half under water, the men quickly cut out pieces of the flesh and tossed them up upon the ice... The five men looked so strange sitting there on the whale. One of them suddenly sprang out. Another's knife had cut a hole through his boot and the water was rushing in. No one having a needle, the man tied the hole up with a string and went to work again (Smith 1997:127).

Figure 8. *Iñupiaq hunters in traditional waterproof seal gut suits crawl on and into a whale to begin cutting and distributing shares. Note that the tail of the whale has already been removed while an individual in a striped canvas parka anchors the whale to shore. Suzanne R. Bernardi photo, circa 1902, courtesy University of Washington Libraries, Special Collections, NA 3233.*

Figure 9. *From* Story of a Whale Hunt: *'catch even a 30 foot pup whale and loses no time in donning his water proof seal skin suit and gets right into the whale's stomach, not through his throat, but through the hole he makes himself with his sharp knife.' Suzanne R. Bernardi photo, circa 1902, courtesy Anchorage Museum of History and Art, Anchorage AK (B97.19.23).*

Figure 10. *An* umiaq *and crew return to Wales. Bernardi notes that this boat had been unsuccessful. Suzanne R. Bernardi photo, circa 1902, courtesy Anchorage Museum of History and Art, Anchorage AK (B96.9.14).*

As the butchering proceeded, each *umialiq* placed a stick in the snow from which he hung a sack of tobacco for his crew. Proudly, says Bernardi, each man took a portion from the bag and filled his pipe while the captain himself displayed his ceremonial costume. He wore a jacket that Bernardi describes as composed of 'many tiny life preservers.' This outfit had been conceived with the capture of his first whale, after which the women of his family cut a piece of skin from the notch in the tail and sewed it carefully into a long pendant to be worn at the back of his neck. After this, she remarks, navels of subsequent whales were attached to the first kill. In a letter to Gordon, Bernardi described the *umialiq's* regalia in somewhat more detail:

> I send you a photo of Eskimo with a whale belt on. Sewed in seal skin is a piece of the skin of every whale he has killed and a white whale skin on the front with a dried head of a duck for medicine charm. These are buried with them (*umialiit*) and this one was dug out by an up to date boy at the Cape. I paid him $2.50 for it—Is it worth $3.50 to you (Bernardi 1906)?

Successful hunters of that era sometimes wore a belt so long, but only during butchering, that it touched the ground. When not in use as ceremonial regalia, this belt was placed in the *umiaq* where it might be eaten if captain and crew drifted away. At the outskirts of this scene, the boundaries of the butchering area were identified with the captain's paddles and seal pokes, all marked with personal identification. Local women then chopped cakes of packed snow, fashioning rough bowls into which they placed the whale's meat, to be taken home by sled. Extended family and friends in neighboring villages traveled to Wales for shares and festivities, just as they do today (Fig. 10).

THE WHALE DANCE OF 1901-1902

As subsistence activities subsided and winter loomed, Wales women began sewing fancy new garments for the coming 'whale dance.' When dancing commenced, each headman was to wear new regalia:[6]

> His costume consisted of a spotted deerskin coat trimmed with wolverine skin, red flannel pants and fancy deerskin knee-length boots. This costume was finished off with a pair of gloves made of white drill cloth or, in some instances, deerskin wrist-length gloves decorated with fancy stitches. A streak of black paint from brow to tip of nose adorns each... face... A circular labret is worn in a hole pierced in his chin when he was just growing into manhood. These labrets... are worn to cover the hole, which is to allow the juice to run out of the body after death, so the devil may not see it and burn them... (Bernardi 1912:12).

Such celebrations and feasting ceased, notes Bernardi, when there had been a death in the family (for a mourning period of one year) or when no whale had been caught. As no whales were captured in 1900, there was no subsequent winter festival. The following year, however, Wales residents invited neighbors that Bernardi mistakenly calls the *Silawamiut* for a reciprocal celebration that Bernardi refers to as the Whale Dance. Since there was no nearby village by this name, the guests may have been the people of *Senoraq* to the east (*Sinaramiut*). The excitement in the village was palpable, she remembers: 'I was unable to understand the suppressed energies of the small boy in school... his arms spasmodically would fly out as if he were punching a bag... Later, I learned that he was mentally reviewing his song and dance for the coming ball that would inaugurate the winter's social season' (Bernardi 1912:12).

During this period, Judge James Wickersham examined one of two 'Eskimo dance houses' at each end of Wales,[7] made a series of

---

[6] Many celebrations were held in Wales' *kasgiit*. In the 1920s, Davey Ningeulook, polar bear hunter from Shishmaref, attended a polar bear dance in the *kasgi* there, where a man danced with 'all his hunting gear and equipment' after killing a bear (Ningeulook 1997).

[7] Herbert Anungazuk of Wales confirms that there were four *kasgriit* in the two historic Wales communities, two in each, and possibly a fifth such

drawings of one of them, and spent many evenings in this 'place of resort for the men.' He also visited private homes in the village. Portions of the entryways of the houses, he notes, were paved with 'the round ends of whale vertebra' but the *kasgiit* were much larger and more elaborate.

> You enter the main hall through an upright door, and having gone down three or four steps find yourself in a large hallway forty feet long and twelve feet wide. The superstructure is made of whale ribs, and arched over with great ice cakes from the beach... At the lower end of this long hall you drop down a step or two more, and enter a low dark passageway at the end of which you find a round hole above your head. Rising you find your head and shoulders in the room, and stepping upon the vertebra [sic] of a whale you can sit on the floor and raise yourself into the 'koz-ge' (Wickersham n.d. 3-4).

Bernardi was invited into a dance house as guests of Norwadluk (Nowadluk) and Egedlenna, friends and employees of the Lopps, during the winter of 1901-1902 for the Whale Dance. This may not have been the same *kasgi* described by Wickersham, as her description of it is not as grand. She crawled through the 35 ft. entryway: '...In the darkness I knew only that we were slowly moving, but could see or feel nothing but the snow walls of the narrowing passage. Not a sound could be heard but the labored breathing' (Bernardi 1912:12). Within the dance house, six drummers sat behind native foods prepared by host-community women. Among these dishes, Bernardi saw two buckets filled with small snowballs which were used to cool the faces of infants and to refresh parched drum skins (Bernardi 1912:12). Arthur Nagozruk, Jr. (Nagozruk), a pupil of Bernardi's who would later became the first Iñupiaq schoolteacher (Smith and Smith 2001:358-361), slipped out periodically to keep her supplied with a tomato can of snowballs. Nevertheless, she fainted from the heat at one point. Young boys who succumbed were hauled out onto the snow to revive.

The entire festival lasted roughly six hours, during which an elder male tended the prominent single seal oil lamp. In family homes, women cared for lamps. Within the *kasgi*, both host and guest females stood or sat down on the floor, while men occupied the seating platform that surrounded the room. One side of the room was reserved for the hosts; the remaining space was occupied by guests and used for dancing. Poles were brought in from elevated rack caches outside to form 'galleries' where youths were, literally, shelved to watch the dances: 'about twenty small boys, like so many sardines,

---

structure. The names of three of these dance houses and associated groups are still known in Wales (Anungazuk, persn. comm. 2002).

[were] packed in their naked glory, their heads peeping out over the edge' (Bernardi 1912:12).

About midway through the celebration, visitors distributed gifts to their hosts throughout the *kasgi* (these items had been cached in the entryway during feasting). Many of the gifts were luxury items and trade goods including strips of seal gut, 'pieces of whalebone mittens,' beaver and deerskins, Iñupiaq-made boots, guns, calico cloth, candy, thread, and sometimes small bottles of liquor. Tobacco, an especially valuable commodity, was apparently not distributed. The feast that followed would feature only the best of local native foods. Guests took uneaten foods home. As gifts were received, a speaker stood in the central pit:

> Perhaps he was telling of the very fine workmanship on a pair of boots. I was quite carried away with his delivery, never having heard finer tones and modulations from any orator. Thousands of dollars worth of gifts were presented that night... It was puzzling until next day one of my neighbors explained to me that it was really an exchange of gifts. If one of your neighbors presented you with a bolt of calico, you could rest assured that you possessed something of equal value which you would be told the next day was expected in exchange. At this one dance of the year one may ask for anything he wants belonging to his neighbor, provided he can give goods of equivalent value in return... (Bernardi 1912:12).

A local shaman announced each song as dancing commenced, and a group of carved driftwood puppets provided entertainment with animated movements. Placed under the seal oil lamp, their shadows must have been greatly magnified, casting lively drama on the floor and walls, contributing forms from the hunt with surreal energy. Bernardi mentions sending Gordon two dolls for the University Museum collection, including one such puppet: 'the half body in old ivory was a charm used by a shaman. The wooden half-doll for automatic dancing figure in dance house' (Bernardi 1905). University Museum records indicate that two dolls were purchased for $1.00, a shaman figure was bought for $2.00, and a carved ivory figure brought $5.50, on February 1, 1906.

Inside the *kasgi*, Bernardi observed a foot-long whale, which was outfitted to spout blades of dry grass from its blowhole while 'sea parrots' (puffins) and tiny hunters moved to the music with their human counterparts. This part of the ceremonies may have been recreational, for she saw one of her 14-year-old students, neither shaman nor elder, manipulating the puppet's strings (Bernardi 1905). The dance, like many festivals, had elements of play including mime and masquerade, but masks were not worn. Participants sang, keeping their eyes downcast on a hole in the floor where each dancer stood. Men danced with their heads hung low, backs to the crowd, while viewers tried to guess who they might be. A woman's long hair would be loosened, then her hair would be thrown over her face at first. After a few motions, husband or brother would aid a woman by arranging the tumbling hair inside her parka hood, this without hampering her rhythm. Then, she could be identified by the crowd (Fig. 11).

Figure 11. *Bernardi was invited into the kasgi in 1902 by Nowadluk, at left, who worked for the Lopps. At sixteen, these young women are posed in their finest spotted reindeer skin parkas. Such romantic images were distributed widely in the north and could have been purchased for the* Story of a Whale Hunt *albums. Photo, circa 1902, courtesy University of Washington Libraries, Special Collections, NA 3209.*

In 1892, Tom Lopp had attended a different type of masked dance in the *kasgi* by invitation of Chief Kokituk, a village leader. There, he recorded a 'roof or ceiling decorated with wooden images— whales, walrus, birds, seals and boat' hanging near a suspended whaling spear. During the festivities, a shaman drummed, bringing forth visions of whaling, while a masked whaling crew of eight men danced. One by one, local shamans, hunters, and herders performed. One man, Oomaligzruk, licked the point of a spear and called on 'stickpaths' from a certain area to assist him. Another, Pootoogook, whom Lopp says was speaking 'an ancient [Siberian] dialect which only a few can understand,' was finally chosen for the ceremony. Two men, Kokituk and Sooksruk, were then selected to spear him. Lopp's notes say:

> General discussion, young men scared, dummy kept on appearing [from central pit]. Called on the Drumapaths and Stickapaths to help them out. Refused to spear him. After much calling, he [Pootoogook] came up and gave their instructions again. (Hands were tied all the time.) Disappeared. In short time dummy appeared at hole again. On fourth appearance Sooksruk speared it. Spear quivered and was drawn down through hole. Thong and bow was attached. Other end thong made fast to timber. Doctors hard at work. Everybody sings for half hour. Draw him up (Smith 1997:133).

When the 'Siberian' Pootoogook surfaced, his back was bleeding from where the spear had apparently entered; blood oozed from his mouth. An ivory spearhead was imbedded in his chest, which fell out as he reeled in and out of the central pit. The masked dances were repeated and the man washed off his chest, which showed no damage. The dramatic performance was over.

CONCLUSION

Formal education introduced the written word to Iñupiaq and other Alaska Native cultures that were, and remain, eloquent in their rich oral knowledge of the past. Along with attempts at a Westernized school year and book learning, Sheldon Jackson's introduction of reindeer to Northwest Alaska and other areas was an insidious plan to convert the local hunters to staid herders, a strategy that was ultimately unsuccessful. In Wales, the Lopps introduced a vibrant tradition of ethnographic arts on paper (later, local artists used skins) and trained students who would become the first Iñupiaq teachers. In 1918, during the global influenza epidemic, Wales lost most of its population.

Suzanne Rognon Bernardi was among this influential group of early teachers in Wales. It is clear that Bernardi recorded Iñupiaq traditions for her own professional advancement, yet we are fortunate that she created numerous copies of *Story of a Whale Hunt*, leaving one of the earliest photographic records of Iñupiaq whaling—one taken from a relatively intimate viewpoint. An adventuresome and perhaps brash young woman, Bernardi wrote about and photographed ethnographic details that would have offended the sensibilities of many Victorian women. Her journals apparently have been lost, her published writing is sparse, and her photographic legacy is small. Yet, Bernardi's photographs and commentary provide an important record of ritual and ceremonial life in a community that was once one of the largest and most powerful centers of Iñupiaq whaling. Hopefully, these materials will be further examined by contemporary Wales' residents in order to interpret and perhaps to re-create some of the traditional whaling activities of their ancestors.

Acknowledgments. This research was supported primarily by a grant from the Alaska Humanities Forum (37-98). Prior research in the region was sponsored by the National Science Foundation (OPP 9631004) and the National Park Service (CA 9910-6-9035). My thanks go to Dianne Brenner, Michael Burwell, Richard Engeman, Nelson H. H. Graburn, Myna Jacobs, Kris Kinsey, Gladi Kulp, Sara Larus-Tolley, Steve Lindbeck, Allen P. McCartney, Barry McWayne, Lincoln and Emily Milligrock, Percy Nayokpuk, Davey and Frieda Ningeulook, Alessandro Pezzati, Laura Samuelson, Jeanne Schaaf, Kathleen Lopp Smith and her husband Verbeck Smith, Rose Speranza, Robert Blair St. George, and Christine Tobar-Dupres. Herbert O. Anungazuk's comments were especially valuable, as is the memory of Edgar Nunageak Ningeulook. Time to write was granted graciously by Joseph Wilder, Director of The Southwest Center, University of Arizona.

REFERENCES

Anungazuk, Herbert
    1999   Personal communication to the author, March 9, 1999.
    2002   Personal communication to the author, June 26, 2002.
Bernardi, Suzanne Rognon
    1901-1902. *Story of a Whale Hunt.* Six photographic albums with ethnographic notes. Locations noted in text.
    1905-1907. Personal correspondence to George Byron Gordon, Director, University Museum. Philadelphia, PA: University of Pennsylvania Museum Archives.
    1912   Whaling with Eskimos of Cape Prince of Wales: Woman Teacher of Local Nativity Writes of the Sport of the Sea and Attendant Weird Customs Among the Aborigines of Alaska. *The Courier-Journal*, October 20, 1912, 1, 12. Louisville, KY.
    1981   'Selections from *Story of a Whale Hunt*,' pp. 134 in: *Alaska Journal*, Vol. 11, August 1981.
Bockstoce, John R.
    1977   *Steam Whaling in the Western Arctic.* New Bedford, MA: New Bedford Whaling Museum and Old Dartmouth Historical Society.
    1986   *Whales, Ice, and Men. The History of Whaling in the Western Arctic.* Seattle: University of Washington Press.

Bogoras, Waldemar.
    1975 [1904-1909] The Chukchee. (The Jesup North Pacific Expedition,
        Franz Boas, ed.). *Memoirs of the American Museum of Natural
        History VII.* New York. (Reprint: AMS Press, New York)
Fair, Susan W.
    1995   'Alaska Native Graphic Arts,' pp. 98-107 in Valerie
        Chaussonnet, ed., *Crossroads Alaska: Native Cultures of
        Alaska and Siberia.* Washington, DC: Arctic Studies Center,
        National Museum of Natural History, Smithsonian Institution.
Fairbanks Daily News-Miner.
    1915   Marriage notice, Suzanne R. Bernardi. May 10, 1915, p. 1.
Gordon, George Byron
    1905-1908 Personal correspondence to Suzanne R. Bernardi.
        Philadelphia: University of Pennsylvania Museum Archives.
Hegarty, Reginald B.
    1959   *Returns of Whaling Vessels Sailing from American
        Ports, 1876-1928.* New Bedford, MA: Old Dartmouth
        Historical Society and Whaling Museum.
Kaplan, Susan A. and Kristin J. Barsness, eds.
    1986   *Raven's Journey: The World of Alaska's Native People.*
        Philadelphia: University Museum.
Kakaruk, John and William Oquilluk.
    1964   The Eagle Wolf Dance. Self-published, © 1964, by Charles V.
        Lucier and William Oquilluk.
King, Eleanor M. and Bryce P. Little
    1986   'George Byron Gordon and the Early Development of
        the University Museum,' pp. 16-53 in Susan A. Kaplan
        and Kristin J. Barsness, eds., *Raven's Journey: The
        World of Alaska's Native People.* Philadelphia:
        University Museum.
Krauss, Michael E.
    1980   Alaska Native Languages: Past, Present, and Future.
        Alaska Native Language Center Research Paper No. 4.,
        University of Alaska Fairbanks.
Lopp, William Thomas, ed.
    1902   Drifted Out to Sea: Na-tax-ite Ninety-six Hours Without Food.
        In: *The Eskimo Bulletin*, 5 (May 1902), 1, 3. Cape Prince of
        Wales, AK.
McKinney, Virginia, ed.
    1981   *Story of a Whale Hunt*, pp. 134-143 in: *The Alaska Journal: A
        1981 Collection.* Anchorage, AK: Alaska Northwest Publishing
        Company.
Milligrock, Lincoln.
    1999   Personal communication to the author. January 30, 1999.
Ningeulook, Davey
    1997   Audiotaped interview, life history with Morris Kiyutelluk and
        Amy Craver. Shishmaref, AK, April 11, 1997. Draft
        transcription by Fred Tocktoo, October 1997. Shishmaref, AK:
        Shishmaref IRA Village Council and Anchorage, AK: National
        Park Service, Cultural Resources Division.
Nome (AK) *Nugget*
    1901a   *Wrecked, Robbed.* October 23, 1901, p.3

1901b   *Missionary Lopp is All Right.* October 30, 1901, p. 3.
1901c   *Teller a Dead One.* September 27, 1901, p. 3.
Phebus, George, Jr.
1995 [1972]. *Alaskan Eskimo Life in the 1890s as Sketched by Native Artists.* (Reprint) Fairbanks, AK: University of Alaska Press.
Qulaut, George Quviq
1998   Keynote Address, p. 20 in J. C. H. King and Henrietta Lidchi, eds., *Imaging the Arctic.* Seattle: University of Washington Press and London: British Museum Press.
Ray, Dorothy Jean
1975   *The Eskimo of Bering Strait, 1650-1898.* Seattle: University of Washington Press.
Simon, Jim and Craig Gerlach
n.d.   *Reindeer Herding, Subsistence, and Alaskan Native Land Use in the Bering Land Bridge National Preserve, Northern Seward Peninsula, Alaska.* Unpublished report prepared for the National Park Service, Alaska Regional Office, Shared Beringian Heritage Program (CA 9700-1-9005).
Smith, Kathleen Lopp, ed.
1995   Suzanne R. Bernardi. Biographical notes for Special Collections, University of Washington Library, Seattle, WA. Compiled from letters of Ellen Kittredge Lopp and Frances Kittredge, Wales, Alaska, 1901-1902.
1997   Reading, Writing, Arithmetic, and Reindeer: A Woman's View of Northwest Alaska, 1892-1902. Unpublished manuscript.
Smith, Kathleen Lopp and Verbeck Smith
2001   *Ice Window: Letters from A Bering Strait Village: 1892-1902.* Fairbanks, AK: University of Alaska Press.
Thornton, Harrison R. and William Thomas Lopp, eds.
1893   *The Eskimo Bulletin* 1. Cape Prince of Wales, AK.
Varnum, W.S.
1902   Log of the Steam Bark *Alexander*, April 3, 1902-November 1, 1902: Old Dartmouth Historical Society-New Bedford Whaling Museum Collection. (141A) New Bedford, MA.
Wickersham, James
n.d(a)   Unpublished drawings, 'Section of the Dance House' Wickersham Collection, Box 54, Folder 3. Alaska State Historical Library, Juneau, AK.
n.d(b)   Kingegan: A Study of the Eskimo. Wickersham Collection, MS 107, Box 54, Folder 7. Alaska State Historical Library, Juneau, AK.

Eskimo Laborers: John Kelly's Commercial Shore Whaling Station, Point Belcher, Alaska, 1891-1892

Mark S. Cassell

Abstract. *Iñupiat Eskimos of North Alaska have been subsistence hunters of bowhead whales for nearly 1000 years. The arrival of the Euroamerican pelagic commercial whaling industry to North Alaskan waters in the 1850s bolstered existing social relations of this traditional whaling society. But in 1888, American whaler and trader Charles Brower hired the Eskimo Poka to help staff his commercial shore whaling operations. This event saw the advent of Eskimo labor in capitalist industry in North Alaska, and within five years the use of Eskimo labor in commercial whaling was omnipresent in the region. The development of an Iñupiat Eskimo labor force in the late 19th and early 20th century both changed and maintained traditional Eskimo whaling. Data gleaned from documentary and archaeological research related to John Kelly's 1891-1892 Point Belcher shore whaling station, a locus of Eskimo labor in commercial shore whaling, help elucidate the process and nature of Eskimo labor development and the indigenous means by which Eskimos used the opportunities created by employment to their own traditional ends.*

INTRODUCTION

Think of the past, dear reader. It was the late winter and early spring on the coast of the Chukchi Sea next to the ancient village of Nunagiak at Point Belcher in North Alaska. About 80 Iñupiat Eskimos were preparing for the coming bowhead whale hunt, and a respected leader was in charge of the operation. This was an age-old practice; since about AD 1000, villagers in coastal North Alaska spent this season readying for spring subsistence whaling under the leadership of Iñupiat *umialiit* (sing. *umialik*, loosely translated as 'whaling captain,' 'boat captain,' 'boss,' 'rich man;' see McLean 1980:70).

But it was not subsistence whaling for which these Eskimos were preparing. The year was 1892, and these whalers, like many Iñupiat in North Alaska, were laborers in the American Western Arctic commercial shore whaling industry. They were employees of American whaler/trader John Kelly, and worked at the shore whaling station that Kelly established at Point Belcher the previous autumn (Fig. 1). They were hunting bowhead whales (*Balaena mysticetus*) for the extraction of baleen, a whale product and an industrial raw material that was profitable on the open market in American financial and manufacturing centers.

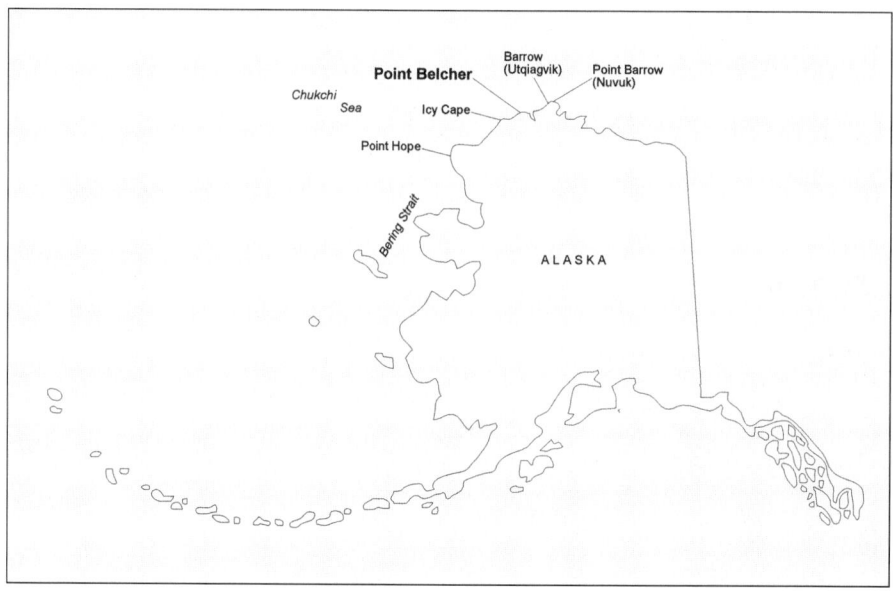

Figure 1. *Location of Point Belcher and other places mentioned in the text.*

The site of Kelly's station represents the result of activities there between September, 1891 and July, 1892. Kelly's whaling station was by no means the only shore station in North Alaska during the Western Arctic fishery; there were dozens. Indeed, Kelly's station was comparatively quite small, and a complete failure in terms of spring whaling success. It is nonetheless archaeologically singular for the people who worked there and what they did, and for its pre– and post–depositional history. Kelly's station may, indeed, be the only site of its kind remaining on the Arctic Alaskan coasts. The story told by the archaeological and related documentary material bears significantly upon the history of the Iñupiat in relation to the commercial whaling industry of the late 19th and early 20th centuries.

In this article, I hope to provide a historical understanding of Iñupiat Eskimo labor in commercial whaling, labor that became virtually universal around the time that Kelly's station was established at Point Belcher. This will provide context for understanding Kelly's station in particular and the material lives of the Iñupiat laborers around the beginning of the 20th century in general.

THE DEVELOPMENT OF IÑUPIAT LABOR IN THE WESTERN ARCTIC COMMERCIAL WHALE FISHERY

Commercial whalers entered Western Arctic waters through Bering Strait in 1848, reaching Point Barrow in 1854 (Bockstoce 1986). While the commercial focus was whale oil in the early decades of the fishery, fashion trends and post-Civil War fossil fuels development led to an emphasis on baleen production. This flexible and fibrous plankton-straining material hanging from the jaws of bowhead whales was commercially profitable as industrial raw material for making corset and umbrella stays and buggy whips (Stevenson 1907; Bockstoce 1986). Prior to the late 1880s, commercial whaling in the Western Arctic was confined to the open sea in ships with primarily Euroamerican crews. Extant *umialiit* traded in baleen with commercial whalers for manufactured goods.[1] At the time, commercial whalers had little use for Iñupiat labor.

In the 1880s, the whaling industry tried to offset reductions in the baleen supply caused by depletions of overhunted bowhead whale populations.[2] Whalers instituted the practice of spring whaling from shore bases to complement summer open sea harvest. Shore whaling was very similar to the Iñupiat method, using small boats to travel the narrow ice leads through which whales pass on their northward migration. Commercial shore whaling was conducted from stations established primarily in extant Iñupiat villages, which were settled initially because of their proximity to ice conditions favorable to spring whaling. The Iñupiat residents, knowledgeable about whaling, offered a prime source for labor to staff whaling stations and operate whaleboats.

---

[1] Among subsistence whalers in North Alaska, those individuals controlling baleen would stand to benefit most from trade with commercial whalers. Sonnenfeld (1957) saw baleen as being fairly equally distributed among crew members before the commercial whaling industry, but that *umialiit*, as leaders of the hunt, began to appropriate larger shares of baleen as commercial visits became more frequent (see also VanStone 1962). Sheehan (1992:188-189, 191-192) noted that archaeological research at Utqiagvik in Barrow revealed marked differences in baleen in both spatial and temporal contexts. While baleen was present in house middens dating prior to commercial whaling, the quantities paled in comparison with areas associated with a contemporaneous *qargi*, which would have been maintained by an *umialik*. Further, strata from house and *qargi* deposits dating to the commercial whaling period were devoid of baleen, indicating that while many Iñupiat likely had access to some baleen for exchange early on, *umialiit* had access to and used much more baleen for their trade, and continued to retain that access. Murdoch (1892:54), writing of the preshore whaling days of the 1881–1883 International Polar Expedition to Point Barrow, observed that 'Of course, men who are rich in whalebone now stay to trade with the ships.' The 'men who are rich in whalebone' would be extant *umialiit*, developing new exchange relations and new prime trading partnerships with commercial whalers prior to large-scale Iñupiat labor development.

[2] Bockstoce (1980:26) conservatively estimated a total bowhead mortality of between 16,000 and 17,000 whales between 1848 and 1890 in the Western Arctic. Over half that number had been killed by 1860.

An understanding of the development of Iñupiat labor considers as its source the nature of Iñupiat society prior to the commercial whaling period (see Cassell 1988b). *Umialiit* had a considerable degree of power in the community. Such persons of power were few, and those without were many. Even in the larger villages, not many individuals were considered *umialiit*. Simpson (1875:237) reported that in 1853 the combined populations of Nuvuk (at Point Barrow) and Utqiagvik (present-day Barrow) amounted to about 536. Both villages together had five *qariyit*, structures associated with whaling activities and each linked to a particular *umialik*. Thus, five individuals held sway over the lives of 500. There is some uncertainty whether this number of *umialiit* represents the number of extended families (e.g., Burch 1975) or a number appropriate to the care and feeding of the village (e.g., Spencer 1959).

What is certain is that *umialik* status was difficult to achieve, and few did achieve it at any given time. Access to knowledge of whaling technique and to the best whale songs and shamans was likely restricted to male sons.[3] Access to materials proffered as recruitment incentives to good harpooners was equally restricted to those with large networks of kin who donated the materials. Part of the material access controlled by *umialiit* centered around the skins needed for *umiak* (skin whaleboat) building which were procured from and sewed by kin. These boats were material underpinnings of whaling, and in controlling access to such primary means of production *umialiit* also controlled access to *umialik* status. Access to some material items was obtained through prime trading partners with which only *umialiit* dealt. Wealth begat wealth and knowledge begat knowledge; the best *umialiit* trained their sons to be good *umialiit*, retained the best whale songs, and continued to build stronger and stronger kin and trading networks. Barring catastrophes in whale population or migration patterns, warfare that killed off whole families, or severely aberrant *umialik* personal

---

[3] The decline of shamanism and subsequent rise of Christianity as a religious focus temporally parallels the fall of the prior *umialiit* and the ascendancy of commercial whalers as controllers of labor. This may relate to power and social control shared by shamans and *umialiit*; the power of one may have been based to some extent upon the power of the other (e.g., Spencer 1977). It has been argued that the arrival of Christian missions did not cause the collapse of shamanism so much as hasten it, that 'shamanism was already an ebbing force when Christianity arrived' (Klausner and Foulks 1982:70; see also Spencer 1959:298). This suggests the possibility of social dissatisfaction with or stress derived from the strong political power held by both shamans and *umialiit*. While assessment of factors in the decline of shamanism in North Alaska is beyond the scope of this paper, the parallels between the historical trajectories of shaman and *umialik* power are intriguing.

behavior, *umialik* status was in large measure restricted to immediate kin.[4] This system tended to perpetuate a power structure in which the few who had power continued to hold power over generations, and those who did not have power never could.

Umialiit provided food and material goods to crew members and their families, and, as an ideal gesture of generosity, to orphans, the elderly, and others who had no kin on whaling crews (e.g., Spencer 1959). In a general sense, however, an individual's material well-being was dependent upon productive alliance with good *umialiit* through participation on a whaling crew. Yet, even participation in a whaling crew was restricted, albeit in a more subtle way. Certain able-bodied men were barred from crew membership for a season based upon social prohibitions against participation when, for example, one's wife had recently died. While an individual and his family were likely provided for in terms of foodstuffs by *umialiit* in such an event, the material benefits obtained from membership on a crew were absent. Consequently, while individuals would find it extremely difficult to materially aspire to *umialik* status, neither could they materially gain at all under certain conditions.

The extent of possibilities for those socially prohibited from whaling are uncertain. Commercial whalers were willing to hire anyone they could; as will be seen below, they practiced no such sanctions against whaling participation. Whatever the case, the nature of power in Iñupiat society suggests why some Iñupiat may have so readily come under the banners of commercial whalers.

The way in which whaler/traders attempted to attract and maintain labor mirrored that of *umialiit*. The commercial operator had to have food resources to permit Iñupiat to forego subsistence hunting during whaling. But while the *umialik* depended upon kin and the previous year's whale harvest to supply food to his crew and their families, the whaler/trader relied on the supply ship, which came every year with provisions from San Francisco. These annual shipments enabled the whaler/trader to provide a relatively steadier and more reliable supply of foodstuffs and other material goods in exchange for labor.[5] This heightened ability to feed and supply crews

---

[4] Such aberrant behavioral situations did occur in the 19th century (and possibly earlier), as Burch (1981) notes for the Point Hope area. One Point Hope *umialik* active during the early days of commercial shore whaling was notorious for his abuses of power, and his behavior was likely a factor in the migrations of Point Hope people to the Barrow area in the late 1880s. He was assassinated by a Point Hope man in 1889.

[5] While the timely arrival in the Arctic of food and supplies from 'outside' occurred more often than not, it was never guaranteed due to hazards from ice and weather conditions.

and their families permitted commercial entry into traditional leadership realms.

Euroamerican foodstuffs entered the Iñupiat world as trade items with the first arrivals of the pelagic (open sea) whalers in the 1850s. As items of trade, these foodstuffs had no real impact on Iñupiat society. The ethnographer John Murdoch (1892:54) of the 1881–1883 International Polar Expedition to Point Barrow wrote that during the early 1880s the Iñupiat:

> have contracted a taste for civilized food, especially hard bread and flour, but this they are unable to obtain for 10 months of the year, and they are thus obliged to adhere to their former habits... They are not absolutely dependent upon the ships for anything except ammunition, and even during the short time the ships are with them they hardly neglect their own pursuits.

There were, however, some Euroamericans who did perceive a desire for trade foods by the Iñupiat, stemming from the inability of *umialiit* to feed the community as a consequence of overhunting by commercial whaling. Lt. P.H. Ray of the International Polar Expedition of 1881-1883 noted with some concern the numerous recent deaths around what is now Barrow:

> ... which bore silent testimony to the fact that famine and disease had quite recently been at work. This is undoubtedly owing to the fact that the food supply is rapidly growing less, and that the great number of whales taken off the coast by the American whaling fleet during the last twenty years has nearly exterminated that valuable animal. That they are decreasing in numbers is well known among the whalemen, and the fact that... there were twenty-four whales taken by the natives [in 1852-1854], while only two were taken during our stay [in 1881-1883], one of them a calf, goes to prove that they will soon be classed among the extinct mammals, and with them will soon pass away many of the people inhabiting this shore... (Ray 1885, in Murdoch 1988: xcix).

Nonetheless, Charles Brower remarked that in 1884 'There was no white man's food that the Eskimos cared for, and it was not until three years later that we could sell them any provisions of any kind' (Brower 1932a:16). The year following those 'three years later' spoken of by Brower saw the initial employment of Iñupiat in commercial shore whaling, and the beginning of the whaler/trader's efforts to undermine the control over labor held by the *umialik*.

Figure 2. *Charles Brower. (By permission of the Brower family via the University of Alaska Press)*

Charles Brower (Fig. 2) started shore whaling in Barrow in 1887 for the San Francisco-based Pacific Steam Whaling Company (PSWC). In the spring of 1887, Brower's Euroamerican crew caught nothing, while village crews took 22 whales.[6] In 1888 he began to organize crews using almost exclusively Iñupiat labor (Bockstoce 1986:233-234). Although the village *umialiit* still maintained sufficient resources to control available labor, Brower was able to recruit some Iñupiat who had been temporarily restricted from village crews. He wrote that:

> In the village I found a native, Poka, who was banned from the native boats that spring, as his wife had died just before whaling, so he was forbidden to work in their boats. Poka was willing to go... and I could get some more crew to go with us (Brower 1932b:87).

If Lt. Ray's description of rapid depopulation by death in the Barrow area is accurate, it is likely that many people were in Poka's situation. The steadfastness of such prohibitions against whaling is uncertain in hard times where many males and females may have perished; such conditions would certainly strain the pool of potential crew members. Disenfranchised under the prior *umialik* power system due to prohibitions against whaling, and therefore unable to gain (or even maintain) materially, persons such as Poka sought and found recourse by switching alliance to other *umialiit*: Euroamericans commercial whalers such as Charles Brower and John Kelly.

Brower left PSWC after the 1888 season and began his own operations; PSWC still maintained a station. While with PSWC, Brower 'had discovered the simplicity of using native crews' (Bockstoce 1986:234). In 1889, he hired two families from Point Hope (Bockstoce 1980:236). In addition, Brower:

> hired two more Eskimos from the village; they were good enough men, and were glad to work for us, but they were taboo in native boats... . If we had not hired them, they could not have worked until the next spring (Brower 1932b: 95).

---

6 After that unsuccessful 1887 season, Euroamericans were never seriously considered for staffing shore stations. Apart from the high skill levels and low maintenance costs associated with the hiring of Iñupiat Eskimos relative to Euroamericans, employing Euroamericans as station labor may have been deemed unadvisable by whaling companies due to the tense labor/management relations extant in the United States around the beginning of the 20th century (e.g., Montgomery 1979, 1989). Labor/management relations in arctic pelagic commercial whaling were as poor as in any industry (e.g., Stone 1983; Williams 1988). Prolonged landfalls such as occurred with the Beaufort Sea overwinterings often led to runaway and other mutinous situations on the part of ships crews (e.g., Bockstoce 1986).

The 1889 season was successful, and in 1890 more commercial, shore-whaling operations were begun at Barrow. That spring, villagers manned 40 boats while commercial operators fitted out ten. Although the whaling season was poor, the combined commercial profits of the baleen and fur trade were substantial (Bockstoce 1986:236).[7]

Brower's increasing influence and control over labor allowed him to mobilize workers for various non-whaling activities, which would maintain and guarantee the subsistence of his employees and their families. One instance occurred in the spring of 1898. That year:

> no whales had been reported, but there was lots of seal, and we wanted to get all we could in case the other meat ran short ... . [S]eals were more numerous than I had ever seen them. My boats killed them by the hundreds. We had the ice-houses full, and then stacked [them] on the ice along the beach (Brower 1932d: 84).

That same year a herd of beluga whales was trapped in a pool of open water. According to Brower (1932d: 84), 'There must have been six hundred killed and saved. I had half as many hauled to the station.' Brower continued to demonstrate his ability to provide food for the community.

Whaler/traders sometimes had occasion to open up their stores of food to the community, and at these times the commercial operators reinforced Native perceptions that Euroamericans behaved like *umialiit*. When an epidemic hit Barrow in 1902, Brower's station 'had to feed the natives during the sickness; if we had not been there, every one would have died from starvation' (Brower 1934:152). Such gestures of generosity were in line with the responsibilities and requirements of *umialik* status.[8]

The commercial shore-whaling season at Point Barrow ran each spring for about six weeks. In order for their Iñupiat crews to survive without hunting during those six weeks of whaling, the commercial operators were required to provide sustenance and other materials as

---

[7] On the fur trade in North Alaska, see Bockstoce (1986) and Libbey and Schneider (1987).

[8] Paternalist tendencies of the industrialist towards employees have often been documented for the 19th century American factory system. Prude (1983:5, 6) has noted that some factory and mill owners used a 'rhetoric of paternalism' to 'justify intrusions into the operatives' family life' and 'could simultaneously assert concern for their workers' well-being and strive to inculcate values encouraging productivity.'

recompense for their labor. As ethnographer and explorer Vilhjalmur Stefansson noted, it was:

> six weeks of fairly easy work at that. For all the rest of the year the men have nothing to do—are their own masters, and can go wherever they like, while their employers must not only pay them a year's wages for six weeks' work, but also furnish them houses to live in ... and rations for the entire year... . The employer supplies them with cloth for garments, and such suitable provisions as flour, tea, beans, rice, and even condensed milk, canned meats and fruit (Stefansson 1922:60).

Figure 3. *Eskimo workers at an unidentified shore whaling station in North Alaska; note Euroamerican in campaign hat, second from left. (By permission of the California Academy of Sciences).*

The commercial whalers controlled trade in such things as ammunition and cloth, and also in productive resources for whaling such as bomb guns, powder, and even whaleboats. While the *umialik* was keeper of the whaleboat and much of the whaling gear in traditional times, the commercial operator held these resources by the late 19th century. Brower recalled that:

> the younger set adopted our whaling-gear, tackles, guns, bombs, and all, even wanting hard bread and tea out on the ice. Tents they would not use ... until after we begun to hire them to whale for the station; then they wanted everything the same as we used (Brower 1932b: 89).

Brower's contemporary, John Kelly, outfitted his whaling crews in Barrow which 'were each supplied with a tent, a stove, and all kinds of food' by 1891 (Brower 1932c:64). By 1894, Brower's crews 'had tents, sleeping bags, stoves, and food of all kinds. The Eskimos were educated now; if they did not have every comfort they would not work' (Brower 1932c:68).

Commercial whalers offered an enormous variety of material goods to crew members and their families, ranging from cloth and needles to rifles and phonographs. Stefansson (1922:60,61) observed that:

> ... these [crew] men get each year as wages about two hundred dollars' worth of supplies. This means that the Point Barrow community leads an easier life than any other community does as a whole in any land I have ever traveled... . The pay-day of the Point Barrow Eskimo comes in the spring, and their employer hands them out rifles, ammunition, cloth, provisions, and various things which the people scarcely know what to do with. So they load them into their skin boats and take them east along the coast [of the Beaufort Sea], to sell them at any point in the Colville or at Flaxman Island.
>
> Whaler and trader Capt. Hartson Bodfish provided a catalog of trade goods carried on his vessel in 1902; and while the goods listed were intended for John Howland Bay in Siberia, he wrote that, 'The lists of trade goods used in the various places do not vary greatly from the one listed' (Bodfish 1936:196).
>
> His vessel carried for trade: 49 rifles, 5 shotguns, 39,000 cartridges, reloading tools, powder, lead, shot, thousands of yards of ticking, drill, denim, calico, flannelette, foot sewing machines, hand sewing machines, needles for same, thread, thimbles, chewing gum, combs, canvas, twine, tobacco, matches, flour, bread, molasses, sugar, tea, baking powder, dried apples, prunes, rice, 3 phonographs, 110 records, phonograph needles, clocks, oak board, boat anchor, brass kettles, primus stoves, dish pans, milk pans, enamel pails, table spoons, serge cloth, coffee pots, canned milk, shovels, tacks, mirrors, scissors, darting-irons, cutting-spades, knives, harmonicas, files, drills, bits, breast drills, planes, hammers, hachets, saws, axes, awls, coat oil, spy glasses, opera glasses, darting bombs, shoulder bombs, boat compasses, boat boards, screw drivers, cigars, beads, caps, suspenders, boys' clothes, leather belts, lady's coat, one 16-foot oar, old sails (these from the ship), whaleboat and gear, paint, paint brushes, playing cards, paper, brooms, and one house, 30 x 20 feet, cut and fitted... and it was only a small part of the entire stock (Bodfish 1936:195).[9]

---

[9] 'The entire cost of this list of trade goods is set down at $6,030.86 ... . In return we received 3553 lbs. whalebone [baleen], 39 white fox skins, 6 deer, 1 seal coat, 1 pair of seal pants' (Bodfish 1936:195, 196). With the 1902 price of baleen at an average of $4.02 per pound, it alone was valued at $14,992.60, approaching a 250% profit on trade goods. Bodfish noted that 'Arctic whalers were trading ships as well as whalers, and it was quite on the cards that a good profit might be made in trade even if very few whales were taken' (Bodfish 1936:191).

Table 1: Iñupiat *Umialiit* and Commercial Whaler Crews and Productivity, 1887-1894. (Data from Marquette and Bockstoce 1980, Bockstoce 1986, and Brower, n.d.).

YEAR	IÑUPIAT *UMIALIIT*			COMMERCIAL WHALERS		
	Number of Crews	Number of Whales	Number of Whales per Crew	Number of Crews	Number of Whales	Number of Whales per Crew
1887[1]	5?	22	4.4	1?	0	0.0
1888[1]	4?	3	0.8	1?	1	1.0
1889[1]	9?	22	2.5	3?	6	2.0
1890[2]	40	3	0.1	10	2	0.2
1891[1]	6?	1	0.2	14?	17	1.2
1892[2,3]	40+	9	0.2	8	5	0.6
1893[4]	?	?	?	?	?	?
1894[5]	?	1	?	28	42	1.5

[1] *Number of crews estimated based upon Marquette and Bockstoce (1980).*
[2] *Number of crews differs between Marquette and Bockstoce (1980) and Bockstoce (1986).*
[3] *Number of whales based upon Brower (n.d.).*
[4] *Marquette and Bockstoce (1980) note a total of 8-10 crews mustered and 11 whales taken, but they do not differentiate between umialiit and commercial operators.*
[5] *Bockstoce (1986) notes 28 commercial crews, while Marquette and Bockstoce (1980) note a total of 28 crews at Barrow without differentiating between umialiit and commercial. Brower (n.d.) recorded native boats as taking one whale that year, so at least one umialiit crew was mustered.*

Brower's 1888 hiring of Poka in Barrow initiated the commoditization of Iñupiat labor and the creation of an Iñupiat industrial labor force in North Alaska. Use of Iñupiat labor spread rapidly at shore whaling stations established in Native villages between Bering Strait and Point Barrow until the collapse of the baleen market around 1914.[10] Within five years of Poka's initial

---

[10] The whale fishery effectively ended with the final printing of the *Whalemen's Shipping List*, the industry trade paper, on 29 December 1914. It read: 'Whalebone: We are unable to quote any sales. At the beginning of the year it was reported that a small quantity of Arctic had been sold for export... There does not seem to be any demand for the large stock on hand...' (cited in Bockstoce 1986:337).

employment, virtually every able-bodied Iñupiat in North Alaska and adjacent environs worked as wage laborers in commercial whaling, receiving pay in the form of foodstuffs and manufactured goods (Fig. 3).[11]

Bowhead whale populations had been decimated by overhunting during the first years of commercial whaling in the Western Arctic. By the last decade of the 19th century, poor shore-whaling seasons were frequent, and good catches were an exception. This was especially so for crews operated by Iñupiat *umialiit*, who fared quite poorly compared to their commercial counterparts in terms of hunting productivity per crew. Table 1 presents data from Barrow comparing numbers of crews mustered and whales taken by Iñupiat *umialiit* and commercial operators from 1887 to 1894. Where data are available and reasonably accurate, hunting productivity in whaling organized by Iñupiat averaged less than 0.6 whales/crew during this time, while whaling organized by commercial operators averaged more than 1.1 whales/crew, or roughly twice that of the former. In appearances alone, this presented Iñupiat with an attractive alternative to extant *umialiit*.

The higher productivity of commercial operators may be in part the result of relative access to or quality of bomb guns, powder, etc., equipment that may have proven more efficient when faced with diminished whale stocks. A greater influence on productivity was likely an increasing inability of village *umialiit* to retain crews who were well versed in whaling and in the need for intra-crew cooperation. Village crews were mustered, but the skill composition of such crews may have been lacking.

The 1889 success of Barrow village whaling crews likely contributed to the 1890 fitting out of 40 crews, with people desiring to cast their lot with successful Iñupiat *umialiit*. Another possible contributing factor to large-scale village crew muster in 1890 may have been a need for *umialiit* to alter their recruitment preferences in order to assert their organizational abilities when faced with increasingly successful Euroamerican competition. With approximately 320 Iñupiat whaling for village crews that year, *umialiit* "were apparently using all available local personnel including very young men and any Iñupiat who visited from other settlements" (Bockstoce 1986:236). This suggests that, like their commercial

---

[11] Eskimos worked for the commercial whaling industry at locales other than shore whaling stations. Between 1889 and 1908, whaling vessels over-wintering at Herschel Island and elsewhere in the Beaufort Sea hired hundreds of Eskimos from the Alaskan coast and interior and from the Mackenzie River Delta as hunters to procure fresh meat to stave off scurvy among sailors (Bockstoce 1986; Cassell 1988a).

counterparts, village *umialiit* began bringing on crew persons without regard to crew skills or kin affiliation. While pragmatic exigencies of crew organization in hard times prior to commercial whaling likely stimulated similar recruitment alternatives, its occurrence in the context of commercial competition seems to signal the erosion of *umialik* power and labor control.

Brower's control over the labor process by the mid-1890s is made clear when he wrote of the 1894 whaling season at Barrow:

> One of Kelley's boats did manage to strike [a whale], and the village boats saved it for him. This was the only whale the natives killed that spring; their ways were of the past, all their gear was of the poorest; and *if they did not work for the station, they were in bad luck* (Brower 1932c:69; emphasis mine).

The commercial whalers, however, 'had not done badly' that year (Brower 1932c: 69), landing 42 whales with as many as 28 crews (Marquette and Bockstoce 1980:12, with references).

With declining whale populations and an inability to attract skilled labor away from the commercial operators, *umialiit* could no longer maintain sufficient resources for the community. But commercial operators like Charles Brower could: to the Iñupiat Eskimo 'of Barrow and environs he was a white *umialik*, a rich man and a whaling captain, who revolutionized whaling and introduced a steady supply of the white man's goods into the community' (Blackman 1989:33). Links between commercial foodstuff provisioning and the ability to attract Iñupiat labor were indicated by one of Brower's contemporaries: 'The presence of [Brower's] station has probably saved the lives of many natives by furnishing food in exchange for work' (Aldrich 1889:110).

At the same time, wage labor led also to a burgeoning of the *umialik* institution in the production sphere by creating material conditions which were favorable to broadscale individual attainment of *umialik* status. Iñupiat wage laborers gained access to their own prime trading partners, the commercial whaler version of an *umialik*, and through the whalers they could gain material access to *umialik* status, the acquisition of which would have been very unlikely under previous social conditions. As noted above, Simpson recorded *qariyit* representing five *umialiit* at Nuvuk and Utqiagvik in 1853. Brower (n.d.) mentioned only two or three *umialiit* during his early years at present-day Barrow. In the early 1900s, however, the named prior *umialiit* in the Barrow area worked for Brower, and a number of new Iñupiat *umialiit* were noted. According to Stefansson (1922:60), 'Some of the Eskimo at Point Barrow now [ca. 1908] carry on whaling on a large scale, maintaining as many as five or six boat crews.' In 1902, one group of Iñupiat employees used their material wealth to purchase the whaling schooner *Penelope* (Bockstoce 1986:328).

The individuals embodying *umialik* status before the development of Iñupiat wage labor did retain an appearance of prestige through an intermediary role, by operating whaling crews for the whaler/trader (e.g., Brower n.d.) and obtaining material goods from commercial whalers which he could pass directly to crews. But access to such goods was largely dependent upon working for commercial whalers. While these *umialiit* controlled production in traditional times, they became a middlemen in exchange during the development of commercial shore whaling. David Riches (1982:142) has observed that:

> in contrast to the traditional period, Canadian Eskimo leadership in the contact period may not have been concerned primarily with productive prowess. Rather, ...leadership activities were concerned more with matters of Eskimo-trader relations.

A similar situation seems to have occurred in North Alaska to some extent, though it may have been limited to *umialiit* considered so prior to commercial shore whaling.

A HABITATION HISTORY OF THE VICINITY OF KELLY'S STATION

Virtually all commercial whaling stations were established in or near extant Iñupiat villages settled originally because of their proximity to good whaling areas. Villages such as present-day Barrow and Point Hope had been occupied for centuries and remain so today. Even if subsequent development and construction did not obliterate all traces of whaling stations in such villages, prior and subsequent material deposition would cloud the resolution of things, which were temporally and spatially associated with station operations.

The specific location of Kelly's station, however, had never been settled (Fig. 4). The adjacent and ancient village of Nunagiak had apparently been abandoned at least a couple decades prior to his arrival at Point Belcher. This situation permits a fine-grained temporal resolution of archaeological deposits at the station, which represent nearly a full year of industrial and domestic activities associated with the commercial shore whaling period. An archaeology of the place has the potential to provide important information on the material conditions of Iñupiat life as laborers in American industry during this crucial era of North Alaskan social history.

The immediate vicinity surrounding the site of Kelly's shore whaling station consists of a relict sand dune feature, sparsely covered by tundra vegetation and running parallel to the active beach of the Chukchi Sea coastline on one side and two elongated brackish freshwater lagoons on the other. The two lagoons, each about 3.2-4 km long, are separated by a narrow strip of tundra across which access is gained to the gently sloping low ridgetops a mile toward the interior. Behind the ridge lies the undulating tundra dotted with thaw lakes and areas of polygonal ground, so characteristic of the permafrost-influenced physical landscape in this part of North Alaska.

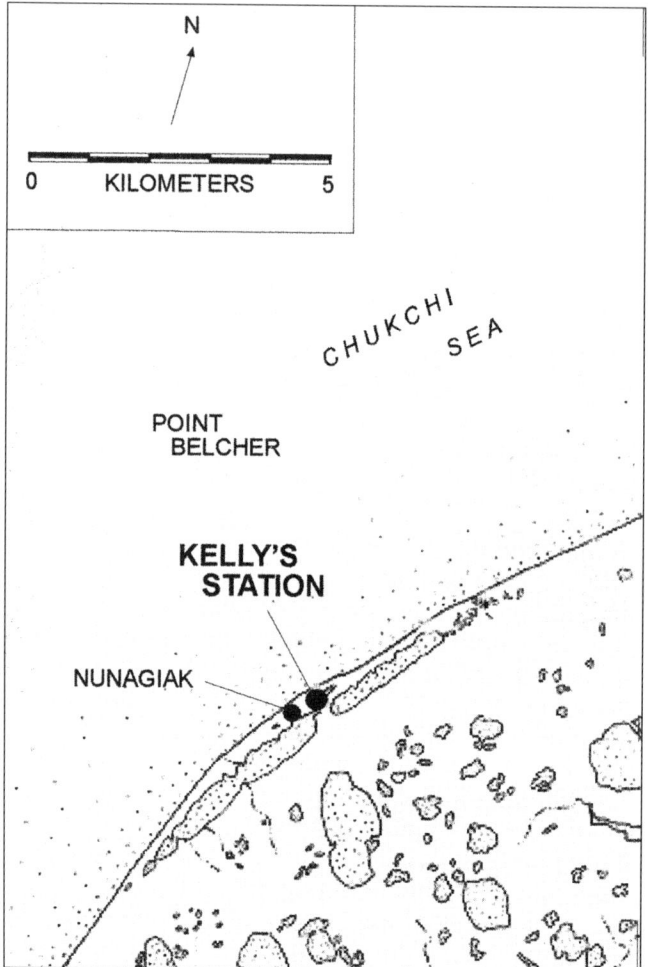

Figure 4. *Location of John Kelly's commercial shore whaling station and Nunagiak at Point Belcher.*

A couple hundred meters south of Kelly's station, near the northern terminus of the southern lagoon, rests the old site of Nunagiak. Nunagiak was likely settled due to its close proximity to near-shore ice-lead formation, which equates with a relative ease of travel to and from the paths of northward-migrating bowheads during the spring. Based upon the archaeological data collected by Ford (1959) in the 1930s, initial occupation of Nunagiak may date to about

AD 1000, a time which ushered in the refinements to technological and social adaptations facilitating large-scale bowhead whaling in North Alaska. In North Alaskan cultural history nomenclature, this initial occupation would be associated with the Birnirk period, although the artifacts are more indicative of the contemporaneous Punuk period of the Bering Strait region to the south.

The village of Nunagiak appears to have been abandoned in 1871. In September of that year, 34 commercial whaling ships were crushed in the pack ice off Point Belcher, and some of the vessels were scavenged by coastal-dwelling Iñupiat. Residents of Nunagiak were among the scavengers. Charles Brower heard from one native about the impact of the scavenging of the wrecked whaleships on Nunagiak:

> In ransacking the wrecked ships the Eskimos came across the medicine chest and as they had been used to buying alcohol from them, they imagined that anything in a bottle was whiskey. They drank everything that was liquid... all bottles were kept and anything in them that had a chance to make drunk come. In trying all that the medicine chests contained many were poisoned, so many dying in one village that it was abandoned, and never used again. This was the village of Nu-na-reah [Nunagiak]... (Brower n.d.: 142, 230).

There are no further records of people living at Nunagiak before the demise of commercial whaling. Apart from his descriptions of Kelly's station, Brower never noted any further habitation in the vicinity of Nunagiak in his travels along this coast during the late 19th or early 20th centuries.[12]

During his archaeological work there in the 1930s, Ford (1959:56) observed that some of the old mounds at Nunagiak were what he considered to be 'recent houses':

---

[12] The locale has no doubt served frequently through time as a temporary camping place, however. The very spot we chose for a campsite during our 1990 fieldwork at Kelly's station (about 250 m from both Nunagiak and the station) was known and described to us by many Wainwright people as the best place to camp in the 30 km or so between Wainwright on the south and Atanik on the north. The area certainly has been (and continues to be) used by Iñupiat for subsistence purposes, if only as a day trip from Wainwright. A driftwood duck blind and shotgun shells were evident in the narrow isthmus separating the two lagoons, which serve as a major flyway for waterfowl as they journey to the gathering place prior to southward mass migrations in the late summer. We used this blind with notable success. Caribou are common in the summer; a Wainwright man hunted there in 1990 while we were encamped. Bearded seals (*ugruk*) and walrus frequent the ice pack, which often drifts within 2 km or so of the shore, and smaller seals abound in the waters and ice floes nearby. Many boats of hunters cruised the shoreline in the summer of 1990.

> The outlines of some of the most recent houses are plainly visible...
> These structures were not abandoned until well after contact with
> the whalers, as is demonstrated by the quantities of barrel hoops,
> old buckets, and objects of metal found in them... The recent
> houses in the Nunagiak mounds average about 12 feet square. Few
> timbers were visible in a surface examination of the house pits, but
> there were enough to suggest that the buildings conformed to the
> pattern of the [traditional] Point Barrow house...[13]

He recorded seven such houses on his map of the Nunagiak site
(Ford 1959: Fig. 19). Of his excavation into one large mound, Ford
(1959:67) noted that 'the refuse at this point was obviously recent,
originating from the several houses built on the top of the mound.' He
only recorded excavation of a few items of what he considered to be
Euroamerican manufacture or materials, the sole objects mentioned
being an iron blade, a bullet, and a trinket box.

There is really just a sense of how recent Ford's 'recent houses'
may be. His identification of seven recent houses approximates the five
or six habitations recorded in the early 1850s. His observation of barrel
hoops, etc. within the houses certainly suggests occupation during the
commercial whaling period; the many shipwrecks off the Alaska north
of Bering Strait throughout the whaling period delivered much nautical
debris to the shore in this area.[14] His association of the recent houses
with the Point Barrow house type indicates human habitation
sometime before changes in house form beginning just prior to the
beginning of the 20th century.

My interest in these recent houses and the timing of their
occupation stems from the possibility, and indeed probability, that
some of Kelly's employees built upon and reoccupied the locations of
houses on the old mounds at Nunagiak. This is despite Brower's
remark that Nunagiak had not been reoccupied after its ca. 1871
abandonment and the absence of observations relating that anyone
living there during his time at Barrow.

---

[13] See Slaughter 1982 on the Point Barrow house type.

[14] While anchored off Point Belcher a few kilometers north of Nunagiak,
Brower (n.d.:230-231) noted that 'all up and down the shore as far as I could
see there was nothing but empty oil casks... . The casks... made ideal fuel to
help keep up steam in the [ship's] boilers. We took all there was in sight of us
that morning, and it was not long before all the ships were doing the same.
These casks that had lain on the shore from 1871 til 1886 were all used that
summer.' Numerous vessel timbers are present on the shoreline today,
reminders of those shipwrecks.

JOHN KELLY'S POINT BELCHER COMMERCIAL SHORE WHALING STATION

Ford's (1959:56) writings of Nunagiak noted that 'About 1/2 mile [*ca.* 1 km] to the northeast of this village, up the beach, are the foundations of what evidently was a building constructed by white men, probably a trading post established here before the site was abandoned.' I wrote to John Bockstoce in 1986, asking if he knew anything about the structure described by Ford. Bockstoce, the foremost historian of the Western Arctic commercial whaling industry and a student of Iñupiat ethnography, had explored the Chukchi and Beaufort Sea coastlines from the water with an eye towards locating places associated with events of the arctic fishery. He wrote back, telling me that the building was John Kelly's whaling station of 1891-1892.

Kelly's station rests near the juncture of the two aforementioned lagoons. Reference was made to the location of Kelly's station in the North Slope Borough's Traditional Land Use Inventory of the Wainwright area (Ivie and Schneider 1978) as Nunagiatchiak: 'Jerome & Kelly had the first houses here; white whaling began. Old whaling settlement.'[15] There is no 'Jerome' known from the time and place apart from this reference, and it is likely that the Iñupiat informant recalled the presence of 'John Kelly' as 'Jerome and Kelly' (David Libbey, pers. comm., 1989).[16]

John Kelly (Fig. 5) had been in North Alaska since 1884, mining low-grade coal at Corwin Bluffs north of Point Hope; Brower described him as a 'bluff, level-headed prospector' (cited in Bockstoce 1986:237). He was employed by the U.S. Census Bureau as a translator for the 1890 federal census in North Alaska, and published brief ethnographic notes and an Iñupiat–English dictionary under federal auspices (Wells and Kelly 1890).

---

[15] The 'Old whaling settlement' is Nunagiak, not Nunagiatchiak (Kelly's station).

[16] Approximately 5 km northeast of the station, however, is a location called Kugalukruak, also identified in the Traditional Land Use Inventory for the Wainwright area: 'Jerome (Jerome & Kelly) old sodhouse ruin.' Ivie and Schneider (1978:177) suggested that 'This may be the place where Jerome [i.e., Kelly] experimented with the Eskimo techniques of spring whaling.' Archaeological investigations at Kugalukruak in 1987 (Cassell 1989) did indicate the remains of a rectangular Euroamerican structure. A subsurface cache of 10 liquor bottles was found inside floor sills and excavated. One bottle from the cache had a maker's mark dating to 1893–1906. Thus, while Kelly may have had something to do with activities at Kugalukruak, the 1893-1906 maker's mark clearly post-dates the 1891–1892 operation of his whaling station.

Figure 5: *John Kelly. (By permission of the Presbyterian Historical Society).*

Kelly had whaled from a Point Hope station in 1889, and in 1890 he was sent to Barrow to run a station for the PSWC. He built a large station, bringing his Point Hope employees with him. In the 1891 spring whaling season, he and his 100 or so Iñupiat employees took 12 whales; Brower's crews took five, and Iñupiat *umialiit* took one. In September 1891, Kelly's Barrow-bound supply ship, steaming from the south, became fast in the ice off Point Belcher, and rather than move everything north by boat and sled, he set up his station there:

> Kelly had a big outfit aboard the steamers. The Company were sending him a two years' outfit. Day after day we lay there, the ice never moving. The first of Sept. came along [and] still the ice was solid. Then there was a little moved off just so the ships could get a few miles north, then all anchored once more... Kelley had all his outfit landed a few miles south of where the ships lay, [and] built a new house saying he would whale there next spring (Brower n.d.:410, 411).

Brower visited Kelly in the December 1891, and his observations during that visit provide the only documentary description of the station and its layout:

Kelley had built a long house of one thickness of lumber, banking it all with snow and sod. It was warm enough with all hands living there. No one had a great deal of room... Kelley kept himself shut in his room all of the time. His house and every place in it was filled with Point Hope people. They mostly lived there, eating at the table (Brower n.d.:417, 418).

Kelly's self-imposed personal isolation at the station was apparently characteristic. When Brower traveled with Kelly in the autumn of 1888 to a village at Cape Lisbourne, he reported that, 'We did not have to take the unoccupied house, as there were three Eskimo families living here. Kelley, however, wanted to keep away from them, saying he thought he was more comfortable by himself' (Brower n.d.:306). Later that autumn, again at Cape Lisbourne, Brower (n.d.:312) wrote that, 'This time I insisted on staying in a native house. Kelley wanted to camp out all night... .'

Brower (n.d.: 418) also observed what he felt were rather unpleasant surroundings inside Kelly's Point Belcher station during his visit, stating that 'I never saw an Eskimo house in the condition that place was.' Brower's Iñupiat wife, Toctoo, seems to have felt the same, as she 'never wanted to visit Kelley again as long as he was in the country' (Brower n.d.:418).

Eventually, Kelly 'hired eighty Eskimo men, many of them starving emigrants from Point Hope, to man about ten boats' (Bockstoce 1986:238). These were Iñupiat who had whaled for Kelly previously, first at Point Hope, and then at Barrow; recall Brower's (n.d.:418) above observation that the station was 'full of Point Hope people."[17] Kelly also took persons who had worked for another commercial whaler, one Peter Bayne. Brower (n.d.:417-418) noted that 'Besides his own men, Kelley had shipped some more of the men that had been [at Icy Cape] with Baine.'

At least two non-Iñupiat employees were with Kelly. During his visit to Kelly in December, 1891, Brower (n.d.:418) observed that, 'He [Kelly] had shipped a new cook, a long colored man from one of the steamers. All night it seemed he was cooking. When he slept I never could find out, unless it was in the afternoon.' Another worked for him earlier at Barrow, but did not last the entirety of the Point Belcher stay with Kelly: 'A month before whaling started [in spring 1892], Ben, a Portuguese working for Kelley, left him and came to [Barrow]. They had some trouble. Kelley discharged him' (Brower n.d.:423).

---

[17] Mary Ann Larson (pers. comm. 2000) appropriately wondered if Kelly's employees were originally from Point Hope or if they may have been from elsewhere and merely worked for Kelly at Point Hope.

Fur trapping, especially fox, had become a major industry in late 19th century North Alaska (e.g., Libbey and Schneider 1987), and Kelly did a brisk fur trade business during the 1891–1892 winter months. Brower (n.d.:416-417) remarked that 'Everyone that [sic] had furs had to go to Belcher and trade with Kelley.' In early December, 1891, 'Tookaloona, an Eskimo from [Barrow] was on his way south to trade. As he was crossing the lagoon at Peard Bay [north of Point Belcher] he found the carcass of a whale that had washed in there sometime in the fall' (Brower n.d.:417). Whale carcasses are magnets for scavenging polar bears, wolves, and foxes, and when Tookaloona:

> told Kelley what he had found Kelley sent a party of his boys there with a lot of traps. During the winter they caught several hundred foxes and a number of wolves... I started for Kelley's to spend Christmas, leaving home on the 18th. We made Belcher without sleep. Kelley's boys were then camped at the carcass. They had a large fire going as we passed. They had not been camped very long... All hands were laying around, some making coffee and others cooking (Brower n.d.:417).

However, Kelly's group caught no whales that spring at Point Belcher, and in the early summer of 1892, he dismantled the station and sold much of its contents to a Portuguese whaleman, Antone:

> Kelley had done nothing at Belcher and was getting all his gear ready to move back to [near Barrow] as soon as there was water enough along the beach. Kelley sold all his fox skins to Antone for a slab of bone [i.e., baleen] each. Then he sold him the house for more bone. Antone was to take it down and move it himself to [Barrow] when the ships arrived. Sometime in July Kelley arrived with his fleet of boats, bringing all he had left with him. He at once opened the old station, starting to buy bone from the Eskimo with stuff left from the season before (Brower n.d.:424-425).

In 1893 Kelly had charge of 100 Iñupiat employees and their family members in Barrow, totaling about 500 people. His Point Hope crew joined him there. In 1894, Kelly staffed 22 crews with nearly 200 employees in Barrow, taking 11 whales (Bockstoce 1986:237-239,240).

ARCHAEOLOGY AT KELLY'S SHORE WHALING STATION

The archaeological site representing Kelly's station is an array of five structures, eight storage racks, and at least 12 trash middens (Cassell n.d.), the result of activities there between September 1891, and July 1892 (Fig. 6). The documentary data indicate that Kelly lived in the main station building with his Iñupiat employees but in a separate room, that additional employees arrived through time at the station, and that fox trapping and processing occurred there. The archaeological data also indicate this. This suggests that while the archaeological record can be seen as serving to confirm the documents, it should also be seen as very useful in understanding sociohistorical dynamics, processes, and events in the absence of documents. This is due to the fine-grained temporal contexts of site deposits and the unusually firm links between the archaeological and documentary data. What follows is a discussion of the overall layout of the station, the midden excavations and the results of those excavations, the links between the documentary and archaeological data, and the implications of data collected for the Kelly's station research to the history of Iñupiat life and labor in the Western Arctic commercial whaling industry in North Alaska during the late 19th-early 20th centuries.

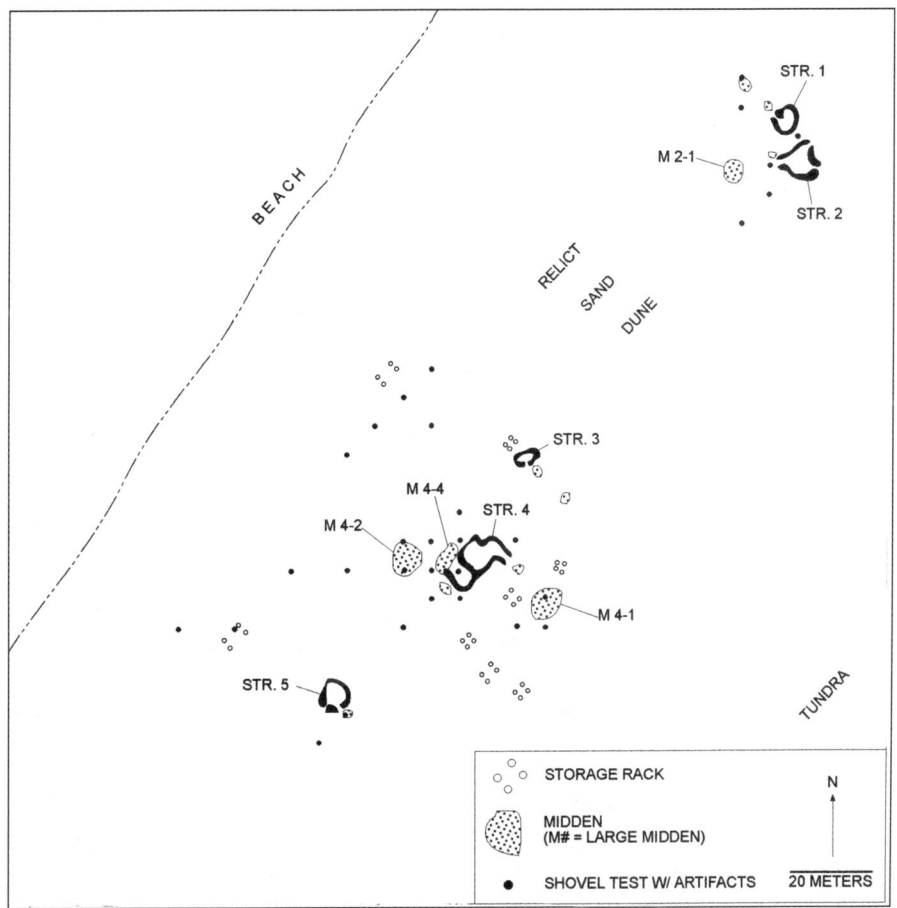

Figure 6: *Site plan of Kelly's commercial shore whaling station.*

Initial knowledge of the history and layout of Kelly's station was gleaned through the historical and archaeological efforts of John Bockstoce (pers. comm. 1986) and Robert Gal (pers. comm. 1986). I visited the site briefly in 1987, not to conduct intrusive excavations, but to familiarize myself with the location and to confirm the previous research. I found what Gal reported in 1983, namely the remains of five structures and an array of storage racks. I also found a small area of sand indurated with seal oil and covered with small brass tacks. The beginning of our 1990 fieldwork consisted of firming up knowledge of the previously documented structure and storage rack remains. We identified a total of eight storage racks; no additional structural remains were discerned at the station in 1990.

Table 2: Tested Middens at Kelly's Whaling Station.

MIDDEN NUMBER	STRUCTURE NUMBER	TOTAL AREA (in m²)	EXCAVATED AREA (in m²)	PERCENT EXCAVATED	RECOVERED ARTIFACTS	% SITE N
1-1	1	2	1	50	21	1
1-2	1	5	2	40	84	2
2-1*	2	16	8	50	595	11
3-1	3	4	2	50	25	1
3-2	3	4	2	50	130	2
4-1*	4	65	18	28	2279	41
4-2	4	45	14	31	1687	31
4-3*	4	2	1	50	43	1
4-4*	4	15	5	33	504	9
5-1	5	1	1	100	20	1
TOTAL	**-**	**159**	**54**	**33**	**5388**	**100**

denotes four largest midden samples, accounting for 92% of recovered artifacts.

All structures were built upon the ground surface, combining driftwood and milled lumber framing and having sod blocks as wall material (sod cutting areas were seen adjacent to sod wall remains at some structures). Of the five structures present, one structure (Structure 4) is clearly of Euroamerican design and construction, rectangular in form, divided into two rectangular portions, and containing within the parallel impressions of floor joists. This appears to be the main station building, referred to by Brower in his December, 1891, visit. Five of the eight storage racks are located immediately adjacent to this main building; all eight racks are within Kelly's purview.

Of the remaining four sod structures, two (Structures 1 and 2) have a distinctive entranceway; Structure 2 also contains an alcove.

The other two structures (Structures 3 and 5) are irregularly elliptical in form, and correspond neither to acknowledged Euroamerican habitation forms nor to the traditional Point Barrow house type. These buildings seem to have been used initially as warehousing or similar special use facilities, with use shifted to occupation as supplies ran down and more employees entered the station from Bayne (and possibly elsewhere), during fox trapping, and before whaling.

It was also important to understand the depositional history of the site, and to this end we excavated more than 500 shovel test pits in and around the station site, covering an area of about 100 x 400 meters.[18] The first 10 m grid of shovel tests showed substantial deposition immediately around the station buildings; deposition away from the buildings was limited to a very few findspots. Subsequent shovel testing was done at 5 m intervals around the findspots.

The shovel test data showed that the highest depositional areas coincided with the building locations, and allowed us to say with some certainty that all deposition was the result of whaling station activities and was confined to 1891–1892. This fine-grained temporal aspect (*sensu* Binford 1978; Hall 1982) of the assemblage removed the potential for depositional background 'noise' stemming from occupation or intensive land use at a given locale occurring before or after the target period. For example, while undeniably vital research, the results of archaeological work into Eskimo/Euroamerican social dynamics at Herschel Island in western Canada (e.g., Friesen 1994a, 1994b) is compromised to some degree by the presence of deposits resulting not only from at least a decade of whaling period land use but also by prior late prehistoric occupation and post-whaling mission and government material presence. The combined archaeological and documentary knowledge that material deposits at Kelly's station resulted solely from less than one year of station activities removed concern for consequences of temporal mixing.

Some deposits found in shovel tests represented trash middens. The middens were characterized by a very high artifact count in tests located at very low rises (not previously noted by us), and had a distinctive crunching sound when probed. Knowledge of the presence of the very low rises and the crunching was used to our advantage: we soon noted a number of additional low rises located between shovel tests that crunched when a pointing trowel gently probed them. The middens, 12 in all, were located immediately adjacent to structural remains; they became the focus of our intensive block unit excavations.

---

18 The site itself, inclusive of all middens, structures, and storage racks, covers approximately 200 x 90 m, and is divided into two areas separated by a small expanse of empty tundra.

Ten of the 12 identified middens were sampled (Table 2). The excavated middens ranged in size from 1 m² to 65 m²; all were less than 0.25 m in depth. The excavation sample percentages for each midden were inversely related to the size, i.e., the small middens were heavily sampled (up to 100%) and the large middens were sampled to a lesser degree (to a minimum of 28%). Nearly 6000 items were collected from the total 54 m² of block midden excavations.[19] The contribution of each midden to the total site artifact sample size reflected the midden size, i.e., the large middens provided many items (up to 41%) and the small middens offered fewer (to approximately 1%). The middens were excavated in natural levels. The levels can be generally characterized as including a sod level on the midden top, underlain by a level of relatively loose and unconsolidated material, and followed by a densely packed level; clean sterile sand lies below all cultural deposits.[20] One can consider the uppermost two levels (sod and loose material) to be a single level, since the sod grew over the underlying material after the station was abandoned in 1892.

Part of the physical distinction between the upper unconsolidated and lower consolidated material may be due to differential post-abandonment effects (e.g., climatic), but artifact assemblage analyses also demonstrate real differences in the nature of deposition between the two levels. When artifact contents are compared by level within individual middens, this indicates changing functions of some station buildings through time.

Additionally, small but spatially discrete pockets of cultural material occurred within some middens, and appear to represent individual deposition events (i.e., dumping a bucket or so of trash onto the midden). The artifact assemblages within each pocket and comparison of assemblages between middens show that activities ancillary to the conduct of commercial whaling occurred at the station, and that more than one group of people were working there.

I want to suggest the likely spatial layout and function of Kelly's station. The place was a complex of structures and facilities used for different purposes, all guided toward the development, maintenance, staffing, and provisioning needs of a shore whaling station. It was a combination warehouse, factory, and company town. Taken together, the archaeological data, including the building types, the layout, and the artifact assemblage, describe a Euroamerican presence with a distinct Iñupiat flavor.

---

[19] One small trench was also excavated, not in a midden but spanning the division between the two rooms in the main station building.

[20] The specifics of level composition varied from midden to midden.

The main station building functioned as Kelly's residence and business office, separated from the quarters of at least some of his Iñupiat employees in the same building. The other structures served as warehousing for a while, until the supplies held in them ran down. As more employees arrived for fur trapping and the impending whaling season, those structures were converted to housing. The storage racks held supplies and provisions off the ground and away from dogs and other scavenging beasts. And despite the spatial gap separating three structures from the other two, the main building provided Kelly his point of observation to the other facilities and structures. The place thus represents a panoptic, in which the central figure, the boss John Kelly, could in large measure survey all aspects of his little domain, including structures, storage racks, and activities; all his Iñupiat employees could be under his scrutiny in the course of their domestic and industrial lives at the station.[21]

Architectural items related to construction in the middens surrounding Structure 4, the main station building (quarters for Kelly and his early employees, as indicated in Brower's observations), were most prevalent in the lower (initial) levels. However, construction items were most prevalent in the upper (later) levels of middens surrounding Structures 1, 2, and 3. This indicates that large–scale construction occurred initially at the main station building (most milled wood fragments came from here), with little refurbishing being conducted in the course of station presence. And while the other structures may have been built at the same time as the main building, considerable alterations to these buildings occurred after they were first built.[22]

Foodstuff remains, comprising nearly half the total artifact sample, included wild country foods and imported varieties. Wild game, especially seal, caribou, and birds, was most abundant. A modicum of bear and walrus was also present, as were fish, mollusks, and eggshells to a lesser extent. Also present were cow and pig bones; this is not surprising given that the initial station supplies were derived from the outside via ships. These domestic fauna were, however, very low in number relative to quantities of wild game. Also, over 500 apricot pits were found in the middens.

Excavated container materials, including tin can and burlap bag fragments, were certainly related to transport and storage of imported

---

[21] Such panoptic interest on the part of management was commonplace in factory and prison layout and design during the burgeoning industrialization of 19th century America (see especially Foucault 1979).

[22] Interestingly, Brower's writings describe only the main structure and not other buildings.

foodstuffs. For example, the apricots may have been preserved in tin cans or dried and stored in bags. The presence of burlap bag fragments point to the likely importation of foodstuffs such as beans into the station larder. Regardless of the relative quantities of wild and imported foodstuffs at Kelly's station, the data demonstrate the joint reliance on country and shipped foods for the station operations.

Faunal materials occur throughout the middens associated with the main station building, but are most frequent in the upper portions of middens associated with Structures 1, 2, and 3. Further, fully 97% of the recovered apricot pits were found throughout three of the four middens associated with the main building. These faunal and floral data demonstrate that the people eating the foodstuffs were present in the main building during the entirety of the station presence, and that Structures 1, 2, and 3 were occupied only later. The extremely high frequency of apricot pits around the main station building relative to middens around Structures 1, 2, and 3 suggests that these latter structures were occupied after the apricot stores had been consumed. It is also possible that apricot presence throughout the main station middens indicates that individuals occupying that building had especial access to apricot resources.

Related to the presence of game foods in the faunal material is, of course, the means for obtaining that game. Approximately 90% of subsistence-related items found in the assemblage primarily concerned rifles and shotguns, including cartridges and shells, gun parts, and powder-can fragments. Less than 10% of the subsistence assemblage were traditional Iñupiat forms, and included wooden net floats, an ivory harpoon head and finger rest, and blunt arrowheads.[23] The subsistence assemblage included equipment useful in taking all the game foods found in the faunal assemblage noted above.

Nearly all the artifacts immediately related to household use were from oil lamps, and included brass lamp parts and chimney glass. The presence of kerosene-can fragments in middens is related to use of these lamps. Coal and coal ash were found throughout the middens, thus providing evidence of heat source.[24] No direct evidence was found

---

[23] Distinctions between Iñupiat and Euroamerican material culture will be examined in more detail below.

[24] Low grade coal is commonly found washed up on beaches in North Alaska. A large coal outcrop south of Point Belcher at Corwin Bluffs was well known to commercial whalers; Kelly himself had mined coal there.

for use of seal oil for light or heat in habitations, although the identification of the oil-indurated sands is suggestive of the possibility.[25]

Personal artifacts, associated with the recreation, dress, and maintenance of individuals included buttons, cloth and leather fragments, combs, toys, knife and watch parts, and sunglass lenses (interestingly, no indications of country skin clothing were present). Most numerous, however, were glass beads (n=795) and kaolin pipe fragments (n=115), together amounting to nearly 97% of the personal items. These latter items, while of Euroamerican manufacture, were intended primarily as trade items for Iñupiat, and replicate the function of material culture types used prior to commercial whaling in the Western Arctic.

Pipes, when sufficiently complete, were found to be universally of the 'TD' variety, embossed 'GLASGOW / MCDOUGALL'; these pipes are commonly found in late 19th century archaeological sites throughout North America. Pipes and pipe fragments were encountered throughout most middens of most structures. This ubiquitous pipe presence may be merely an indicator of Iñupiat tobacco use during the run of the station.

The 795 glass beads from Kelly's station, however, tell other stories. They were used as decorative materials for clothing or other accoutrements; decorative styles could identify family village residence, or signal an alliance with trading partners who could provide certain varieties of ornamentation to the wearer.

Fifty-three bead types were derived from the sample excavated at Kelly's station, based upon size, form, and color. The beads ranged in size from seed beads (<2 mm) to beads 15 mm in diameter, in form from flat to tube to oval, and including blues, reds, green, yellows, and colorless. While beads were encountered in each midden and each level, they were most frequent in Midden 4-1, outside the front (east) of the main building, where fully 87% of the beads occurred. Within Midden 4-1, the lower level contained 83% of all beads. And within this level was found a number of discrete pockets of deposition from which were recovered five clusters or bundles of beads. These five bead bundles held over 60% of the total site bead assemblage, with bundles containing from 12 to 205 beads. Little else was found in the pockets; the beads comprised 74% of the recovered artifacts.

The sheer quantities of beads in these bundles is striking in itself. They would appear to represent individual events of bead

---

25 The oil-indurated sands may have been associated either with Kelly's employees trying out seal oil, with their burning of oil-soaked casks derived from the 1871 shipwrecks, or with activities of the shipwreck survivors themselves. The brass tacks amidst the sands are indicative of nautical origins.

disposal. Why were so many beads deposited within the narrow time range needed to create a discrete level within a midden created in just nine months? One can perhaps explain the deposition of 12 or 15 or 20 beads in any one event as part of the tasks. But one wonders why as many as 205 beads would be disposed of in one event. Were beads so easy to obtain that one could just throw so many of them away? Beads comprised fully 85% of the personal category items that amounted to just 17% of the site artifacts; they were clearly commonplace in the material lives of the Iñupiat workers. Could access to resources such as beads be limited to the extent that some could have many and others few? The identification of bundles containing frequencies of beads so far exceeding those found singularly (i.e., not in bundles) in the other middens is certainly suggestive. If the bundles do indeed represent individual dumpings, then someone apparently had access to many beads. Or could the signaling aspect of ornamentation indicate shifts from one prime trading partner to another during the commercial whaling period? Brower's observations indicate this occurred with the transfer of Bayne's employees to Kelly.

A clue to the resolution of this quandary can be found in the variability of bead types represented in the assemblage when comparing beads within and between middens. Of the 53 defined beads types, the bead bundles in Midden 4-1, Level 3 contain 11 types unique to those bundles; they were recovered nowhere else. Inclusion of the beads recovered individually in Midden 4-1 shows 22 types unique to that midden. Middens other than 4-1 have 13 unique types.

The bead data suggest a few interlinked possibilities. First, the two bundles holding 200 and 205 beads in Midden 4-1, Level 3 stem from deposition by Point Hope individuals who arrived initially with Kelly and had been with him in previous years. Kelly was their trading partner; the bead types unique to those deposits were bead types preferred by Kelly as trade items. The presence of so many suggest these individuals had been with Kelly long enough or had been sufficiently skillful to have acquired considerably more desired material items than had others working for Kelly. The deposition of bead types not found in Midden 4-1 but unique to other middens may coincide with the arrival of persons who had access to commercial whaler/traders with bead types differing from those offered by Kelly, such as Peter Bayne.

Chert flakes and scrapers also occurred in the middens across the site. While chert debitage was present throughout the midden levels, finished scrapers were present only in the lower levels, indicating the use of scrapers relatively early in the station's existence. Five of the 16 recovered scrapers were found in the debris pockets representing specific deposition events; four of those five were in pockets associated with bead bundles from Midden 4-1. These scrapers were used in the preparation of fox hides, and represent a period of intensive fur procurement described by Brower as occurring at Kelly's station. The recovered scrapers are for the most part intact but dull from use; rechipping the edge would likely make them too sharp for the delicate fox skins (Robert Gal, pers. comm. 1992). The deposition of so many scrapers over such a short period of time is very unusual in prehistoric North Alaskan archaeological sites, and indicates not only the extent of the fur trade at Point Belcher but also the unique industrial labor social conditions in which the scrapers were used and discarded by the Eskimo employees (Cassell 2003).[26]

Finally, I wish to compare the recovered traditional Iñupiat material forms with those of Euroamerican origin. This addresses a broader issue of the degree of Iñupiat material acculturation during the commercial whaling period by the turn of the 20th century. This is useful because of the fine-grained temporal aspect of Kelly's station deposits and the existence of prior research on this topic.

Table 3: Iñupiat-Affiliated Items, Kelly's Station.

TYPE	N=	TYPE	N=
Antler, cut	1	Lithic	53
Bead, glass	795	Object, anthropomorphic	6
Blunt, cartridge/wood	2	Pipe, kaolin*	115
Bone, worked	6	Sled runner, whale bone	4
Drum handle, wood	1	Stake, wood	6
Finger rest, ivory	1	Toy bow, wood	1
Fishing float, wood	9	Ulu, iron	1
Harpoon, ivory	1	Whetstone	1
Ivory, worked	11	Wire loop	1
Leather dog harness	1		
		TOTAL 19 TYPES	**1016**

Beads and pipes are included here as Iñupiat-affiliated types for two reasons. First, although they are of Euroamerican manufacture, they were primarily used as trade items to Iñupiat. Second, they conform to indigenous pre-commercial whaling material forms.

John Kilmarx (1986) attempted to understand the nature of Euroamerican documentary description of material culture used by

---

[26] Euroamerican observers in North Alaska after the commercial whaling period noted that the chert end scraper was among the last enduring components of traditional Eskimo material culture. Ethnographer Diamond Jenness wrote that chert scrapers were 'more adapted to the purposes to which they are applied than any implement which we can supply. It is interesting to note that while the implement used for stretching and softening skins is generally an iron ferrule set in a short wood handle, the actual scraper, which is set on a similar handle, is still commonly made of flint' (Jenness 1918:93; see also Anderson and Eels 1935:131).

North Alaskan Iñupiat around the turn of the 20th century. He compared categories of material culture used by Iñupiat as described in the documents with artifacts derived from the Utqiagvik Archaeology Project excavations in Barrow. He found that Euroamerican observers greatly overemphasized the frequency of traditional Iñupiat material culture categories, to the near exclusion of introduced Euroamerican material culture. The archaeological data from the Barrow excavations indicated a much greater Euroamerican material presence than would have been expected from the documentary data.

Table 4. Euroamerican Items, Kelly's Station.

TYPE	N=	TYPE	N=
Bag, burlap	9	Leather	8
Band, rubber	1	Marble, ceramic	1
Bar, iron	3	Nail / spike	765
Bottle glass	56	Object, iron	1
Bracket, iron	1	Object, jet	1
Bullet, lead	2	Pencil, graphite	1
Bushing, iron	1	Platter, ironstone	9
Button	7	Pug, wood	2
Can	114	Powder can	5
Cap, brass	1	Primer cap	8
Cartridge, rifle/shotgun	123	Shank, brass	1
Chalk	1	Sheet metal	15
Cloth	3	Sleeve, metal	2
Comb	2	Stopper, glass	1
File, iron	1	Stove part, iron	1
Fishook, iron	1	Strap, iron	3
Fixture, iron	2	Sunglass lens	2
Flatware, iron	3	Tableware, ceramic	13
Gunpart	3	Thimble	2
Handle, wood	1	Unid glass	1
Hinge	2	Valve, iron	1
Holystone	2	Vessel, iron	6
Hoop iron	15	Washer, brass	1
Hose, rubber	1	Watch face cover	1
Knife	3	Window glass	106
Lamp	172	Wire	11
		TOTAL 53 TYPES	1499

Kelly's station and turn-of-the-century Barrow can be seen as similar in terms of ethnic population composition (almost entirely Iñupiat but with a few Euroamericans) and context of everyday life (heavily involved in the commercial whaling industry). Consequently, it seems appropriate to examine relative material culture type frequencies of Iñupiat and Euroamericans from the Kelly's station excavations.

Approximately 44% (n=2515) of the assemblage from Kelly's station were items to which manufacture, form, or context of use could be attributed with some certainty to an Iñupiat or Euroamerican affiliation. These are presented above in Tables 3 and 4. The relevant artifact types are sufficiently broad to provide a conservative estimate of affiliation content for the assemblage. The data show that the total quantity of Iñupiat–affiliated artifacts does approach those affiliated with Euroamericans (1016 *vs.* 1499).[27] However, the number of material types associated with Iñupiat are far exceeded by those of Euroamericans (19 *vs.* 53). This supports Kilmarx's findings that despite the vastly greater numbers of Iñupiat relative to Euroamericans, Euroamerican material culture had substantially permeated the lives of Iñupiat. At least in locations where those groups mixed about in the context of commercial whaling, the introduction and use of Euroamerican materials was substantial.

DISCUSSION

The results of archaeological research at Kelly's station provide numerous links to the documentary knowledge of the station. But the archaeological information also yielded significant information about the daily lives of the people who lived there, information perhaps not available elsewhere. This is important because, as noted above, there may be no other shore whaling station found, and if they are found they would not likely have the fine-grained temporal qualities that are exhibited at Kelly's station.

The building forms, station layout, and the bulk of the artifacts identify the site as primarily Euroamerican, while the traditional Iñupiat artifact types show that Iñupiat were there as active participants in station activities. The panoptic potential of the place suggests the possibilities of Kelly's conduct of his industrial pursuits.

Foodstuffs evident from the faunal and floral remains show a primary reliance on wild game but also partial reliance on imported foods. The artifacts related to subsistence acquisition run the gamut of technical means necessary to obtain the game species found; the container goods reflect the transport and storage of the imported varieties.

Brower wrote that while most of Kelly's employees had been with him at Point Hope and had arrived with him at his Point Belcher station, some of the employees came from Peter Bayne's operation. The architectural-related materials at the lower (initial) midden deposits represent the early construction of the station buildings, while those in middens associated with structures away from the

---

[27] This total is due in large measure to the presence of glass beads and pipe fragments in the Iñupiat category; removal of these yields only 106 Iñupiat items.

main building seem to be related to refurbishing those structures upon arrival of other employees. The bead varieties also suggest the arrival of folks after the initial settlement of the station. The presence of unique bead types between middens indicates the likelihood that different groups of people had access to different types of beads.

The bead assemblage shows further that some of Kelly's employees had greater access to resources than others, and that Kelly was a prime trading partner. This may be a local material manifestation of the process by which some previously disenfranchised Iñupiat attained *umialik* status in the commercial whaling period through material acquisition from commercial whalers acting as prime trading partners.

Documentary data do not indicate the population structure of Iñupiat working and/or living at the station. Brower's writings frequently indicated that he supported whole families, and that Iñupiat women and men worked for him. We know that Kelly supported families of employees in Point Hope and in Barrow. The recovery of a toy wood bow and small anthropomorphic carvings at Kelly's station suggests the presence of children, which further suggests the presence of families there.

The presence of chert scrapers demonstrates the existence of hide working activities at the site. Since no evidence of skin clothing was recovered at the station, one can only follow the archaeological leads to Brower's report of the intensive fox trade conducted by Kelly, and conclude that the scrapers were used virtually solely in the fur trade.

CONCLUSIONS

It is apparent that the archaeological data from Kelly's 1891-1892 Point Belcher shore whaling station not only complement the documentary data but go beyond them. Archaeological and documentary research yield independent databases, useful both in their own right and in conjunction with other independent databases. The documentary data provide vital information about specific names, places, and events. But archaeology does more than just fill in gaps in the documentary record. Because of its emphasis on material culture, archaeology can address aspects of social and historical processes in ways that documents cannot; it is removed from the observer biases inherent in documents. It yields data that lead to questions, answers, and interpretations that the documents simply cannot reach.

One issue raised by the results of archaeology at Kelly's station concerns the nature of social change with the coming of commercial whaling, of industrial labor development, and of enormous quantities and varieties of Euroamerican material culture. The presence of the traditional chert scrapers is instructive in this regard. It is clear that while Euroamerican material culture permeated the lives of Iñupiat by the late 19th and early 20th centuries, some vital material aspects of traditional Iñupiat lifeways continued. Of related interest here is the possibility that some of Kelly's employees lived at the old site of Nunagiak, re-excavating and re-using the semi-subterranean living areas, an age-old practice continued.[28] This time, however, they used milled lumber in the reconstruction of the old houses. Like the manufacture and use of traditional chert tools in commercial industry, the possible re-use of the Nunagiak houses by Kelly's employees during the operation of the station indicates a synchretic Iñupiat adaptation to shifting social and material environments.

Commercial whaling and the development of an Iñupiat wage labor force for and by the industry was a landmark process in the social history of the Iñupiat in North Alaska because of the links between the events of today and those of that not-too-distant past. It marked the formal Euroamerican exploitation of arctic natural and human resources, an exploitation nowadays seen in petroleum-related extraction and employment. The remuneration for Iñupiat wage labor engendered by participation in the whaling industry (and continuing into the present day) led from an old system of power relations to a new system, from one in which the few established *umialiit* had material power and opportunities, to one where the material power base was much more broadly distributed. The new system engendered different entrepreneurial freedoms, and begat opportunities for individual access to *umialik* status. For the Barrow area, for example, recall Simpson's recording five *umialiit* for a population of about 500 persons in the early 1850s, or approximately one *umialik* for every 100 people. A population estimate for Barrow in the late 1980s described 40 *umialiit* for over 4000 individuals, but less than one half of the population were Iñupiat (David Libbey, pers. comm. 1987). This translates into 40 *umialiit* representing about 2000 persons, or about one *umialik* for every 50 Iñupiat Eskimos. This is roughly double the figure for the 1850s, prior to Iñupiat labor in commercial whaling.

With reference to the decimation of bowhead populations by commercial whaling, Lt. P.H. Ray (1885, in Murdoch 1988:xcix) wrote in 1885 that the Iñupiat in North Alaska 'are slow to take up with an innovation, and they do not really adapt themselves to the new condition of affairs which the loss of this great food supply has brought about.' Ray was clearly speaking a bit prematurely. In an anthropological sense, Eskimos in Canada and Alaska have been viewed as the prime example of human adaptation to the physical environment, and as George Wenzel (1991) and Norman Chance

---

[28] This intriguing idea was offered by Albert A. Dekin, Jr. (pers. comm. 1999).

(1990) have pointed out, they are equally adaptive to changing social environments. The strength and vitality of Iñupiat culture today in the face of a rapidly changing world is indicative of their adaptive capabilities.

North Alaskan Iñupiat Eskimos today are active participants in the public and private commercial economic sectors and in Western forms of bureaucracy. They are ardent followers of Western religious faiths. They are Americans when they want to be or have to be. For all that, they are no less Iñupiat Eskimo than they were prior to the arrival of commercial whalers; indeed they are entirely Iñupiat and more.[29] The late Waldo Bodfish, Sr., Iñupiat elder and son of commercial whaler Hartson Bodfish, has said, 'I am Eskimo. I don't want to change my nationality and I like to speak Eskimo' (Bodfish 1991:1). As his son Dempsey Bodfish told me in 1990, 'We are Eskimos. We are hunters.'

Acknowledgments. I wish to thank Bruce Mattioli for his professionalism, companionship, and camaraderie during the 1990 fieldwork at Kelly's station. LouAnn Wurst was of great assistance, durability, and friendship during the initial 1987 reconnaissance project. My gratitude is extended to the Bureau of Land Management and the U.S. Department of Fish and Wildlife for logistics and supplies support; especial thanks go to John Cook and Chuck Diters of those respective agencies. Bob Gal provided valuable suggestions after reviewing an earlier version of this paper. I very much appreciate the editorial skills of Allen McCartney and the anonymous reviewers of this article. Finally, whatever success was achieved during the archaeological fieldwork at Point Belcher was due in large measure to the hospitality of the people of Wainwright, Alaska, and especially to Dempsey and Anna Mae Bodfish of that fine community. Anna Mae has since departed this earth, and the void is tangible.

REFERENCES

Aldrich, Herbert L.
    1889   *Arctic Alaska and Siberia; or, Eight Months with the Arctic Whalemen.* Chicago: Rand, McNally and Company.
Anderson, H. Dewey and Walter C. Eels
    1935   *Alaska Natives: A Survey of their Sociological and Educational Status.* Stanford: Stanford University Press.
Binford, Lewis R.
    1978   *Nunamiut Ethnoarchaeology.* New York: Academic Press.

---

[29] This is reminiscent of McFee's (1968) '150% man.'

Blackman, Margaret B.
  1989   *Sadie Brower Neakok, an Inupiaq Woman.* Seattle: University of
          Washington Press.
Bockstoce, John R.
  1980   A Preliminary Estimate of the Reduction of the Western Arctic
          Bowhead Whale Population by the Pelagic Whaling Industry,
          1848-1915. *Marine Fisheries Review* 42(9-10): 20-27.
  1986   *Whales, Ice, and Men: The History of Whaling in the Western
          Arctic.* Seattle: University of Washington Press.
Bodfish, Hartson H.
  1936   *Chasing the Bowhead.* Cambridge: Harvard University Press.
Bodfish, Waldo, Sr.
  1991   *Kusiq: An Eskimo Life History from the Arctic Coast of Alaska.*
          Fairbanks: University of Alaska Press.
Brower, Charles D.
  n.d.   *The Northernmost American: An Autobiography.* Unpublished
          manuscript in the Stefansson Collection. Hanover, NH:
          Dartmouth College Library.
  1932a   My Arctic Outpost. *The Blue Book* 54(3): 6-19.
  1932b   My Arctic Outpost. *The Blue Book* 54(6): 86-101.
  1932c   My Arctic Outpost.'*The Blue Book* 55(1): 60-75.
  1932d   My Arctic Outpost. *The Blue Book* 55(2): 76-87.
  1934   Land of the Long Night. *The Blue Book* (58:4): 146-160.
  1942   *Fifty Years Below Zero.* New York: Dodd, Mead & Co.
Burch, Ernest S., Jr.
  1975   *Eskimo Kinsmen: Changing Family Relationships in Northwest
          Alaska.* American Ethnological Society Monograph No. 59. St.
          Paul, MN: West Publishing Co.
  1981   *Traditional Hunters of Point Hope, Alaska: 1800-1875.* Barrow,
          AK: North Slope Borough.
Chance, Norman A.
  1990   *The Iñupiat and Arctic Alaska: An Ethnography of Development.*
          Fort Worth: Holt, Reinhart, and Winston.
Cassell, Mark S.
  1988a   *Ethnohistory, Native Labor, and Commercial Whaling in the
          Beaufort Sea, 1889-1910.* Master's thesis, Department of
          Anthropology, State University of New York-Binghamton.
  1988b   Farmers of the Northern Ice: Relations of Production in the
          Traditional North Alaskan Iñupiat Whale Hunt. *Research in
          Economic Anthropology* 10: 89-116.
  1989   *An Archaeological Reconnaissance of Commercial Whaling Sites
          in the Vicinity of Point Belcher, Alaska.* Prepared for the Bureau
          of Land Management, Fairbanks District Office.
  2000   'If They Did Not Work for the Station, They Were in Bad Luck':
          *Commercial Shore Whaling and Iñupiat Eskimo Labor in Late
          19th/Early 20th Century North Alaska.* Doctoral dissertation,
          Department of Anthropology, State University of New York-
          Binghamton. Ann ArbourL University Microfilms.
  2003   'Flints and Foxes: Chert Scrapers and the Fur Industry in Late
          19th and Early 20th Century North Alaska,' pp. 151-164 in C.
          Cobb ed., *Stone Tool Traditions in the Contact Era.* Tuscaloosa:
          University of Alabama Press.

Ford, James A.
    1959   Eskimo Prehistory in the Vicinity of Point Barrow, Alaska.
            *Anthropological Papers of the American Museum of Natural
            History* 47(1).
Foucault, Michel
    1979   *Discipline and Punishment: The Birth of the Prison.* New York:
            Vintage.
Friesen, T. Max.
    1994a   'The Qikiqtaruk Archaeology Project, 1990-92: Preliminary
            Results of Archaeological Investigations on Herschel Island,
            Northern Yukon Territory,' pp. 61-83 in J.L. Pilon ed., *Bridges
            Across Time: The NOGAP Archaeology Project.* Canadian
            Archaeological Association Occasional Paper No. 2.
    1994b   *Archaeological Approaches to Intersocietal Contact: Inuvialuit-
            Euroamerican Interaction on Herschel Island, Yukon Territory.*
            Paper presented to the Society for American Archaeology,
            Pittsburgh.
Hall, Edwin S., Jr.
    1982   The Potential Significance of Small, Single Component
            Archaeological Sites for Eliciting the Culture History of
            Northern Alaska. *Anthropological Papers of the University of
            Alaska* 20(1-2): 7-13.
Ivie, Pam and William Schneider
    1978   *Wainwright: Land Use Values Through Time in the Wainwright
            Area.* Barrow, AK: North Slope Borough, and Fairbanks:
            Cooperative Park Studies Units (National Park Service.
Jenness, Diamond
    1918   The Eskimos of Northern Alaska: A Study in the Effect of
            Civilization. *The Geographical Review* 5(2): 89-101.
Kilmarx, John N.
    1986   Archaeological and Ethnohistoric Evidence for Material
            Acculturation in Barrow, Alaska. *Etudes/Inuit/Studies* 10(1-2):
            203-231.
Klausner, Samuel Z., and Edward F. Foulks
    1982   *Eskimo Capitalists: Oil, Politics, and Alcohol.* Ottawa, NJ:
            Allenheld, Osmun Publishing.
Libbey, David and William Schneider
    1987   'Fur Trapping on Alaska's North Slope,' pp. 335-358 in B.G.
            Trigger, T. Morantz, and L. Dechene eds., *Le Castor Fait Tout:
            Selected Papers of the Fifth North American Fur Trade
            Conference, 1985.* Montreal: Lake St. Louis Historical Society.
Marquette, Wilman M., and John R. Bockstoce
    1980   Historical Shore-based Catch of Bowhead Whales in the
            Bering, Chukchi, and Beaufort Seas. *Marine Fisheries Review*
            42(9-10): 5-19.
McFee, Malcolm
    1968   The 150% Man, a Product of Blackfeet Acculturation. *American
            Anthropologist* 70(6): 1096-1103.
McLean, Edna A.
    1980   *Abridged Iñupiaq and English Dictionary.* Fairbanks: University
            of Alaska Native Language Center, and Barrow, AK: North
            Slope Borough Iñupiat Language Commission.

Montgomery, David
1979   *Worker's Control in America: Studies in the History of Work,*
        Technology, and Labor Struggles. Cambridge, UK: Cambridge
        University Press.
1989   *The Fall of the House of Labor: The Workplace, the State, and*
        American Labor Activism, 1865-1925. Cambridge, UK:
        Cambridge University Press.

Murdoch, John
1892   'Ethnological Results of the Point Barrow Expedition,' pp. 19-
        441 in: *9th Annual Report of the Bureau of American Ethnology*
        for the Years 1887-1888, 19-441. Washington, D.C.:
        Smithsonian Institution Press.
1892. 1988. *Ethnological Results of the Point Barrow Expedition.*
        Reprint of 1892 edition, W.W. Fitzhugh (ed.). Washington,
        D.C.: Smithsonian Institution Press.

Prude, Jonathan
1983   'The Social System of Early New England Textile Mills: A Case
        Study, 1812-40,' pp. 1-36 in M.H. Frisch and D.J. Walkowitz,
        eds., *Working Class America: Essays on Labor, Community, and*
        American Society. Urbana: University of Illinois Press.

Ray, Patrick Henry
1885   *Report of the International Polar Year Expedition to Point*
        Barrow, Alaska. Washington, D.C.: Government Printing
        Office.

Riches, David
1982   *Northern Nomadic Hunter-Gatherers: A Humanistic Approach.*
        New York: Academic Press.

Sheehan, Glenn W.
1992   *Proto-historic Social Organization of the Coastal Whaling*
        Communities of North and Northwest Alaska. Doctoral
        dissertation, Department of Anthropology, Bryn Mawr College.

Simpson, John
1875   'Observations on the Western Eskimo, and the Country They
        Inhabit: From Notes Taken During Two Years at Point Barrow,'
        pp. 233-275 in *A Selection of Papers on Arctic Geography and*
        Ethnology. London: Royal Geographical Society.

Slaughter, Dale C.
1982   The Point Barrow Type House: Analysis of Archaeological
        Examples from Siraagruk and other Sites in Northern Alaska.
        Anthropological Papers of the University of Alaska 20(1-2): 141-
        158.

Sonnenfeld, John
1957   *Changes in Subsistence Among the Barrow Eskimo.* Doctoral
        dissertation, Department of Geography, Johns Hopkins
        University.

Spencer, Robert F.
1959   *The North Alaskan Eskimo: A Study in Ecology and Society.*
        Bureau of American Ethnology Bulletin 171. Washington, D.C:
        Government Printing Office.
1977   'Shamanism in Northwestern North America,' pp. 351-364in
        R.D. Fogelson and R.N. Adams eds., *The Anthropology of*

Power: Ethnographic Studies from Asia, Oceania, and the New World. New York: Academic Press.

Stefansson, Vilhjalmur
    1922   *My Life with the Eskimo.* New York: MacMillan Co.

Stevenson, Charles H.
    1907   *Whalebone: Its Production and Utilization.* Department of Commerce and Labor, Bureau of Fisheries Document No. 626. Washington, D.C.: Government Printing Office

Stone, Thomas
    1983   Atomistic Order and Frontier Violence: Miners and Whalemen in the 19th Century Yukon. *Ethnology* 22(4): 327-339.

VanStone, James W.
    1962   *Point Hope: An Eskimo Village in Transition.* Seattle: University of Washington Press.

Wells, R. and J.W. Kelly
    1890   English-Eskimo and Eskimo-English Vocabularies, Preceded by Ethnographical Memoranda Concerning the Arctic Eskimos of Alaska and Siberia. *Bureau of Education Circular of Information* No.2. Washington, D.C.: Government Printing Office.

Wenzel, George
    1991   *Animal Rights, Human Rights: Ecology, Economy, and Ideology in the Canadian Arctic.* Toronto: University of Toronto Press.

Williams, George O.
    1988   Share Croppers at Sea: The Whaler's 'Lay' and Events in the Arctic, 1905-1907. *Labor History* 29(1): 32-55.

Whaling: Indigenous Ways to the Present

Herbert O. Anungazuk

To many who are unfamiliar with the world of the Iñupiat[1], it is a dark, unforgiving world. Even for those who know and must live on the coastal plain, it is, at times, forbidding and cold. It is then that you are most helpless, and most insecure. It is when you feel most unprepared. The land and the sea will show you its wrath if you cannot read what it tells you. The land may not, at times, reveal that it is the faithful bosom and the life's blood to a very special group of people, and it has been the bosom and the lifeline to the people, since the first dawn. The land can appear very unforgiving in the eyes of other shades, but not in the eyes of its people. The land and the sea are one, as they share the same environment. The sea brims with life once a year, or so it seems, and she can readily show her wrath, but fortunately for us, she always shares with us what she will do.

There is constant life in the oceans and the seas, and the sea is also the life's blood of the sea mammals, birds, and fish, and they will readily disappear into its vastness of the land and sea, as they too must move in search of their prey. When the animals begin their movement is when we wait for them, and the land is enforced with place names which tells us what you may find there. The sea itself is a being, and our people, over millennia, placed their names upon her. If there were not life in the sea, the people would not live near the migratory path of animals that provide for their well-being. The coastal plain would not carry a theme of abandoned villages seen today in the form of earthen pits, seemingly empty when seen from passing aircraft. The indentations are seemingly mere dimples upon the land, as they are evidence of former semi-subterranean homes of the people now claimed by the shadows of time. We are a part of the universe because we have lived within our realm, unchanged, and without damaging the delicate land, since dawn immemorial.

Our world was already a very special place, long before the newcomers began showing intense interest upon our land. It is a place that has provided for the spiritual and physical well-being of the people who learned to rely on its renewable resources. We too, like others in far-off lands, taught others who came to us, to survive as we have

---

[1] Iñupiat: A derivative of Inuit. The Iñupiat inhabits sub-arctic and arctic regions in Alaska, from Norton Sound to the Canadian border. The term also identifies the language spoken by the people, and is used as a plural for Iñupiaq.

survived, because there was, 'no other way.' Everything that the people ever needed was from the land and the sea; our ways are natural. The Creator provided for our well being very well.

The ancient hunter learned to strive for total perfection in the environment he shared with his prey. He knew he was not made to survive the cold waters, and the possibilities of being thrown into the element of his prey could become a final reality if he was not careful. The ancient hunter did not realize that the generations who succeeded him would encounter challenges other than the natural conditions that profoundly tested his success. Living in extreme conditions is to live a life of caution, and to live cautiously is continually expressed, and the hunter and his crew scan the horizon for any signs of change. The responsibility placed upon the hunter as someone who must provide for the community is most important. The family, the elders, the children, and those who cannot survive without the help of the hunter must be fed. The hunter is taught the ways of survival by his elders, just as the ways of survival were taught to them.

Some of the truly good parts of indigenous life are viewed with disdain by the newcomers, and the rituals and ceremonies that are involved in fulfilling a complex culture, and not recognized and respected, has caused a breakdown of the innermost being of our culture. It is sad that those who wish to regain the special rituals that must be done before hunting do not have the support of the people who fear the tone of resentment and jealousies of those with other beliefs. Multiple beliefs invaded the lives of northern people very recently and their influences are strong and firmly entrenched in many communities. Some beliefs are so new that those who teach them have not yet adjusted their ways with ours. True, there is a specific line between people, but some members of other societies have not yet seen or leaned that difference. Other ways have taught a deep indifference to indigenous people that has changed deeply the common understanding about our daily lives. Northern man has always had beliefs. That our Creator did not come to us, in form, may be why we do not have a name for him as other beliefs do, yet His presence is felt each and every day. With many among the indigenous community, He is simply *Silaam Inua*, the 'Man of the Sky' or *Munaaqshighuut*, the 'One Who Cares For Us.' And, equally simply, there are 'good' and 'bad' forces that are both very powerful, and a person must differentiate between the two. 'Good' is described in many ways in the spectrum of humanity, but within the indigenous community, it penetrates into another life and into the animal kingdom. It is a measure of showing respect, and it is with no wonder that in other societies, 'good' has won many levels of description, which in many ways, some of the descriptions of the term lacks merit and truth.

The importance of being a part of a family is no longer recognized in the manner as applied with indigenous genealogy. The family is important, as the unit enforces the social standing of everyone. The family is a crucial element of any society, but social order in respect to applying dominance between people has done what it can to weaken the indigenous family unit for many decades. Genealogy must remain an important factor. Many people, keeping in tune with the winds of change, are now separated by greater distances, which often separates families for long periods. Important songs and dances have sunk, as into the sea, or are in a moribund state in many communities through the land, but in gatherings of Nations song and dance continue to provide a special enlightenment, a special happiness. Dance, as a part of celebration, continues to marvel the indigenous world, and has done so since dawn immemorial.

Our languages and dialects define our culture, our heritage, and our being, and our ways explain our standing in perfect mode with our universe. Foreign tongues arrived and overrode the ancient tongues of the land in the diplomatic front, and this was done without any undue consultation with original people. This trend continued into very recent times. Our well-being is decided in a language that was foreign in the land very recently. A horrific history can be heard, but remains to be told by the elders of the land on their encounters of speaking the language in the presence of government teachers, some of whom were clerical representatives of religious organizations. The degree of abuse encountered by the children of much earlier times have never been revealed in full because no one has asked the question fully, although minute revelations of this period have been revealed. Issues of importance to indigenous people can create utter confusion because certain words important to the diplomatic front are not found in indigenous languages. Our languages represent our total being, and they are a part of a world of subsisting off of the land and the sea, and yet, foreign languages were brought to us without proper representation by those who came upon the land of the people.

Education still has a 'Dick, Jane, and Sally tune' to it, but like the times, how it is being applied has changed. Many people have grasped the reins of education with an enlightened goal to receive all that is available in a world that was, at one time, full of barriers. The doors of education that were formerly restricted by race, creed, and certainly not according to choice, or enjoyment, have sprung wide open; however, a seemingly impenetrable barrier remains in the lack of funding support for many people.

The rituals that are practiced to show respect to harvest continued to lie dormant for generations, but our relationship with universities and museums, and the return of ancestral heirlooms is assisting us in retrieving some of the ways of acknowledging the spirit of the whale, or other mammals. The sharing of this special knowledge is forthcoming from the elders, as many are coming forward to assist in unraveling many questions that exist between man and the animal kingdom. The elders are revealing to the novice hunters what they had heard from the elders before them when they, too, were children and novice hunters.

Time is a confidential being, but at times it has been forced to reveal its silent secrets. To learn from the elders are now special moments to many descendants of the ancient hunter.

The land shows signs of ancient residency and the names of the land tell you the story of our existence and our special relationship with the land. The names reveal danger or places to harvest food to your heart's content. Some names will honor a fallen warrior, and in some places even a fallen foe is memorialized because our warriors found favor in him before he was slain. Yes, the land continually beckons you, but can it be that in just a few generations the land may no longer beckon its silent invitation to the descendant of the ancient hunter? We are in an age when the names of the land have disappeared from the stories of the hunter. To the hunter, it was the land that unfolded before them each time the story of a hunt is told. The sea also has its names and they tell your position even if you are in search of the sea mammal under the fog, at certain times of the season. The fog does not stop the hunter; we continue to find our prey by the sounds of their voices.

The seal and other sea mammal species have provided very well for the people, when the harvest of the whale was prevented by excessive wind or ice conditions. Wind and ice conditions are justifiable factors that restrict our will to launch after a whale. It was not every year that a whale was harvested. The weather is a factor that can decide on the outcome of survival during a cold winter when the hunter is prevented from hunting and is not able to go in search of the whale and other sea mammals. A successful whale harvest can add to the larders of many people, and a full cache means the people will pass through the cold winter safely with adequate food supplies until the time of fresh meat returns.

What guides the whales within reach of the whalemen is not known. Is it the points of land, or capes that guide the whale on their northward migration? Or, is it the position of the setting of the sun that guides the whale and other sea mammal species? Celestial forms that are assumed to guide migratory bird species are beginning to disappear during this period, as the evening night leaves northern latitudes. Some landforms may possibly be used as vantage points, but they are not always visible. Are the ocean currents at play in the migration of sea mammals? Or, is it some unseen force, never to be known by the descendants of the ancient hunter that guides our prey to their summers feeding ground far beyond the reach of many hunters? The elders say that the mammals themselves have a trail they follow; but what is it?

The hunters take advantage of the stillness of the spring weather, as it is this period when the seal and the walrus bask on ice floes, resting, during their northward journey to the summer feeding grounds. Although the sea mammal is ever vigilant, the hunters have learned to mimic the movement of the seal so that he can move closer to ensure a successful harvest. The hunter must ever sharpen his skill by learning the movement of his prey as he moves closer. The period when the rays of the rising sun push the wind gently over the waters is the time when the animals are at peace and man and other predators search the ice for animals that would certainly be basking under the warm sun. The animals sleep, but this is not an idle period for them as the current carries the ice northward at a phenomenal pace. Sleep becomes a necessity for the hunter, but it is good weather that gives the hunter his best advantage, and the animals are within his reach for only a short period. The hunter returns, again and again, to the hunting grounds until the distant mountains and hills, or the bells of the walrus tell him that he must return home. The bells of the walrus are a testament to the hunter that the tempest is coming and that they must return to the safety of their homes. A storm can mean that what has been gathered before the storm may not meet the needs of the people for the coming winter. Tranquil waters can become churning waters very quickly, and in the effort to return to safety, the hunters must at times cast valuables overboard to lighten their craft; otherwise, the men themselves would be lost to the sea.

If winds bring change, the wind has certainly brought change into the lives of northern people in just more than a century. The insistence of the missionary condoned to change the ethics of northern man, or the cries of the explorer that riches can be easily acquired drove madness to many, and a mass migration from outside of the traditional boundaries occurred overnight. Changes came swiftly from that period, and although people with foreign concepts struggled in transforming their beliefs into the minds of the original inhabitants, they succeeded in planting a foothold in many communities.

Every hunter remains responsible for others within his Nation, but the values of sharing and the willingness to share is being harmed with regulatory measures forced upon the people. Where once it was natural forces alone that regulated his success as a hunter, rules and regulations enacted by unseen dominance far away prompts families to receive charitable alms from outside of the normal realm. Western foods do prevent starvation, but they do not have the nutritional values to meet the needs of the people, and the health of the people has shown a questionable trend. Why do many of the people suffer from life-threatening ailments that were never before suffered by our ancestors? We can feel the strength enter our bodies when we feed on the dark red meat of the sea mammals. Many indigenous people who live far from home are fortunate that we had extended our universe through the generous practices of sharing the meat harvested by the hunter with others that moved away from their homes of origin. Our families are our lifelines of survival away from home, and the trait of sharing must not be removed from our ways.

The mammals of the arctic seas and coastal plain ever vigilant, and they too have been told by their kind that the hunter is vigilant, and that he has the edge over any adversary. Although the hunter does not have the size and the strength of his prey, he has the stamina, patience, and the ability to foresee unseen movement that outclasses his prey. Every single day is a rude awakening in the life of the hunter, his family, and his children in their daily encounters with alien control that is applied upon them from outside of their original realm. The tranquil season of spring is a time of expectation to the hunter. It is a time when many species of sea mammals, and bird species of the land and the sea, are on the move, and it is this time that the people move to ancient harvest areas in expectation of replenishing food stores. A spring season of successful harvest provides fresh meat to the table after a winter of subsisting on food resources that were put away the previous year. Henceforth, this period is known as *Nutaaqtugviq,* or 'the moon when you can once again consume fresh meat.' A smile from the elders provides the most pleasing moment to the hunter. These special moments make the hunter realize that he has fulfilled an unwritten promise that was expected of him as a provider for the community. Today, that promise is contested daily by domineering principles that are set by unseen and unheard decisions far away. In spite of many barriers, the hunter enjoys a special place in the hearts and minds of the people.

It is our relationship with the realness of our universe that continually receives undue attention from other ways. When at one time we harvested all of what we needed in the belief that more shall come to us, we now meet regulations that, at times, seriously offset the balance of nature. The animal with the genetic superiority over his kind that are needed to bring forth new generations are used to grace the walls of the sportsmen, and this measure has taxed the survivability of many species: mammal, bird, and/or fish. From the time of our ancestors, the people relied upon the land and the sea for their sustenance. Time has certainly changed in a short period to a point where the original people of the land must pursue change or get left behind. Being a hunter/gatherer is a very serious lifeway. All hunters are shown the ways of the sea, and the ways of the land of their domicile. It is not an enlightening feeling when you enter into the hunting territory of another to feel the helplessness of not knowing what to do, or not knowing what may be there for you. You must have someone to guide you, else, you will be purged you're your lack of understanding of what the land or the sea holds for you. It is also not an enlightening thought to understand that when the sea mammals are gone, the descendants of the ancient hunter may not be far from following the sea mammals, birds and fish into oblivion. The land and the sea is bleak and barren in the eyes of many others, but they have supported many generations of our people, since dawn immemorial.